CREATING ALTERNATIVE FUTURES

CREATING ALTERNATIVE FUTURES

The End of Economics

Hazel Henderson

Foreword by E. F. Schumacher

A BERKLEY WINDHOVER BOOK
published by
BERKLEY PUBLISHING CORPORATION

Published by arrangement with the author's agent

Berkley Publishing Corporation
200 Madison Avenue
New York, N.Y. 10016

SBN 425-03715-0

*BERKLEY WINDHOVER BOOKS are published by
Berkley Publishing Corporation
200 Madison Avenue
New York, N.Y. 10016*

BERKLEY WINDHOVER BOOK ® TM 1062206

Printed in the United States of America

Berkley Windhover Edition, FEBRUARY, 1978

SECOND PRINTING

ACKNOWLEDGEMENTS AND CREDITS

The author wishes to acknowledge and give credit for permission to reprint her previously published articles and for use of the illustrations in this volume, as follows:

Business and Society Review, copyright, 1976, by Warren Gorham and Lamont, Inc. 210 South St. Boston, Mass. (all rights reserved) for the illustrations by Huffaker and others on pages 10, 62, 92, 208, 236, and 296; Public Media Center of San Francisco for the illustrations on pages 198, 204, 284 and 360; Diane Schatz and *RAIN* magazine, Portland, Oregon, *Co-Evolution Quarterly,* Sausalito, California, Mr. Sim Van Der Ryn, State Architect, Sacramento, California, and the Office of Energy Research and Planning, Governor's Office, State of Oregon, for Diane Schatz's illustrations on pages 40, 44, 51, 56, 146 and 380; the American Federation of State, County and Municipal Employees, Washington, D.C., for the illustration on page 278; Harold Brammeier, Jr. St. Louis, Mo. for the cartoon on page 268; Pat Oliphant and Paul Conrad and the Los Angeles Times Syndicate for the cartoons on pages 136 and 352; and to *The New Yorker* and artists Booth, Drucker, Lorenz, Mirachi, Reilly and Richter for the cartoons on pages 154, 157, 173, 202, 242, 276 and 304. For permission to reprint or adapt my articles, credit is due to the following publications: *Liberal Education, The Financial Analysts Journal, Co-Evolution Quarterly,* the *Harvard Business Review,* (copyright 1971 and 1973, by the President and Fellows of Harvard College, all rights reserved), *Nutrition Action, Saturday Review, Technological Forecasting and Social Change, Planning Review, Business and Society Review, Human Resource Management, Management Review, MBA Magazine, Columbia Journalism Review, The Annals* of the American Academy of Political and Social Science, *Just Economics, Public Administration Review, The Futurist, Earthrise,* and the *International Social Science Journal* of UNESCO in Paris, France.

To Dorothy Mustard, my Mother and wisest teacher,
and to Ali Henderson, my beloved daughter—
my links in the great chain of evolution.

CONTENTS

LIST OF FIGURES

FOREWORD

I greatly welcome this collection of Hazel Henderson's essays. As they appeared over the years, one by one, in various journals, every time I found them illuminating the landscape of our society like a flash of lightning, causing me intense delight but also anxiety that somehow I might miss the next flash.

To have them in one handy volume provides more illumination than some of us will be able to take in a hurry. But never mind. Our eyes will get used to the light; our minds will stretch and widen so as to gain—or regain—the *capaciousness* required for doing justice to reality. To me (and I can speak only for myself) Mrs. Henderson's essays, every one of them, have more "reality" than almost any other writings on societal problems I know. How is this possible? What are her credentials?

There aren't any. She has never attended college. She describes herself as "not only . . . relatively unschooled but also unchurched . . . not institutionalized . . . [and] lastly . . . a foreigner and a newly naturalized citizen." Maybe this accounts for the amazing *freedom* of her thinking, a freedom matched by courage and power.

Maybe many people are born like that and then lose everything in the process of being educated—paralysis by analysis, trained incapacity—the result of, among other things, the systematic rejection of any kind of metaphysical, theological, or philosophical training. Hazel Henderson is a self-trained metaphysician (or theologian or philosopher—whichever name you may find least objectionable), and this enables her to "measure" the various sciences, disciplines, ideologies, and myths—above all, to "measure" economics, which she not unjustly describes as "our reigning sophistry." She is thus able to see, and does not hesitate to say, that, in all too many cases, "the emperor has no clothes."

She says, in effect, to the establishment:

> Your theories are mostly make-believe. You yourself don't really believe them. You are becoming schizophrenic. You are trapped in the gilded cages of your institutions; but individuals are learning faster than institutions. So watch out. What holds you in your cages? Why not get out? And if you cannot physically liberate yourself, can't you at least free your thinking? You don't have to *admire* the naked emperor's clothes.

But what am I doing paraphrasing or interpreting Hazel Henderson! I am sorry to have taken your time and delayed you. A splendid meal has been prepared for you—go and enjoy it.

Caterham, Surrey, England —E. F. SCHUMACHER
1st June 1977

1st June 1977

On Sept. 4, 1977, the many millions of admirers of Dr. E. F. Schumacher were saddened to learn of his death in Switzerland. The world has lost a calm voice of sanity and humanity. His best memorial will be in the continuation of his work by the growing global movement for the wiser, more humanly and ecologically-appropriate use of technology. At heart, Fritz Schumacher was a metaphysician, and those who wish to know this beautiful man more deeply should read his newly-published book, A Guide for the Perplexed.

Hazel Henderson

CREATING ALTERNATIVE FUTURES

INTRODUCTION

In April, 1977, President Jimmy Carter spoke on nationwide TV and, for the first time, officially described the American energy crisis. He called it a quiet crisis, an insidious crisis, in that there were no dramatic events, such as the bombing of Pearl Harbor, to alert us to the danger. And yet, Jimmy Carter, 39th President of the United States, cited for the first time in our 200-year history, facts and figures indicating that, at last, the American energy-resource cornucopia was running dry, and that our spree of wasteful consumption and production was drawing to a close. The President warned that we now had no choice but to tighten our belts, and that the transition to living within our energy means would require nothing short of the moral equivalent of war.

Since then, although there has been much debate over the time frame needed for action, the means to be employed and the precision of the forecasts President Carter cited, and although other projections have been advanced by numerous think tanks—from the even gloomier Club of Rome reports to the euphoric onward-and-upward extrapolations of Herman Kahn's Hudson Institute—there has been very little disagreement from any quarter about the need for immediate, massive efforts at energy conservation. And since energy is the basis for all industrial activities and material transformations, this augurs a major transition in our economy, technology and, more importantly, our values.

All of the aspects of this transformation are the subject of this volume, which was written over the past dozen years, as I tried to understand, predict, interpret to others and to organize politically for the changes I was sure must come as the mass consumption joyride of Americans collided with global population/resource realities. In 1964 I began my own self-education (there were few college courses then in these matters) concerning this needed transformation of our economy and technology. At the same time, I began to express my active concern when I joined with some other worried citizens and mothers of small children in New York City to form an organization called Citizens for Clean Air. I soon learned that if the air was to remain breathable and the environment life-sustaining for my infant daughter during her lifetime, I and other citizens would have to commit ourselves to a process of learning about the complex, interdependent, urban, industrial societies in which we lived and about the basic assumptions on which their technical and economic systems were founded. I also discovered in studying ecology, economics, corporate behavior, mass media and our sociopolitical system that many of the traditional assumptions about economic growth, resource exploitation and competition were literally killing us.

My developing consciousness of global ecological and human interdependence grew with the burgeoning environmental movement during the nineteen sixties and seventies. Ecological models have the advantage of providing larger contexts and more inclusive matrices which subsume economics and other pseudo-rigorous, narrow forecasting techniques, the shortcomings of which I have reviewed in Part One, "The End of Economics." It is for this reason, I believe, that ecological and more systemic models have been more predictive of the crucial events of the past decade, which economists and technical forecasters have often viewed as "acts of God" or "unpredictable" happenings, for example, the new vagaries of the world's weather, overfished oceans, floods, the change into deserts of formerly fertile land, dying lakes, the threatened atmospheric ozone layer, and our rapidly depleting energy and mineral reserves. Even major "surprises" in geopolitical events could have been inferred, if not actually predicted, using ecological models and theories which highlighted the worsening population/resource ratio, which was bound to heighten global resource conflicts sooner or later. Seen in this light, the

Organization of Petroleum Exporting Countries (OPEC) cartel might have come as less of a "shock," as well as the Third World countries' resource strategies and demands for a New Economic World Order, the politicizing of the United Nations and its major conferences on environment, food, population, habitat and planetary resource-allocation, like that involved in the law of the sea. As the reliance on global price mechanisms blinded economists and other forecasters, the ecologists and pragmatic ethicists clearly saw the interdependent global village emerging, with all the attendant stresses and strains on formerly sovereign, independent nation-states as they sought vainly to adjust.

My sense of foreboding and alienation from the dominant, mass-consumption, competitive, individualistic, "free" market U.S. value system grew during the nineteen sixties and seventies. In my travels around this country speaking to hundreds of groups, from business, labor, professional and college audiences to consumers, environmental and other citizens' organizations, I sensed the growing cognitive dissonance between citizens and their business and government leaders. Even as late as April, 1976, then Vice-President Nelson Rockefeller was still claiming at a meeting of the Club of Rome:

> More growth is essential if all the millions of Americans are to have the opportunity to improve their quality of life. . . . It is naive, indeed dangerous to assert, as some do, that the industrialized nations of the world must support the underdeveloped nations of the globe through massive and long-term foreign aid and in goods and services and massive grants of capital.[1]

Meanwhile, about a year earlier, in September, 1975, a Harris poll showed how far ahead of their leaders Americans already were in dealing with the realities of global interdependence. Harris noted that an 85%-90% majority felt that "most government leaders are afraid to tell it like it is—that is to tell the public the hard truth about inflation, energy and other subjects," while a three-to-one majority endorsed the statement that "the trouble with most leaders is that they don't understand that people want better quality of almost everything they have rather than more quantity." The poll also showed that Americans were aware of the consequences of U.S. overconsumption, as 6% of the world's population (now 5%) consuming 40% of the world's resources. Seventy-four percent said that this used up our own resources and those of others abroad, thereby making products and raw materials scarce and driving up prices. And by an almost three-to-one majority, Harris found that Americans believed that this course is morally wrong, while a 77%-78% majority opted for changes in their own lifestyles, such as eating less meat, eliminating annual model changes in their automobiles, rejecting fashion and wearing old clothes and reducing the amount of advertising urging people to buy more products.

Today, I sense that, at last, the logjam of official silence and conventional wisdom is being breached. Leaders in business and government and other areas of our society are being pressured by the new ranks of citizen leaders and activists that stepped into the vacuum. The Carter Administration shows signs that it is reconceptualizing our situation, as does the U.S. Congress, which in a six-week period of 1977, for example, introduced some 36 bills to encourage energy and materials conservation and the shift to regenerative resource methods and technologies, as described in this book.

Never has a blossoming of pluralistic leadership from the grass roots been more sorely needed, since such times of change must call forth multiple leaders rather than rely on formal, centralized control which is cut off from adequate feedback. Luckily, our creaking, but still functioning, political system does allow these new leaders' voices to be heard, as they rally Americans to their greatest challenge: the peaceful accommodation of our society to the inevitable social and economic transition that is already upon us, and the orderly recycling of our values, our culture and ourselves. I also discovered that it was no use coming up with fancy hypotheses, theories and "headtrips." Mine had to be tested in the real world. This meant, "Could one find enough other citizens to whom these ideas sounded plausible and explanatory so as to be able to organize another citizens' group, network or public-interest activity around them?" To me, political scientists should never be awarded their Ph.D.'s until they have demonstrated that, at least, they can organize a respectable bake sale!

Therefore, as my concepts about the bankruptcy of economics developed, I was at the same time constantly testing these ideas, not only with iconoclastic economists and debunkers from other disciplines, but also by starting, or helping others start, a series of public-interest organizations. After my initial involvement with Citizens for Clean Air from 1964 to 1967, I served as an adviser to New York's Earth Day organizers in 1969 and 1970 and still advise Environmental Action, the wonderful group that grew out of Earth Day. I worked with the Campaign to Make General Motors Responsible in 1968 and 1969, a pioneering effort to seat three outside "watchdog" public directors on the corporate board. I have served on the board of the Council on Economic Priorities, of New York and San Francisco, from 1970, and am as proud as ever of its innovative research studies comparing the social performance of American corporations and of its physically diminutive but intellectually heavyweight founder, Alice Tepper Marlin. I became a speaker for the Conservation Foundation program developed by Byron Kennard, which fostered nationwide citizen involvement in setting air-quality standards in 1967 and 1968, and I helped Mr. Kennard organize the first national Conference on Public Transportation in Washington in 1971. At the 1969 annual meeting of the National Association of Business Economists, I proposed in a speech the development of "advocacy," public-interest economics, so as to help economists become aware that they are not scientists, and I urged that economists already sensitized to this fact volunteer their services to groups impacted by economic projects or decisions, but unable to hire economists, to quantify the dis-economies, dis-services and dis-amenities that they might suffer. Such economists, aware of the limits of their discipline, could also critique those cost/benefit analyses that were intellectually fraudulent, discounted the future, or concealed social conflicts by averaging out the costs and benefits per capita, making it unclear which groups would reap the benefits and which groups would bear the costs. Today many citizens groups understand that economists are not much different than lawyers and that a cost/benefit analysis is not unlike a legal brief in support of whatever economic action the writer is paid to justify to the public. So in 1972, Byron Kennard, by then my closest public-interest collaborator, and I proposed and helped found, with Allen Ferguson, the Public Interest Economics Foundation, which matches volunteer economists to those citizens groups needing help in clarifying the other side of the picture than the one usually painted by the promoter:

the social, human and environmental costs of economic activities and decisions. My abiding interest in the social impact of large corporations led me to co-chair with Prof. Prakash Sethi of the University of Dallas and Kirk Hanson the Second National Symposium on Corporate Social Policy for the National Affiliation of Concerned Business Students held at the University of Chicago in 1974.[2]

Many of my organizing activities were failures, such as my premature effort in 1967 to launch a National Citizens Committee for a Guaranteed Income, modeled on the ideas of its inventor, Robert Theobald; an equally abortive effort to set up a women's executive talent bank in 1972; and my unsuccessful struggle in 1975 to save the small, underfunded Center for Growth Alternatives, some of whose ideas I had helped develop, together with Sydney Howe, Tom McCall, former Governor of Oregon, and Prof. Georg Treichel of California and the Public Media Center of San Francisco, into a public-service advertising campaign with the theme "WE CAN'T GROW ON LIKE THIS" (see p. 360). These ideas are even more relevant today. My views on the need to head off future environmental and social problems by focusing on the policies directing the scientific community and technological innovation led me naturally to the emerging field of technology assessment and to my appointment, in 1973, to membership on the Advisory Council of the U.S. Congress Office of Technology Assessment (OTA), now a separate science and technology policy research organization serving the needs of Congress in assessing the impacts of existing technologies and likely impacts of future technological and scientific decisions.

Since then, I have tried to assure greater public participation in such science and technological policy making and to promote the ideas for more appropriately scaled technology of my friend the late E. F. Schumacher, author of the best-selling *Small is Beautiful* and *A Guide for the Perplexed.* I helped found a coalition of environmental groups in 1975, Environmentalists for Full Employment, designed to draw attention to the fact that an environmentally sound economy must, by definition, be a full-employment economy. We must now conserve our natural resources by more fully using the talents of all of our people. We must therefore now run our economy with a leaner mix of capital, energy and materials and a richer mix of labor and human resources.

Indeed, I have learned that it is almost impossible to be a thinking, fully functioning human being in a complex society without doing politics—not necessarily the old politics of geography, but issue politics, citizen-movement politics, public-interest-group politics, corporate-stockholder "proxy-fight" politics, media-access politics, and finally, "networking" politics linking concerned citizens all over the U.S. and the world around emerging planetary awareness that only ecological sanity, social justice and new forms of social, individual and technological development can save human societies from disaster.

Millions of Americans like me have been struggling with this cataclysmic personal and social agenda: by rethinking their lives, their jobs, their relationships and the functioning of their local communities, as well as their state, national and now single global community. They are creating alternative institutions, communities, lifestyles, and safety nets and small lifeboats, rather than fighting over deck space on the old "Titanic"—our declining mass-consumption-based industrial society. This book is a personal view of this great transition and the conceptual shift it requires, which hinges on the problem of transcending traditional economic thought and practice, what I have called "The End of Economics," the title of Part One. Part Two, "Creating Alternative Futures," reflects my years of activism, as I sought to validate my social hypotheses and integrate my thinking and doing. It describes some of the ways in which we might re-vision and re-make our futures. Up to now, I have been busier doing this revolution than writing about it. The past decade reviewed in these essays was for me also a rich, exhausting, exhilarating, deeply moving period. Now it is time to take stock. I owe heavy intellectual debts to all those whose activist and theoretical work is mentioned, many of whom I am privileged to count as my friends. My errors of omission and commission, of course, are solely my own. My deepest thanks are owed to Carter F. Henderson, my partner and co-director with me of our own small group, the Princeton Center for Alternative Futures, Inc., and also my oldest friend, without whose support over the past 20 years I could not have grown at all. I remember sociologist Harold D. Lasswell asking me once at a party in what discipline I had been trained. I replied, "None, I am just a human being trying to act sensibly under the current circumstances of existence." That is the best any of us can do.

---

[1]Ervin Laszlo, et al., *Goals for Mankind,* Dutton, New York, 1977, pp. 38-40

[2]Symposium published as, George Rohrlich, ed., *Environmental Management,* Ballinger Press, Cambridge, Mass., 1976

PART ONE

The End Of Economics

CHAPTER ONE

Recycling Our Culture

I have confessed that I have never attended college, and that not only am I relatively unschooled but also unchurched. Instead, my mother taught me her brand of pantheism which, in retrospect, was a fairly useful basis on which I developed my particular ecological model of reality. I must also note that I am not institutionalized, but operate as an individual in an institutionalized society—at least, during my working life. Since, as Bertram Gross has pointed out, institutions are devices for screening out reality, this gives me a view of society unmediated by many of the organizational filtering devices that color the perceptions of most of my fellow Americans. Lastly, I am a foreigner and a newly naturalized citizen. Therefore, any relevance that my thoughts might have may lie in these somewhat detached conditions of my existence.

Let us try to suspend for a brief moment our culture-bound and geo-specific consciousnesses and recognize ourselves for what we are: a group of humanoids, heir to a particularly lush segment of this blue planet which we have arbitrarily designated the United States of America. Now we have celebrated our 200th birthday, it is fitting that we examine the ideologies, paradigms and myths that have sustained us through the past two centuries of our growth and development. We need to assess which of them reflect that which is eternal in the human condition, in our continually evolving interaction with our ecosystem, and which of them have served a transient purpose and should now be jettisoned as excess baggage, as we continue our journey into our third century.

I suggest that the first ideology that we need to assess anew is what my Japanese friends refer to as the "curse of individualism": the legacy of John Locke, which was so vital to the small band of our ancestors in conquering a wilderness and forging a nation. In today's overcrowded, urbanized, interdependent America, individualism as an ideology creates needless conflict and exacerbates loneliness and alienation. John Dickinson, a delegate to the Continental Congress, noted as far back as 1768: "A people is travelling fast to destruction, when individuals consider their interests as distinct from those of the public. Such notions are fatal to their country and to themselves."[1] And political theorist, Joseph Mazzini, pointed out in 1835 in referring to the ideas of the French Revolution, that declarations of human *rights* alone would not build a society, since they did not take account of our social interdependence.[2]

Similarly, our Declaration of Independence, being a document forged in rebellion, asserts human individual *rights*—a historic achievement in social philosophy, and entirely appropriate as a goal for a small new nation of farmers and entrepreneurs faced with an almost empty continent. But in our complex modern society, individual rights must now be balanced with the concept of individual responsibilities, and we may also require a Declaration of Interdependence.

Jonas Salk, in his book, *The Survival of the Wisest,* provides us with some useful imagery which may help set our current scene of cultural confusion in a useful context. Salk points out that in the life-cycle of any biological species in a finite environment, growth follows the now typical S-curve. During the first phase of the curve the behavior pattern is characteristically that of maximizing growth through vigorous competition and ecosystem colonization and exploitation. As the curve reaches its fulcrum point, a new phase is entered where the past behavior patterns are no longer rewarded. It is as if the species, like Alice, had gone through the looking glass and, on the other side, growth gives way to differentiation and maintenance, competition to cooperation, and exploitation of the ecosystem is transformed into restoration and recycling. The implications for economic theory are devastating, as we shall see.

This "Alice through the looking glass" phenomenon may explain why so many of the pronouncements of our politicians, businessmen and other leaders are now viewed by many aware citizens as a mirror-image of what is needed: for example, continued attempts at increasing industrial growth, cutting social spending in the federal budget without questioning military, space, highway building and other expenditures, or trying to force-feed private consumption of materials and energy-intensive goods, while admitting that we are facing a new era of capital, energy and materials scarcity.

Such contradictions are viewed with increasing levels of cognitive dissonance, since in our media-rich society, individuals are learning faster than institutions and are more open to feedback, while most of our large public and private institutions operate with ten-year average time-lags and still focus on goals that may no longer be appropriate. Their leaders are necessarily programed into the same time-lags, and the larger the institution the less flexible and responsive to new demands, and the more insulated its leaders are from apprehending new conditions.

Worse, the higher the institutional level, the greater the level of unreality. For example, while the leaders of Western nations intoned banalities at the economic conferences in London and earlier at Rambouillet, France, at another lower level meeting, experts from the same group of countries had talked realistically about the collapse of Keynesian policies and the new intractabilities of structural inflation and growing scarcities.[3]

Several recent polls in the United States demonstrate this growing lag between the goals and paradigms of our leaders and the average individual's perceptions of current conditions. Harris Polls, Roper Reports, and Opinion Research all testify to the drastic decline between 1959 and 1973 of confidence in our business institutions (53 per cent from 75 per cent).[4] The Gallup Poll found in 1975 that for the first time Americans are realistically coming to terms with the prospect that their own futures will be less bright than once imagined.[5] Hart Research found in 1975 that 33 per cent of Americans believe that our capitalistic economic system is now on the decline; 41 per cent favored making major changes in our economy and applying policies so far untried and 56 per cent said that they would support a presidential candidate who favored employee control of U.S. corporations.[6] And Opinion Research found in 1975 that despite recession and unemployment and rising fuel costs, 60 per cent do not believe that we should cut back on environmental control programs, even if they must pay even higher prices.[7]

Another measure of our cultural shift and the resulting confusion for business leaders is a poll of the attitudes of corporate executives conducted by George Cabot Lodge and William F. Martin (*Harvard Business Review,* December 1975). Of these executives, 70 per cent preferred the old ideologies of Lockean individualism, private property and free enterprise, but 73 per cent acknowledged that, although they preferred these values, they thought that by 1985 collective models of problem solving would have supplanted them, and, furthermore, 60 per cent thought that the more collective value orientation would be more effective in finding solutions.[8] The truth is that in a highly complex, interdependent, technological society, individual freedom, when armed with polluting, disruptive technologies, now destroys the freedom and amenities of others. Further, as system theorist Todd LaPorte points out, the market is no longer a valid arbiter of choices having "indivisible social consequences."[9]

Indeed, a majority of individuals now share beliefs that untrammeled corporate freedoms allow big business to run the government, export capital and jobs, despoil the environment and waste resources, and as the 1974 Yankelovich study confirms, 66 per cent also believe that inflation is caused by business seeking higher profits. However, these beliefs have not yet destroyed Americans' faith in private ownership and enterprise. They might be reassured by the words of Thomas Jefferson in 1814. "I hope we shall crush in its birth the aristocracy of our moneyed corporations, which dare already to challenge our government to a trial of strength and bid defiance to the laws of our country."[10]

Luckily, many Americans can still distinguish between the rights to property ownership for the purposes of self-sufficiency and the unlimited rights to accumulate property which may become oppressive to others, as in the case of gigantic corporations, which are clearly not private, involve little entrepreneurship, inculcate dependence and conformity, avoidance of risk and responsibility, and which display a bureaucratic dedication to survival and growth at all costs, whether publicly or privately born. We might remember Samuel Webster's words in 1777, "Let monopolies and all kinds and degrees of oppression be carefully guarded against,"[11] and Thomas Jefferson's warning in 1816, "I sincerely believe, with you, that banking establishments are more dangerous than standing armies."[12]

And yet we see in Chapters 10 and 17 that the debate about "big government" deregulation and "the need to lift the shackles off the backs of business" is as deluded as ever. This does not discriminate between big and small businesses and voids the real choices and trade-offs involved that neither big business nor big government want to face. The inevitable axiom of complex, industrial societies is that each order of magnitude of technological mastery and managerial control *inevitably* calls forth an equal order of magnitude of government coordination and control. For example, if the nation's pharmaceutical industries mass-market thousands of patent medicines and mood-altering drugs, the FDA will be forced to undertake (with tax dollars) the task of testing them, and vast efforts mounted by government to combat resulting drug-addiction and crime.

All this demonstrates the inadequacy of the Cartesian world-view which has held sway in our minds for 300 years, and which permits us to view the world in terms of such unreal dichotomies as those of "public" and "private" sectors, goods and services. This view leads us to ignore the link between private profits and the mounting public costs they engender. It leads us to believe that we can "afford" oversize private cars, thousands of brands of patent medicines, and billion-dollar industries devoted to pet foods and cosmetics, while we cannot "afford" nurses, teachers, police, fire and sanitation services in our cities.

We will see how this narrow logic has caused us to over-reward competition while ignoring cooperation and all the cohesive activities that bind the society together, which we have relegated to the status of unremunerated activities, to be performed by women. It has caused us to overvalue property rights while undervaluing amenity rights and to overvalue individual freedom over community needs. Psychologist Robert Ornstein notes that this type of linear, sequential, quantitative, reductionist cognition is a function of the left hemisphere of the human brain. The right brain hemisphere processes information in spatial, simultaneous modes, and is the source of intuitive, imaginative modes of cognition, such as the great hypotheses of science. Both modes of cognition are equally important: the brilliant intuitive leap of a good hypothesis and the careful processes of its validation or rejection.

Nowhere is our culture's overdose of left-brain, Cartesian cognition breaking down faster than in the Tower of Babel it has created in academia, where reality has for so long been carved up into neat little disciplinary boxes. The spontaneous reassertion of right-brain cognition, perhaps as an almost biological-level survival response, is producing new yearnings for reintegration of head and heart, mind and body, and a rich new yeast of intellectual insights as well. I am fortunate enough to receive many fascinating papers containing such insights, sent to me by academics who are trying to transcend their original disciplines but whose efforts at new syntheses are rejected by reductionist journals and suppressed by disciplinary territorial imperatives. They have discovered that not only academia, but our whole society, is biased in favor of analysis while punishing equally necessary attempts at synthesis.

For example, a fascinating book manuscript was sent to me recently by a physician, Dr. C.A. Hilgartner, who has come to understand, out of his deep knowledge of human physiology, that our Western semantic structures, and indeed traditional mathematics, are flawed by the same nonintegrated, binary, Cartesian worldview. In our semantic structures this is evidenced in our verb/noun, observer/observed, subjective/objective speech conventions, which Werner Heisenberg's Uncertainty Principle in physics has already invalidated. In mathematics, Hilgartner notes the same fatal flaw of subjective/objective dichotomizing, evidenced in the operant (subjective) symbols: + (plus), – (minus), × (multiply) and ÷ (divide), and the number symbols (objects) which are passively manipulated.[13]

Such crude intellectual conventions can no longer map the seamless, interacting totality of which we humans and our perceptions are a part. Taxonomy and method can become the enemy of thought, and our greatest intellectuals present us with paradoxes: whether a Heisenberg in physics, a Kurt Godel in mathematics, or a Nicholas Georgescu-Roegen or an Oskar Morgenstern in economics. Such genuine intellectuals take the methods and logical constructs of a discipline and lead its hapless practitioners up the primrose path of impeccable adherence to these existing paradigms, and then zap them with the limits to their application and recognition of the abyss beyond.

The ferment Heisenberg caused in physics is now leading to new efforts by Wheeler, Everett, Capra and Wigner and a host of audacious young physicists to write the observer back into the equation—an overdue recognition in the most basic of the "hard" sciences that, in a very real sense, reality is what we pay attention to. In fact, the humanoid is a perceiving/differentiating device of limited range whose focus inevitably distorts the visioning of the totality. Indeed, perhaps original sin is nothing more than differentiating, out of which grows individuation and the hubristic concept of free will which causes us such communal grief. Out of more holistic insights we may discover a different view of probability theory, rooted in the understanding that "randomness" and "disorder" are only measures of human ignorance. While the recognition of peripheral vision, by the device of "probabilities," was an imaginative leap, perhaps we may also embrace the possibility that those "probabilities" actually exist, even though we are not paying *attention* to them, as the many-worlds-interpretation in quantum physics suggests.

Not only is the Cartesian paradigm bankrupt, but it is not an exaggeration to assert that our culture itself is collapsing, as have so many others before. In spite of the great psychic pain of such collapse of a major belief system, there is a yin-yang rebirth ready to flower. As Thomas Kuhn points out in his *Structure of Scientific Revolutions,* a major paradigm shift leads to a major cultural shift.[14] We see heroic efforts to reintegrate perceptions in the struggle to create more holistic research methods and models, embracing more variables. The most imaginative researchers all complain of experiencing "boundary problems" in their new understanding of ecological insights and consequent realization of the arbitrariness of human categorization schemes. We see these new problems emerging in efforts to develop methods for environmental-impact statements and technology assessment.

Ironically, the usual rationale for decision makers to purchase such research is to reduce uncertainty. Now such new, inclusive forms of research only *increase* uncertainty for the poor administrator or executive—specifying more carefully what is still not known. Thus decision making in centralized, bureaucratic institutions is becoming more palsied and inept, while delay is excoriated by those with vested interests in "getting on with the job." Those with holistic awareness are relieved that the megamachine is slowing down a little, believing that if decision makers now obviously do not know what they are doing, then at least it is well that they do it more slowly. The existential problem of the bureaucrat or administrator in such conflicts is obvious: "What does one *do* when one gets into the office in the morning?" It is impossible to admit to oneself that it might be better *not* to come to the office, or to drop out and reassess one's existence, and whether one's lifestyle has not become a gilded cage.

The most aware and sensitive among us are already swept by such personal doubts, or identity loss, feelings of meaninglessness in their careers and even a sense of moral schizophrenia as they become aware of the social costs their institutional activities engender. Management people and scientifically trained professionals are caught in many such role conflicts and are questioning their allegiance to the goals of their employers. Everywhere I go in this country, such questions abound: "Am I going insane, or is the culture insane?" My answer is that it is the culture that is insane, and that their seemingly private perceptions of the dissonance are shared by millions of aware Americans.

But political and corporate leaders who will sometimes privately admit that they share such doubts about our goals and values maintain the conspiracy of silence, reasoning: "How do I break it to my stockholders or my constituents?" Political change is possible only when such private perceptions are widely shared and finally confirmed as a new social reality. When a critical mass is attained, the political process begins and eventually ratifies the cultural change which has taken place. I have had the good fortune to be intimately involved in such processes, in the development of ecological consciousness since the early sixties; in the growing manifestation of female social wisdom, and in the growth of the movement for corporate accountability. Thus, the most vital function of social movements is as psychological support structures for developing value shifts that enable a society to adapt peacefully to new conditions.

The rise of citizen movements for peace, social, racial and sexual equality, consumer and environmental protection are grounded in the knowledge that information is the basic currency of political decisions. They seek with their public interest research efforts to restructure information, politicize its modulation and amplification channels and create new social awareness and insight that may lead to political and cultural change. They have successfully politicized the sciences and the professions and shown the extent to which most research is commissioned as ammunition for political manipulation. Economics, our reigning sophistry, has been revealed as normative to its core and, happily, dares not continue to parade as a scientific discipline. Such words, formerly unchallenged, as "progress," "efficiency," "productivity" and even "profits" are now part of the tug of war of symbols and the destruction and refashioning of the language. As the paradigm shifts occur, we see that they do not necessarily create new information, but repattern existing information so as to render it more elegant, explanatory and efficient.

New conceptual "peg-boards" are now a prerequisite. The image of the true reality of our situation televised from space, the blue planet, is the most useful and widely available, since it transcends the problem of illiteracy. Another image which is becoming more important to me as I struggle with the vision of a decentralized, communitarian society based on humane, organic technology which will *not* recreate factionalism and parochial viewpoints is the image of the hologram. Vastly increased communications may be a key to achieving this vision. The image of the hologram is that of an information system where every bit contains the program of the whole. It is a key metaphor for our time.[15]

A political system based on this model might at last permit the anarchists' dream, where the state could finally wither away. Since people (analogous to the bits in a hologram) would have incorporated into their individual consciousnesses an understanding of the whole system, i.e., the mental image of the blue planet and all it represents, socially appropriate behavior would be internalized, without the need for external controls. If all persons were so "programed," they would instantly visualize the extended chains of causality flowing from their actions, however delayed or displaced. At last we might recreate on a planetary scale the social sanction system of the medieval or tribal village where aggressive actions brought swift feedback and retribution. Certainly our fractionated, disordered consciousnesses must somehow be harmonized if we are to survive, since they are the basic source of conflicts and tensions that we objectify by compulsively manipulating each other and retooling our environment, rather than retooling ourselves.

Yet much of this retooling of ourselves is proceeding. Some manifestations are very strange, in the flight into new cults, religions and conversion experiences. Much is pragmatic, such as the safety nets being devised by networks of people to help each other through the identity crisis, the citizens' organizations, the alternative institutions and groups testing their capabilities anew in attempting greater self-sufficiency and group coherence. Naturally, most of this new learning is occurring outside existing structures although much is now occurring within old institutions as individuals try to stretch their constraints.

Not surprisingly, during such a period of cultural experimentation, academic institutions, the custodians of the old culture, are particularly suspect. In a rich, media-saturated society, education is bound to move away from old forms and becomes an individual and small-group enterprise, where the whole society and all its dimensions of experience are used as one vast, metaphysical university. Meanwhile, the ferment within academia is beginning to produce fragile flowers in the form of interdisciplinary programs and other experiments, nurtured from within by courageous individuals with new visions. The pressures are also coming from without, due to the changing perceptions of our citizens, and are part of the necessary questioning of all authority figures intoning old platitudes, and the vital ridiculing of the inadequate formulations of our fragmented disciplines. However painfully, academic hypocrisies, territorial jealousies, intellectual and financial vested interests, as well as the commitment to the "value-free objectivity" of the sciences and other myths, must continue to be exposed.

Only in this way can our citizens continue to learn the big lessons: how the consent of the governed is too often engineered by impounding and distorting information; how intellectuals have too often become the servants of the powerful and help control the allocation of resources by mystification; how professionals corner the market on specific knowledge in order to maximize their income and influence; how business leaders manipulate preferences and cultural norms through advertising and endowments, and how political leaders too often govern by capitalizing on ignorance.

Citizen movements are a good measure of the extent to which such insights are now illuminating the perceptions and behavior of Americans. They represent in a real sense, social learning and are, I judge, much more effective than any formal, passive programs of adult education. On another level they constitute vital feedback to our body politic about the deficiencies of all our linear, Cartesian policies, because they spontaneously organize around all of the dis-economies, dis-services and dis-amenities which such policies create as unanticipated, second-order consequences.

All of this bad news being trumpeted by less-than-popular messengers such as Ralph Nader, Betty Friedan, Cesar Chavez, Gloria Steinem, Jesse Jackson, Margaret Mead, David Brower, Paul Ehrlich, Lester Brown, Jay Forrester and other activists like myself is now needed to draw attention to the mounting social costs of our current linear preoccupation with maximizing industrial growth as measured by the Gross National Product (GNP) which, incomprehensibly, *adds* these social costs as positive contributions to production and wealth. As Ralph Nader has said, "Every time there is an automobile accident the GNP goes up." Similarly, the social and environmental costs of growth: the cleaning up after the wastes of production and consumption, the maintaining of adequate supplies of clean air and water, the caring for the increasing numbers of human casualties of massive incomprehensible technology and inhumanly scaled organizations, the mediating of conflicts, the controlling of crime, addiction and other pathology and generally maintaining a fragile "social homeostasis"—all are counted in the GNP as positive production.

We will explore how these social-cost components of the GNP are now the only part that is rising, and that we now may have reached the point in our society of an evolutionary cul-de-sac, which I have described as *The Entropy State.* In such a society, due to its unmodelable, unmanageable complexity and interdependence, social costs begin rising exponentially and exceed actual production. Such a society has already drifted to a soft landing in a steady state, but its still rising GNP and increasing rates of inflation mask its declining condition.

As a modest start, I suggest that if only one paradigm change could be instituted by legislative decree, it should be the institution of a crash program of researching and documenting these flagrantly visible and quantifiable social costs. As I have suggested to the staff of the Joint Economic Committee, we should then start building a social-cost model of the U.S. economy in much the same style that Jay Forrester has constructed his enormously useful models for the Club of Rome. Such a social-cost model of the U.S. economy would provide us with a mirror-image of the GNP, and would be no more difficult to develop than those for the Club of Rome. One could look at our economy by industrial sectors and begin stating relationships between these "private" sectors and the mounting social costs they are engendering in the "public" sector.

For example, it is fairly easy to assign to tobacco companies a reasonable portion of the medical costs associated with lung and respiratory ailments, as well as the costs of absenteeism. Similarly, reasonable calculations can be made as to the portion of the social costs of alcoholism and charge these to the distillers, or the costs not born by the producers and consumers of polyvinyl chloride or aerosol containers but now suspected to be mounting sharply. Much of these efforts to document social costs have been accomplished by citizen groups, such as the Council on Economic Priorities, on whose board I serve, which has pioneered the careful comparative analysis of corporate social performance in areas such as environmental protection, consumer and minority rights and military contractors and their foreign operations.[16] We will have to start paying economists to do such studies, since there is no market incentive for collecting such data.

Once social cost data begin to accumulate, they will eventually find their way into GNP calculations. Meanwhile, we might take a leaf from the Japanese, who have already begun the task of reformulating their GNP to a new indicator, Net National Welfare, which will deduct these social costs. Another needed change will be to value housework, volunteer work and leisure time, which GNP ignores, and to stop treating money spent on education as thrown down a rathole rather than as an investment in vital human resources. As physical resources become even scarcer, investing in human resources may prove to be our best strategy. Increasing the skills and knowledge, and hopefully even wisdom, of our citizens is one form of growth not limited by the dismal laws of physics, and its exponential growth might be our best chance of survival.

At an even deeper level, many now understand that we are rapidly exhausting the limits of empty technique and our dominant modes of instrumental rationality and materialism. There is now a new hope that the fruitless dialectic between capitalism and communism will be exposed as irrelevant, since both systems are based on materialism, technique and narrow rationalism. The two dominant societies representing these so-called opposing value systems, the U.S. and the U.S.S.R., are beginning to appear rather similar in their major contours: both are dedicated to industrial growth and technology with increasing centralism and bureaucratic control, whether their chief institutions are nominally designated private (as in the case of multinational corporations) or public. There is, in effect, little to choose between centrally planned, bureaucratic socialist economies and the emerging bureaucratic state capitalism of the U.S. and other Western economies. Both neoclassical "free market" and Marxist economics consign social values, morals, art and consciousness to dependent status. While Marxists view such froth as merely the superstructure on a materialistic base, neoclassical economics relegates this aspect of human existence to the limbo of the private and unquantifiable, and therefore merely ignores it.

But the spiritual and emotional dimensions of humans will not be denied, and we are now witnessing what might be viewed in Freudian terms as the return of the repressed on a societal scale. The human soul is determined to find meaning and cannot live by bread alone. The predominantly Cartesian, masculine-oriented, objective style is now too limited a plane for the expression of our new multidimensional awareness. Intuitive "body-wisdom" is coming to our rescue in the spontaneous growth of new organizational forms, networks, rap groups, cooperatives and all the other manifestations of the human potential movement.

In Part Two, I hope to show that this same body-wisdom is visible in the growing opposition to massive, big-bang, capital-intensive technology which only serves to concentrate power, wealth and knowledge in fewer and fewer hands, at the expense of making most of us more stupid, more dependent and poorer. It is manifest in the turning away from the charade of national politics, with its vast abstractions based on statistical idiocies, which hypnotize bureaucrats and politicians into the self-delusion that people are white rats, and that their mass manipulation can be accomplished with a stroke of the functionary's pen. Body-wisdom is operating to prevent obfuscation of the issues by intellectual mercenaries. The social costs of our economic system have now risen above the threshold of sensory awareness: we smell the dirty air and water, see the waste and pollution, hear the rising noise levels and sense the growing social disorder and breakdown.

All this is leading to spontaneous efforts to create alternative futures and decentralize our society and its technological means. Organic change is now emanating from small groups at the local level and from reassertions of state and regional initiatives. Similar movements are afoot in Europe, as the Scottish and the Welsh demand relief from the domination of London, while the Basques, Catalans, Ukrainians and other ethnic groups fight for greater self-determination. This need to redress overcentralization is manifesting itself in demands for greater citizen participation in science policy and in the decision making of corporations alike. The divine right of property is increasingly viewed as being as oppressive as was the divine right of kings. Demands for worker control and self-management are growing in this country as well as in Europe.

Lastly, we in the United States are developing a more humble and tragic view of ourselves and our nation's role in the world. Hubristic, machismo nationalism is crumbling as we sample the psychic relief in store for us when we relinquish efforts to police the world and to keep up with the Joneses. As far back as 1937, psychologist Karen Horney cited the pressures on Americans of their industrial, competitive, materialistic society. She noted that three basic value conflicts had arisen: aggressiveness grown so pronounced that it could no longer be reconciled with Christian brotherhood; desire for material goods so vigorously stimulated that it can never be satisfied; and expectations of untrammeled freedom soaring so high that they cannot be squared with the multitudes of restrictions and responsibilities that confine us all. And as we begin to deal with the external and legitimate demands for a new economic world order, we are beginning to realize that having now created a globally interdependent economy, we must develop the "software" to operate it cooperatively.

The new interest in searching for extraterrestrial life is a psychological improvisation we have fashioned to assist us in our new survival quest as well as a legitimate subject of scientific enquiry. Likewise, the new explorations of power of the human mind are a survival tool to help us project ourselves out of modes of thinking and being that are now evolutionarily blocked. The dialectics have shifted to the higher system levels of the planetary and the interplanetary. Perhaps another useful image is that of this planet as a gigantic Skinner box, with all the positive and negative reinforcers programed in for us. If we learn how to operate in it we shall be rewarded by survival; if not, the planet will simply return to an equilibrium state by eliminating us. The linear extensions of the old instrumental rationality are manifested in the space colony proposals of Gerard O'Neill, Kraft Ehricke and others. The metaphysical mode is now represented by the new explorations of mental and spiritual dimensions in the writings of Theodore Roszak and William Irwin Thompson[17] and the quest for cosmic awareness, rather than physical travel in tin-lizzie space ships.

After 300 years of "enlightenment," we are dealing realistically again with the basic human dilemma: a consciousness capable of wandering in time and space among planets, solar systems and galaxies, but trapped in a body composed of a few dollars' worth of chemicals which will disintegrate in a few short years. Death: in our frantic efforts to deny it, we rush around scratching "Kilroy Was Here" on ourselves, each other and the world around us. We commission endless research to rationalize our fears, hopes and desires and to shield ourselves from the knowledge of what we must do. As Ernest Becker makes so clear in *Denial of Death,* we do not need any more research; we know what we must do.[18] We have always known what we must do: face ourselves, accept death and our existential agony and learn to *live* in daily awareness and *delight* in the mystery and wonder of our consciousness and the cosmos.

---

[1] *Voices of the American Revolution,* The Peoples Bicentennial Commission (Bantam Books, 1975), p. 147

[2] *Manas,* Vol. XXVIII, No. 46 (Nov. 1975), p. 1

[3] Bilderberg Meetings, Cesme Conference, 25-26 April, 1975, Cesme, Turkey

[4] *Bell Magazine* (Jan.-Feb. 1975), p. 10

[5] *New York Times* 26 Oct., 1975, p. 1

[6] *The Progressive,* Oct. 1975, p. 13

[7] Opinion Research Corporation, Report to Management (Princeton, N.J.: August, 1975)

[8] George Cabot Lodge and William F. Martin, "Our Society in 1985: Business May Not Like It," *Harvard Business Review* (Dec. 1975)

[9] Todd LaPorte, *Organized Social Complexity* (Princeton, N.J.: Princeton University Press, 1975), p. 19

[10] *Voices of the American Revolution, op. cit.* pp. 151, 154.

[11] Ibid., p. 153

[12] Ibid.

[13] C.A. Hilgartner, M.D., "The Method in the Madness of Western Man" (Unpublished MSS.)

[14] Thomas Kuhn, *The Structure of Scientific Revolutions* (University of Chicago Press, 1962)

[15] See for example, Iztak Bentov, *Stalking the Wild Pendulum.* E.P. Dutton, 1977

[16] The Council on Economic Priorities, 84 Fifth Avenue, New York, N.Y. 10011 (publications list available on request)

[17] See for example, Theodore Roszak, *Where the Wasteland Ends* (Doubleday, 1972), and William Irwin Thompson, *At the Edge of History* (Harper & Row, 1971)

[18] Ernest Becker, *The Denial of Death* (Free Press, 1975) (paperback edition)

Reprinted from *Liberal Education,* the Bulletin of the Association of American Colleges, May, 1976.

CHAPTER TWO

The Exhaustion of Economic Logic

The first part of this book zeros in on economics, a pseudoscience whose inappropriate concepts, language and methods are now impeding the needed public debate about *what* is valuable under changing conditions.

We are already encountering social and conceptual "limits to growth"—well ahead of actual depletion of specific material resources. Our conceptual crisis involves the limitations of economics in mapping the immense structural changes that have characterized the technological developments of industrialization since its beginnings in 18th-century England, as described by Adam Smith in his masterful opus *An Enquiry into the Nature and Causes of the Wealth of Nations,* 1776. His equilibrium model of supply and demand still underlies our economic policy making. Meanwhile, the ideas of John Maynard Keynes in his *General Theory of Employment, Interest and Money,* published in 1936, have been misunderstood, as we shall explore. This has given rise to today's confusion among economists, who tinker endlessly with our disequilibrium economy, while still visualizing it as a fluid, equilibrium system which can be managed with the simple hydraulics of aggregate supply and demand. Their obsolete conceptual models now map a vanished system, monitor the wrong variables, generating many statistical illusions. All mature industrial economies are in a process of transition from their maximizing of material production, consumption and "throughput," based on nonrenewable resources, to economies based on minimizing materials throughput, more recycling and product durability and the use of renewable resources, and managed for sustained-yield productivity. Therefore, our most urgent task is to remap our economy, account for its structural evolution and redesign our models and indicators more in accordance with today's realities. A key proposition is that this task is interdisciplinary, and, given the lag in economics, insights from other disciplines such as general systems theory, thermodynamics, game theory, biology, anthropology and psychology, must now be called upon by all economic policy units in government. Economics is not a science, and economic policy is now too important to be left to the economists.

Vain efforts to restimulate the economies of the U.S., West Europe, Canada and Japan are failing. The stop-gap floating exchange rates instituted in 1973 to shore up the world's financial system are in deep trouble. These countries' economists in their respective confusion, are advising sporadic policies, resulting in gyrations in exchange rates and declining growth of world trade, from 11½% rate in 1976 to 6% in 1977. *Business Week* noted recently that the only alternative to floating rates would seem to be a total breakup of the world economy, with nations hiding behind protectionist policies not seen since the 1930s. The nature of the industrial revolution has been its growing structure and organizational, technological scale—culminating in today's globe-girdling multi-national enterprises and interlinked trading nations. The logic of this industrial system is now exhausted.

A sure symptom of conceptual crisis in any discipline is the proliferation of apparent paradoxes. Today, paradoxes abound, in economics.

□ The paradox that advancing technological innovation in a free society systematically destroys the conditions required for free markets to function, and destroys the conditions required for voters in a democratic society to master sufficient technical information to exercise well-informed votes. The inherent complexities of some advanced technologies, *e.g.*, nuclear power, cannot be fully mastered by Senators, Congresspeople, or even the President, let alone the average voter. Therefore, such technologies become *inherently* totalitarian. Worse, their very scale requires social investment and taxpayer subsidies at the same time as it precludes full participation and representation in the direction of technological innovation.

□ The paradox that in mature, industrial societies with highly complex technologies, free-market, *laissez-faire* policies become unworkable, while (at the same time that this private-choice system is eroding) we have not yet devised public-choice systems adequate to manage the complexity we have created, and we clearly have not yet learned how to plan. Facing this paradox squarely will be necessary before we can proceed with the task of devising a "Third Way."

□ Paradoxes in economics are now signaling the collapse of its traditional models. The most glaring of these anomalies is the Phillips Curve formulation of a supposed trade-off between unemployment and inflation. At the recent London summit meeting, leaders of the industrial democracies at last faced up to the need for new approaches. It is now possible to prove that the Phillips Curve is inoperative and that there are many other sources of inflation beyond wage costs. In fact two new sources of inflation are now best understood from beyond the disciplinary view of economics, as we shall discuss.

The first arises from the unmodelable, unmanageable levels of complexity of our society and the soaring unanticipated social costs it is now generating and which culminate in a metalevel trade-off between specialization and division of labor, on the one hand, and the soaring social costs and general transaction costs of maintaining coordination, on the other. Rather than the much vaunted "post-industrial state" of Daniel Bell, I describe this syndrome in Chapter 5 as "The Entropy State," where the heralded tertiary, knowledge-based, service sector Bell envisions is really nothing more than the growing "social-cost sector."

The second new source of inflation, described in Chapter 9, is rooted in our declining resource base and the worsening population/resource ratio on the planet. We must now cycle ever more capital back into the process of extracting energy and raw materials from ever more degraded and inaccessible resource deposits, with ever declining net yields. The theory of continual substitution is

Fig. 1 Time and System Scale Coordinates for Determining "Efficiency"

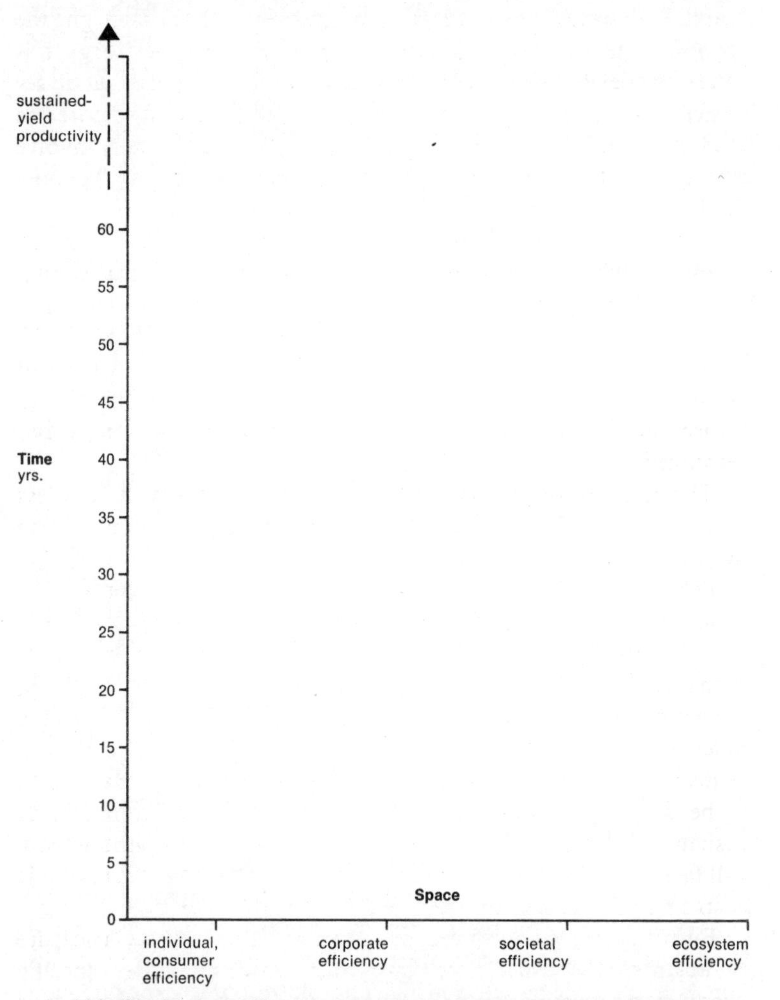

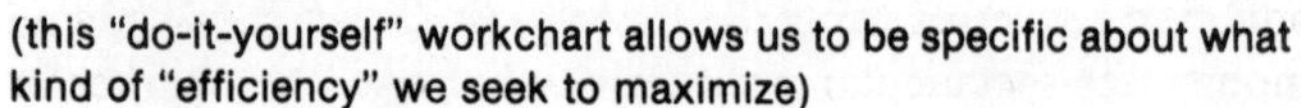
(this "do-it-yourself" workchart allows us to be specific about what kind of "efficiency" we seek to maximize)

over-optimistic and does not deal with simultaneous rates of depletion across a whole range of resources, thus reducing substitution options. This type of declining productivity is beginning to manifest itself as a "capital shortage" and exerts a multiplier effect through the economy. Since it also involves declining productivity of energy, it is better modeled using thermodynamics and net-energy analyses rather than traditional economics. These problems underscore the inadequacy of our measurer of "productivity," which usually involve measuring output per employee-hour, or labor productivity. We now need measures of capital productivity and energy productivity to correct this overemphasis on labor productivity and pervasive drift to excessive capital intensity that it, together with tax credits for capital investments, has created. We must now corroborate with new indicators the overuse of capital and energy that our old statistics and policies still encourage and subsidize, as well as the rising efficiency of labor in hundreds of processes, which is still masked by linear projections of past labor costs relative to past cheap energy and resource inputs.

☐ The paradox of greater micro-efficiency in production, but less social efficiency and individual-consumer efficiency, now is leading to widespread social alienation. This indicates an inadequate modeling of "efficiency" criteria, since efficiency is a meaningless, subjective concept unless time horizons and system levels are specified. Figure 1 is offered as a corrected model of efficiency, where such coordinates are provided to clarify efficiency criteria. Similarly, the term "ephemeralization," or "doing more with less," is vague unless submitted to similar criteria. The essence of the matter in both terms is "Efficiency for whom?" For example, "efficiency" is assumed to be the goal of increases in "productivity," but it cannot be as casually assumed, as it is today, that such increases in productivity will be shared on an average *per capita* basis, nor that the inevitable costs and dislocations incurred will burden us all fairly.

Furthermore, we now need an additional "productivity" measure to augment the usual micro approach which examines specific production processes using the *labor*-productivity measure and thus demonstrates spectacular productivity increases per worker in such *capital-* and *energy-intensive* processes, while overlooking that many workers are shaken out of the bottom and join the ranks of the structurally-unemployed, while *their* productivity falls to below zero, and they show up on the social cost side of the economy as welfare recipients. Another important example of our curiously inaccurate view is that we do not bother to assign economic value to work performed by volunteers or in households, and yet according to economist Scott Burns in his book, *Home, Inc.,* the total amount of work done by men and women in the household, would equal, in monetary terms, the entire amount paid out in wages and salaries by every corporation in the U.S.

☐ The paradox of increasing production and economic growth coexisting with structural unemployment and a significant and stubborn proportion of our population below the poverty line. This paradox relates to the obsolete modeling of the production process as if individual input factors (capital, land, labor) could be specifically related to their proportional share of output, and thus yield an objective formula for distributing the fruits of production. Yet in a technologically complex society, production becomes a similarly complex, social process, where such neat casual relationships of inputs and outputs can no longer be established, and therefore yield no clear formula for fair distribution. Therefore, not only are our models inadequate for analyzing the relative productivity of the various factors of production, but they are no longer useful in determining equitable private distribution, nor in designing public-sector-transfer programs, or in assessing technological development and public works projects by traditional cost/benefit techniques of averaging costs and benefits per capita.

In the intervening decade since structural unemployment and hard-core poverty were first addressed by the President's Commission of Automation, Employment and Economic Progress in 1966, little has been achieved conceptually in remapping our society. In hindsight, we can see two erroneous assumptions made by the Commission: it confirmed that although automation and the drift to capital-intensity did create structural unemployment, it assumed that essentially "perfect" labor markets would redeploy workers with little disruption, and that any workers who remained unemployed would be absorbed by a continually growing economy. Today, we are less sanguine as we try to address the new worldwide disease of "stagflation."

Today's choices are no longer the simple choices of yesteryear. They involve higher technological stakes and graver human risks than ever before. These new trade-offs involve not simple choices between energy options of coal, solar or nuclear, between transportation options of autos and mass transit, or from the usual menu of public and private goods and services. These metalevel trade-offs involve choices between the societal specialization and division of labor *versus* its social and transaction costs; between centralization and decentralization of production and population; between capital and energy-intensity versus labor-intensity with a much more complex reckoning of externalities and societal impacts. Since rationality now dictates that we conserve our scarce and costly capital and natural resources, we must now fully utilize our human resources. We must run our economy on a leaner mixture of capital and a richer mixture of labor. I shall explore all these issues in Part One.

Such a resource-conserving, full-employment, less inflationary economy would, of course, be an environmentally benign economy also. The new choices we are now called upon to make consciously in our own generation, are usually made by other biological species through eons of evolution and genetic changes.

As in genetics, timing is all: if adaptation to change is too rapid, this may only mal-adapt us for the subsequent changes we must face. The imminent paradox is that nothing fails like success. We may have exhausted the evolutionary potential in our GNP-measured industrialization path, and the next adaptation will be in a new dimension for which new measuring rods will be needed. Perhaps now is the time to recognize that the real factors of production are energy, matter and knowledge, and that the output is human beings.

Based on invited testimony before the Joint Economic Committee, 94th Congress, Washington, D.C., Nov. 18, 1976.

CHAPTER THREE

The Finite Pie: The Limits of Traditional Economics in Making Resource Decisions

Today the discipline of economics and its practice as the basic tool used in allocating resources is being challenged on many fronts, by scientists from other disciplines and by an increasingly skeptical public. The current mismanagement of our economy calls into question the basic concepts of neoclassical economics and later Keynesian variations. I shall review the problems encountered in economics, now clearly a subsystem discipline, which has been expanded in a vain attempt to embrace phenomena which its concepts are inadequate to explain. By and large, most economists have tended to ignore those social and environmental variables that do not fit into their theoretical models, such as questions concerning the distribution of wealth and income which is too often accepted as a given, or ways in which the concepts of the "free market" and the "all-knowing, ever-rational consumer" are distorted by the wielding of institutional power, by the manipulation of information, by the speed-up of technological change and by those human needs that lie beyond the marketplace. Economics and its modern tools, such as the cost/benefit analysis, have now begun to obscure social and moral choices and prevent a vital, new, national debate about what is valuable. Today, business cycles themselves are created by economists, rather than the market, as they alternately inflate and deflate the economy. Such aggregate demand management cannot address the structural problems of our complex, mature economy, where only vestiges of such free markets remain.

There are, of course, some economists, notably, Kenneth

Boulding, Kenneth Galbraith, Gunnar Myrdal, Barbara Ward, Robert Heilbroner, Adolph Lowe, Gardner Means and Nicholas Georgescu-Roegen[1] and others, who have kept such questions alive. However, the anomalies economists cannot address are now painfully visible, whether in global inflation, pollution or the unwanted side-effects of economic development, such as social disruption, cancerous urbanization, soaring infrastructure costs, unemployment and mal-distribution of income and wealth. Indeed, many Third World nations now question the advisibility of trying to imitate the capital-intensive development of the West, as typified by Walt W. Rostow in *The Stages of Economic Growth.* Many are now looking to China as a more viable model, because its labor-intensive system uses the human resources that are abundant in all countries, and does not require the surrender of national autonomy, which often becomes the price of foreign capital. The Chinese stress that they do not maximize "efficiency," in Western terms, but rather see it as one goal to be optimised in relation to others, such as decentralized population, domestic production, discouragement of elitism and equalizing income distribution. Obviously this kind of economy, which substitutes exhortation for incentives, and utilizes the energy of its own people in mutual non-mechanized service to each other, is a pragmatic response to the lack of capital to seed economic growth any other way; but it must also result in a resource-conserving, and therefore more environmentally-benign economy than a capital-intensive one.

Much of the new questioning of the goals of economic development has fallen into the re-hashing of the communism versus capitalism dialectics of the last century. The Chinese denounce capitalism as the root of environmental problems. The U.S.S.R. after initially taking the same position, has now acknowledged its own environmental problems and collaborates with the U.S. on the bi-lateral committee now set up to explore solutions to these mutual problems. Many reject dogmatic environmental arguments against capitalism and point to government-directed investments in many centrally planned economies, such as power generation, steel and auto production and many extractive industries which create problems in the same way that they do in capitalistic settings. Furthermore, many less developed countries without noticeably capitalist leanings, proclaim their willingness to capitalize their relatively clean environments in their understandable drive for economic growth. However, the now-famous Founex Report prepared by experts from developing countries for the 1972 U.N. Environment Conference raised the newer issues. "In the past, there has been a tendency to equate the development goal with the more narrowly-conceived objective of economic growth as measured by rises in Gross National Product. It is usually recognized today that high rates of growth do not guarantee the easing of urgent social and human problems. Indeed, in many countries high growth rates have been accompanied by increasing unemployment, rising disparities in income, both between groups and between regions, and the deterioration of social and cultural conditions." In their 1974 book, *Economic Growth and Social Equity in Developing Countries,* economists Irma Adelman and Cynthia Taft Morris reached essentially the same conclusion.

All these new issues challenge prevailing economic policies in most industrial countries. How economists address these issues will determine their future usefulness, and whether the current drift toward irrelevant reductionism in the vain quest for "scientific objectivity" can be reversed, so as to permit integration of the new variables, whether the behavior of oil sheiks, multinationals or ecosystems, into their models.

Let us focus on the priorities by which a nation determines the allocation of its resources. These are a product of many factors: its myths and traditions, its cultural assumptions of "value," its stock of knowledge, its assessments of risks, costs and benefits within various contexts of space and time, the availability of land, material and human resources, as well as the mix of public and private decision mechanisms by which its citizens' needs and priorities are shaped, articulated and implemented with sufficient general satisfaction to contain dissent at manageable proportions. Under such a general description of most nations' systems for allocating resources, is subsumed the relative value weightings between individual autonomy and societal goals, and the various centralized and decentralized configurations of power they produce. Many industrial nations in the West have opted for a greater degree of reliance on market mechanisms of allocation, on the assumption that they optimize individual autonomy while approximating shared societal goals. Other industrial nations have followed the lead of the U.S.S.R. and prefer centralized political mechanisms for resource allocation, on the assumption that overall social goals are optimized which simultaneously approximate individual needs. However, the two largest, most advanced models of these two differing value-systems, the U.S. and the U.S.S.R., are beginning to appear very similar in several of their major contours, for example, in their dedication to ecologically-unassessed growth, technological determinism and their increased dominance by bureaucracies, whether officially designated as "public" or "private."

A brief comparison of the environmental merits of these two major resource allocating systems is necessary because there are increasing convictions among resource economists and thermodynamicists that environmental degradation is an index of an economy's inefficiency in utilizing resources; while many social critics in market-oriented economies contend that overall efficiency and general welfare can be improved by shifting resources from the private to the public sectors of an economy. John Kenneth Galbraith, in his book, *The Affluent Society,* focused widespread attention on the public amenity problems developing in the U.S. through over-reliance on market mechanisms to allocate resources. We now see in many other "overdeveloped" countries, how overheated consumption by an affluent stratum produces the excessive resource consumption, depletion, waste, obsolescence and pollution which Galbraith had described. He pinpointed the role of advertising in overheating such consumption in order to keep expanding the private sector production of goods on which the major reliance for employment had come to rest. Other critics in the 1960s offered solutions to this purchasing power dilemma, such as Robert Theobald, Milton Friedman and James Tobin, who proposed new distribution devices to guarantee minimum incomes to satisfy more basic unmet needs, and to prevent these distortions in production patterns. Theobald accurately predicted that advanced, technological economies would be socially unstable and inflationary, because consumption must be continually increased, while capital-intensive production would require less and less labor input. While many service industries have grown to take up some of the slack, today, unemployment and simultaneous inflation are our two most serious problems; thus invalidating economists' traditional concept known as the Phillips Curve, which postulates a no-longer operative tradeoff between these two curses of mature, industrial economies. The issue of whether a technologically-advanced economy produces both structural unemployment and structural inflation has finally surfaced, after its successful submergence by Keynesians and their policies of general stimulation through tax cuts, easing credit, incentives for capital investment, and retraining programs for "unemployables" in the hope that if skills were increased, jobs would somehow materialize.

Such anomalies must now be vigorously debated, especially since capital itself is now in short supply and many of our most pressing needs lie in the public sector. Market-oriented economies cannot deal effectively with these needs until potential consumers of these public goods and services aggregate themselves politically, and develop sufficient power to shift public funds into underpinning these new "markets" for mass-transit, education, health care, parks, and water-treatment facilities, as well as long-term investments to research and develop non-polluting, renewable energy sources, such as solar and wind power. Not to be overlooked when massive public works projects are proposed to cure our recessions, all these public sector goods, services and investments create vital, rather than make-work jobs. Not only does over-reliance on private production and consumption of material goods unnecessarily waste resources, but it cannot be relied upon as a major source of employment in an advanced economy without other strategies to distribute purchasing power. In addition, Kenneth Boulding has pointed out that economic welfare constitutes *using,* rather than *using up* resources; the enjoyment of the stock of wealth, rather than the throughput of production, consumption and waste. Market economies, with their emphasis on private property rights, encourage such accelerated throughput, because they assume that ownership confers the right to use up, rather than merely use resources.

However, the more centrally-planned economies seem to exhibit similar ranges of environmental problems, not caused by market decisions, but by bureaucratic ignorance or deliberate central decision making that sacrifices the environment to economic goals.

In addition, socialistic economies have other problems uniquely their own, particularly in finding incentives more thrilling than "plan-fulfillment" to substitute for the individual profit motive and reduce the need for costly unpopular bureaucratic regulation. Indeed, in Eastern Europe and the U.S.S.R. we now see the age-old human motive of profit slipping in again through the back door, whether as individual productivity rewards, workers' councils or in the form of royalties in deals with Western corporations. Advanced technological societies, programed by whatever set of economic assumptions, all suffer from bureaucratic giantism, technological determinism, human alienation and environmental degradation. Marxian, socialistic and Western-style utopias all rely heavily on technological abundance, seemingly unconstrained by resource-depletion.

The new convergence in advanced economics of problems of inflation, pollution, resource-depletion together with human alienation, unemployment and maldistribution, is forcing new assessments of our almost subconscious labor-oriented theories of value. Such an anthropocentric emphasis on our own human inputs to value is understandable. All economic activity is human, and it is to be expected that economic policy discussions in democratic societies stress labor's input to the production process relative to the objective role of land, resources and capital in determining value. Indeed, in the early stages of the industrial revolution, the role of these objective factors was limited, compared with the vast amounts of human toil required to produce commodities. Marx went so far as to attribute virtually all value in commodities to the labor factor. Although as technology advanced, economists have assigned increasing weighting to land and capital factors of production, their orientation toward labor inputs to value is illustrated by persistent use of concepts such as "man-hours" and "labor productivity," even though this latter term most often refers to additional *capital* placed at the disposal of the worker.

This emphasis on labor inputs to value, even in advanced, capital-intensive economies, became politically-necessary to mask the fact that jobs were becoming a distribution device of major proportions. For example, in their current plight many industries use as a rationale for federal assistance, not their primary function as supplying needed goods, but that of providing jobs. If we were to acknowledge that in

Fig. 2 Energy Flow Through the Biosphere

Transition report, 1975, Office of Energy Research and Planning, Governor's Office, State of Oregon

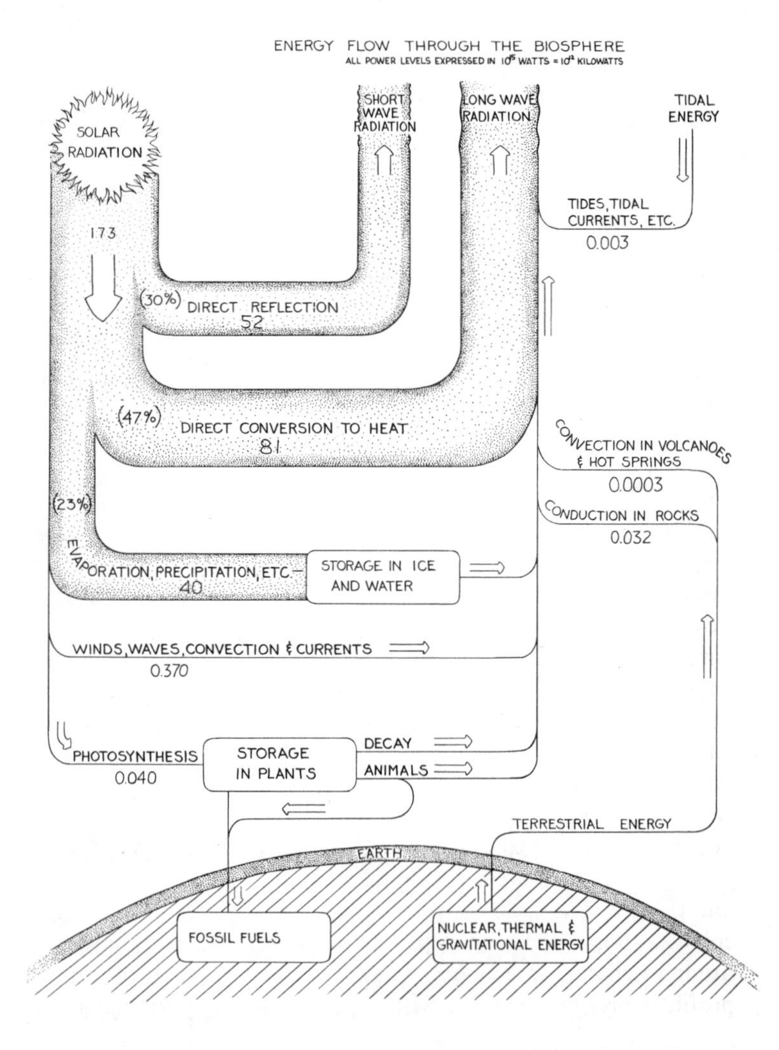

many highly-automated industries capital creates wealth unattended by anything more human than a humanly-programmed computer, we would also have to deal squarely with the need to create institutions for distributing wealth, so that the increasing welter of goods can be consumed by those who still have unsatiated needs for them. This in turn would undermine many current assumptions in market economies concerning property rights, and that only work or contributions to production entitle the right to an income (except in cases of capital-ownership, age or disability). Furthermore, our emphasis on labor inputs still shortchanges nature's contribution to production at a time when natural resources are becoming scarcer in relation to human populations. Therefore we must not only reverse our former notions of "efficiency," but also abandon attempts to neatly quantify the relative inputs to production provided by labor, land, capital and knowledge and recognize the increasingly social nature of production in advanced economies.

Since the planet's resources are finite and its processes are bound by the laws of physics, the 1st Law of Conservation, which states that matter can neither be created or destroyed, and the 2nd Law, the Entropy Law of gradual disordering and decay, the basic requirements of economies operating as subsystems within it must eventually be "steady-state" economies, with constantly maintained stocks of people and physical resources. If economic growth of material wealth must be constrained at some point in time, however distant, then human development must find another dimension. Luckily, knowledge development and, hopefully, wisdom is unfettered by the dismal laws of physics and is still wide open for evolutionary progress. A steady-state economy can no longer rely on employment in the production of energy and resource intensive goods as its major distribution device, but must gear its production and distribution strategies to a sustained-yield system based on renewable resources. Its theories of value must embrace the subjective, changing goals of people, the role of information and human knowledge and the limits of the physical resources of the planet and its daily energy income from the sun. The issues raised by the Club of Rome concerning the ecological and psychological limits to growth will require a major paradigm change in economics, as we reexamine such concepts as "profit," "productivity," "efficiency," "utility," "maximizing" and "progress." None of these concepts has any meaning unless the frame of reference is made clear, and boundaries in space and time horizons clearly specified. We must know the answers to such questions as "profit for whom?"; "efficiency at what system level?"; "maximizing in what time frame?", for such terms to be precise, and to avoid the multiple crises of suboptimization that their fuzzy use by economists, politicians and businessmen has unwittingly created. (See Fig. 1)

In its dedication to scantily defined "progress," we now see that the Keynesian enterprise of pumping up whole economies to ameliorate structural pockets of unemployment and mask distributional inequities, has now become too costly in raising rates of both inflation and resource depletion. The easy assumptions that an ever-expanding pie would provide increasing portions to the poor, no longer offers the comforting rationale whereby the world's affluent justify inequities as essential to the formation of new capital for investment. Economists and businessmen with intellectual and financial investments in the growth syndrome, can no longer defend it on the grounds that it is the only way to improving the lot of the poor and providing the "resources" to clean up the environment. There is now too much evidence that growth does not often trickle down to the poor in the prescribed Keynesian manner and using our current form of flawed, excessively polluting production to create the "resources" to clean up its results leaves us with a trade-off. Yet businessmen without prior noticeable commitment to the poor, suddenly display hearts newly bleeding with concern for them. Their crocodile tears at the prospect of the dispossessed being denied hopes for increasing private consumption, to which they must aspire if private-sector prerogatives are to be preserved, are new "red-herring issues" to obscure the need for reassessment of the *nature* and *direction* of growth. The new growth debate is uncovering all the value assumptions it has relied on, and forcing us to examine whether growth of consumption in the private sector, however harmful its neighborhood effects, is the only form of growth. Of course, we are obliged to admit that it is not, and that growth could be channeled into the many public service areas of our economy mentioned previously: mass transit, health care, education and research into new energy-conversion system and recycling with minimal environmental impact. But such a consciously controlled readjustment would require internalizing the social costs of private production and consumption, diverting private resources through taxation, prioritizing investment and allocating credit; measures which businessmen and many capital-owning citizens still vehemently oppose.

Indeed, we must ask whether in an age of increasing complexity, without vastly more information between buyers and sellers, the simple aggregation of micro-decisions in the market adds up to anything more than the macro-chaos described by biologist Garrett Hardin in his now-famous treatise, *The Tragedy of the Commons* (see p. 76). Problems of commonly owned "free goods" such as air, water and oceans, where everybody's business becomes nobody's business, are some of the knottiest theoretical questions of how we are to make social choices in the areas where market choices fail. Herman Daly addressed the dilemma in his 1974 book, *Toward A Steady State Economy,* and states that for a society to achieve a political economy of biophysical equilibrium and nonmaterial, moral growth will require radical institutional changes and a paradigm shift in economic theory. Daly suggests that three institutions are needed for a steady-state economy with constant stocks of people and capital maintained at a low rate of throughput; aimed at providing macro-stability while allowing for micro-variability, to combine the macro-static with the micro-dynamic. Daly endorses Boulding's earlier plan for issuing each individual at birth a license to have as many children as corresponds to the rate of replacement fertility. The licenses could then be bought and sold on the free market. Secondly, he argues for transferable resource-depletion quotas, based on estimates of reserves and the state of technology, to be auctioned off annually by government, and thirdly, a distributive institution limiting the degree of inequality in wealth and income.

Somber proposals such as Daly's may be considered impractical, or "social engineering," and yet the concepts of the "steady-state economists" are beginning to gain a hearing. Most favor theories of value based on entropy, such as Boulding, who states in his essay "The Economics of the Coming Spaceship Earth" that the economic process consists of segregating entropy, where increasingly improbable structures of low relative entropy are created at the expense of

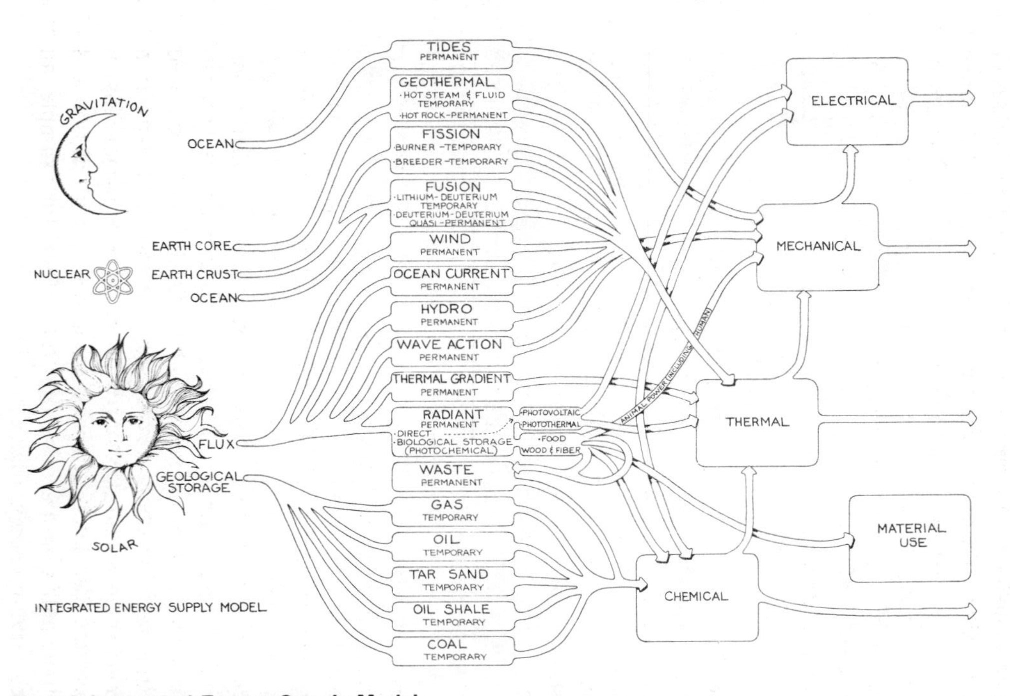

Fig. 3 Integrated Energy Supply Model.

Transition report, 1975, Office of Energy Research and Planning, Governor's Office, State of Oregon

higher entropy level wastes somewhere else. Nicholas Georgescu-Roegen in *The Entropy Law and the Economic Process* traces entropy theories of economics back to German physicist, G. Helm, who in 1887 argued that money constitutes the economic equivalent of low entropy. Georgescu-Roegen pierces the fallacy that economic processes are analogous to the mechanical Newtonian processes of locomotion. Because economic processes also produce qualitative changes, usually associated with higher entropy levels, he believes that they also elude "arithmomorphic schematization" and, therefore, economics with its "arithmomania" ignores them. Basically, the problem is that although resources (matter) may be recycled, it can only be done with inputs of energy, and energy use not only creates inevitable loss (generally heat), but it cannot be recycled. For instance, in most advanced countries, services are becoming major constituents of their economies, including communications (which often replaces the need for more energy-intensive transportation), movies, TV, insurance, health care, education and research, whether performed in the public or private sectors. Even though these services are less entropic than heavy industries, we cannot forget that they rest on a base of extraction and production which pollutes and depletes resources, although they share the chameleon quality of appearing to be environmentally benign at the point of delivery. Even pollution control and recycling services, such as electrostatic precipitators and waste-water treatment processes, use a good deal of energy and resources in operation and manufacturing. In fact Georgescu-Roegen states flatly that all economic processes use up a greater amount of low entropy than is represented by the low entropy resulting in the finished product, and that in entropy terms most recycling is equally fruitless. This is why he and the other "steady-state" economists stress that the real payoffs are in *durability,* which reduces this unnecessary flow of production—consumption—waste—recycling to the lowest level achievable. Therefore, we need very careful simulations of entire economic processes from extraction to refining, to manufacture, to consumption, to waste, to recycling, in order to assess their relative efficiencies in resource utilization and concomitant pollution and depletion rates. (See, for example, an integrated view of energy, Fig. 3.)

Georgescu-Roegen's entropy theory of value cites as separate, additional factors of production natural chemical processes, rainfall and solar radiation, which are usually subsumed under the factor of land, as free gifts of nature. Since some would view this as double counting, he adds that land, far from being inert, as in Ricardo's definition, is an agent of production in that it contains the chemical processes, catches the rainfall and the solar radiation, which is the only income, or fund source of energy available for the performance of all planetary processes from photosynthesis (the most basic and vital) to our economic activities. (See Fig. 2.) The energy "capital" stored in the earth's crust as fossil fuels is a rapidly depleting stock of fossilized solar energy collected in the past by photosynthesis which is nonrenewable. The chief difference in the process of agriculture as opposed to the process of industry is that traditional agriculture must rely on utilizing the unchanging rate of flow of solar energy, while industry can mine the stocks of stored energy in the earth's crust, at least while they last, at its own determined rates. Georgescu-Roegen's book analyses many current input-output models of economic processes in light of his entropy theories and cites the omission in all such dynamic models of the representation of production *processes,* rather than merely the production of commodities, as well as other critiques. His theory further challenges the assumption that the increase in "labor productivity" resulting from capital input is only limited by economic costs of additional mechanization and depreciation, rather than any ultimate limits of how much matter/energy nature can put at our disposal. Such inadequacies of economics give credence to self-defeating strategies, such as that proposed by Henry Kissinger, to place a floor under oil prices to make it "profitable" to develop shale, tar sands and coal liquefaction, in spite of their dismal payoff in real net energy terms. Fig. 4 illustrates the brief Fossil Fuel Age in the context of human history.

A shift toward entropy theories of value would require that "profit" be redefined to mean only the creation of real wealth, rather than referring to private or public gain which excessively discounts the future, or is won at the expense of social or environmental exploitation. Similarly, we would recognize that the concept of maximizing profit or utility is imprecise until qualified by a time dimension. Such realistic profits would include improvements in energy-conversion ratios and better resource management, and recycling geared to using the solar energy income available in nature's processes rather than further depleting energy "capital" in the earth's crust. As more externalities are included in the price of products, we may find that many consumer items' profitability will evaporate and these goods will disappear from the market. This is already happening as manufacturers such as Alcoa discontinue production of aluminum foil, and other goods requiring large inputs of energy/matter, such as high-powered cars, are being replaced by smaller models and the new boom in bicycles.

Or take the question of the unalloyed desirability of capital investment itself, which is used to justify much inequality of distribution. Under what circumstances are capital investments socially and environmentally destructive; and since we must and will continue our economic activities, how can we reduce their resource-depletion rates and restrain the often arbitrary and irrational investments of increasingly scarce capital. Economists, hypnotized by their elegant equilibrium model of free market supply and demand, cannot readily handle the possibilities of absolute scarcity on the supply side. We must also question the concept of "productivity," another value-laden term, which economists seek to "maximize" by raising the level of capital invested in each working person or the machines they use. Raising agricultural "productivity," for example, by mechanization and application of fertilizers and pesticides can often produce social costs, such as the income inequities engendered by the "green revolution," and environmental costs in breeding insecticide-resistant pests, runoffs of fertilizer-polluted water, destroying more stable and resilient forms of agriculture and rapid soil depletion. There are also some limits to investments in machinery and automation beyond which workers rebel at the increasing robotization of their jobs and begin sabotaging the production process, as has occurred in plants in the U.S. Many useful and profitable functions cannot use much capital investment, such as private tutoring, or producing works of art or custom, hand-crafted goods; and they provide workers with psychic pleasure often envied by workers in capital-intensive industries. Economist E.F.

Schumacher, in *Small is Beautiful*, points out the culture-bound nature of economics in his chapter on Buddhist economics, which, based on the concept of "right livelihood," would define labor as an *output* of production rather than an input, and valuable for its own sake. Schumacher also stresses the need for intermediate, labor-intensive technology to meet developing countries' requirements for rural employment, decentralization and political stability, substituting the Western economists' dedication to market value with the concept of use value.

All this suggests the extent to which economic theories have fallen behind the welter of changes wrought by technological innovation. All these new issues lead to a reexamination of human cultural notions of "value." For example, we in the U.S. tend to overvalue and overreward competitive activities, which can only exist within an equivalent field of cooperation and social cohesion. At the same time, we undervalue all these cooperative activities which hold the society together, such as child nurture and the vast array of services lovingly performed in the voluntary sector, and for the provision of which women bear an unfair burden of the opportunity costs.

Therefore, in the last analysis, we must zero in on the normative nature of economics and how economists' often subconscious value assumptions weight their analyses. Economics also attempts to deal with humans' subjective perceptions of value as well as the objective realities concerning the actual values of the complex matter/energy exchanges which maintain the viability of our global habitat. Kenneth Boulding and Barbara Ward were among the first to perceive that Spaceship Earth and its natural cycles powered by the sun, contain information on the values of these matter and energy exchanges in the biosphere, and that economics must repair to the physical and biological sciences to obtain this essential baseline data for the accuracy of its own models. Unfortunately, human perceptions of value, i.e., prices, with which economists deal, are notoriously inaccurate because they are based on (1) our subjective, imperfect observations of the objective world and our resulting unrealistic expectations of the availability of its resources, and (2) our subjective evaluation of what is important to us, or "valuable." If our assessments of value are either arbitrary, or erroneous, as they usually are, then our primary tool for studying their relative exchange values: economics, must be similarly flawed. Indeed, if prices reflected accurately the true survival values of humans, then why would tobacco be expensive while air is more than merely cheap, but is actually free? The arbitrary nature of human expectations is familiar to all who have studied the behavior of stock exchange prices. In addition, there are often serious lag times between the reports of scientists on, for example, increasing pollution-related eutrophication of lakes or acid rainfall, and the incorporation of such data into economists' reports to bankers and investors or policy makers, on how they may affect prices.

However, prices still have much useful potential for allocating resources in all situations where buyers and sellers still meet each other with equal power, and have faster information on true costs, so that lags in response and price correction are reduced. As Gunnar Myrdal has stated, "We can begin to fill that empty box in our diagrams marked 'externalities,'[2] so as to calculate as far as possible the social costs of production so that they too can be accurately reflected in prices. In this way more accurate pricing can still function as an alternative to bureaucracy." In the same vein, Myrdal contends that organized citizens and consumers can function as a countervailing check on the power of public and private institutions, as is evidenced in the U.S. by the rise of the movements for consumer and environmental protection and the direct confrontation of corporations by boycotts, the use of proxy machinery, and the politicizing of company annual meetings and institutional investment policies. Many externalities can be calculated or reasonably approximated, so as to bring us closer to determining true value added, rather than immediate but evanescent gains won only at the expense of social and environmental exploitation. Such improved calculations of what market economies call "profit" and state-directed economies call "economic growth," would vastly improve all resource-allocation decisions. But in market economies particularly, the quantification of these externalities has been shortchanged or overlooked, because the majority of economists are employed by private interest groups or the empire-building public agencies that often cater to them, for the purpose of preparing biased and sometimes blatantly fraudulent cost/benefit analyses in advocacy of their profit-making or bureaucratic-aggrandizing projects. Even academic economists in both capitalistic and socialistic economies tend to be influenced by the prevailing political pressures and cultural assumptions of their societies. They also ignore the indisputable fact that the Fossil Fuel Age, powering all industrial economies, is coming to an end. Therefore many economic analyses suffer from unacknowledged biases, and overestimate immediate benefits, while underestimating more elusive social and environmental costs, whose impact may be borne by the society in general or a group within it, another nation, or succeeding generations. We need to enrich the public debate by critiquing the often frankly promotional cost/benefit analyses used to promote both public and private projects. Costs and benefits are usually averaged out per capita, which conceals who will bear the costs, in perhaps neighborhood despoliation or loss of jobs, and who will reap the benefits; the contracts, bond-issue business, profits and new jobs. The Public Interest Economics Foundation has a roster of some 500 volunteer economists willing to perform such economic analyses for groups who could not otherwise afford economic expertise to buttress their case, either in courts or legislatures, such as citizens groups working for environmental protection, social justice or other volunteer causes. This new branch of "public interest economics," is analogous to similar movements that have been established in law and the sciences, as well as in the accounting profession, which recently set up its own National Association of Accountants in Public Interest.

In some cases, the mere collection of data and its dissemination in the most effective channels can create pressure for change. New York's Council on Economic Priorities, for example, has broadened the traditional concepts of security analysis to cover the social and environmental performance of corporations. The Council's reports and in-depth studies count among subscribers a growing number of brokerage houses, banks, mutual funds and other institutional investors, as well as socially concerned stockholders and citizens. It publishes comparative information on the social impact of corporations in various industries in the area of environment, minority rights, military contracting, consumer

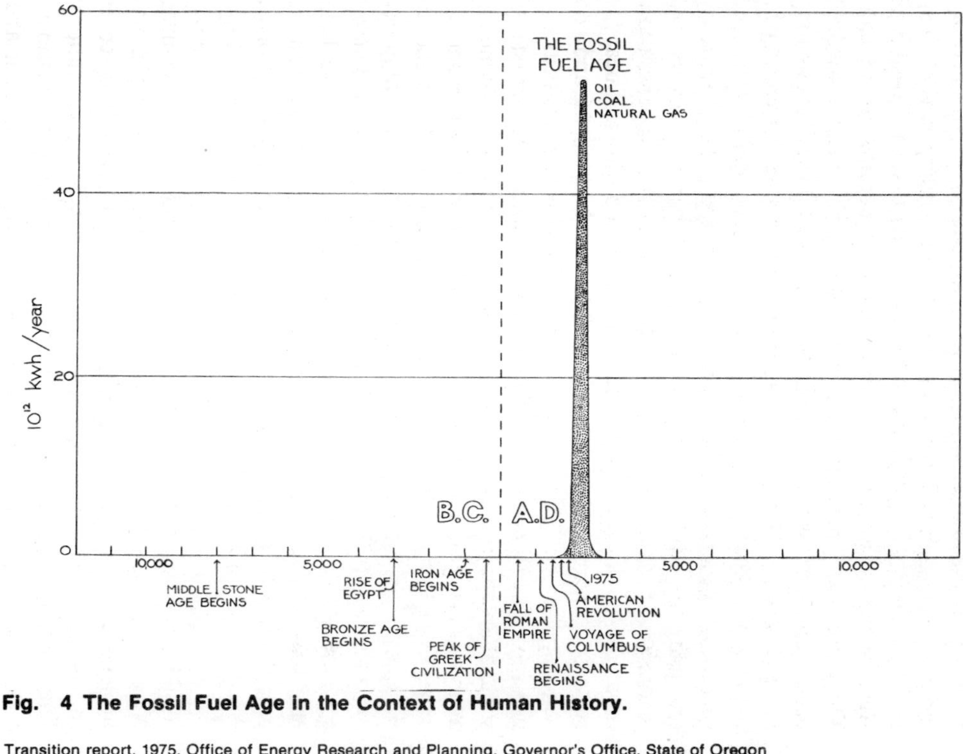

Fig. 4 The Fossil Fuel Age in the Context of Human History.

Transition report, 1975, Office of Energy Research and Planning, Governor's Office, State of Oregon (after M. King Hubbert)

protection, political influence and foreign investments. The growing political power of these multinational corporations which now threatens national sovereignty and world monetary stability, confirms the need for this type of analysis. In addition, there are now enough U.S. investors to provide a market for these reports, as stockholders see the desirability of having portfolios that do not contradict their personal values. In response to these new stockholder pressures on their members' clients, the American Institute of Certified Public Accountants is attempting to develop social auditing methods for corporations. One fruitful avenue growing out of their own experience would seem to be that of expanding the familiar concept of "goodwill," which however unquantifiable, is routinely capitalized on hundreds of company balance sheets. It should also be possible to refine calculations of short- and long-term profit so as to elucidate the time dimensions which always qualify maximizing behavior.

Much new and useful work on modelling externalities is now in progress, by such economists as Wassily Leontief, and those working at Resources for the Future, including Allen V. Kneese, Talbot Page and John Krutilla, as well as Charles Cicchetti of the University of Wisconsin. Hirofumi Uzawa of Tokyo University advocates an annual deduction from GNP analogous to the capital consumption adjustment that now distinguishes Gross National Product from Net National Product. The new deduction allows for the depletion of natural resources: the consumption of the irreplaceable original capital of the planet. On the assumption that industrialized nations are exhausting resources more rapidly than nature can renew them, each year Uzawa's deductions will increase. In the U.S., Thomas Juster sets forth a more realistic set of criteria for restructuring our own GNP, which include in the assets: knowledge, skills and talents, physical environment and socio-political assets, which appears in the 50th Annual Report of the National Bureau of Economic Research. Resource economists, including Allen V. Kneese, argue for effluent and emission taxes as the *most* efficient way to control pollution through the market mechanism. Yet transaction costs also occur, and effluent taxes are more likely to be decided by the political power of corporate lobbying than the objective market. Neither can such taxes deal with toxic substances which must be prohibited, or irreversible changes. Similarly, the subsidy method also discounts true social costs of pollution, particularly the new pollution-control bonds, which are tax-exempt to encourage corporate spending on environmental improvement; but are proving to be little more than another tax loophole.

Efforts to simulate nature's closed-loop energy cycles are described by Howard T. Odum in *Environment, Power and Society.*[3] Odum's "value-system," calculated and converted from kilocalories to dollars, enables a cost/benefit analysis to credit the chemical exchange work performed by a host ecosystem of a proposed economic activity at the same rate that humans would have been paid for comparable work. This invisible and unaccounted activity performed by natural systems includes, for example, absorbing carbon dioxide from combustion and replacing oxygen that all such processes use, or converting industrial wastes and sewage back into fuel or fertilizers. Until such ecosystem activities are included as costs of production, environmental activists will be bargaining from weakness. Policy proposals growing out of such work as Odum's include such new devices as an amortization tax, as proposed by thirty-three British scientists in the now-famous "Blueprint for Survival" published in *The Ecologist* in January, 1972. The amortization tax would penalize throwaway goods and obsolescent products, while encouraging with the least tax those items most durable.

One study in Illinois concerns the relative costs in total energy of refilling returnable beverage bottles versus the collection, destruction and refabrication of throwaways. Findings confirmed fears that, ecologically-speaking, recycling centers are little more than public relations tools. The study by Bruce Hannon (*Environment,* March, 1972) found that throwaway bottles consume 3.11 times the energy of returnables and that, in the State of Illinois, a complete conversion back to returnables would also save consumers some $71 million annually.

Similarly, a consulting firm in Florida prepares total-energy cost/benefit analyses for its clients on the relative merits of different methods of heating and cooling buildings. For each system, whether using gas, electricity or oil, the firm estimates the relative quantities of sulfur oxides, nitrogen oxides and particulates discharged to the environment. After reviewing such three-dimensional cost/benefit data, the State's school system became more interested in cross-ventilation and increased tree planting than in air conditioning. Another consulting firm in Germany has developed a decision model for use in determining the best mix of fuels to supply an urban area, taking into account topography, meteorology and sources of energy, which incorporates similar environmental criteria.

Still another very useful analysis by R. Stephen Berry, published in the *Bulletin of the Atomic Scientists*, evaluates the processes in the production/scrap cycle of automobiles to pinpoint hidden energy subsidies. Berry estimates that the largest energy and thermodynamic-potential savings can be achieved in basic methods of metal recovery and fabrication which could, in principle, reduce the thermodynamic costs of autos by factors of five or ten or more. By comparison, extending the life of the vehicle could realize thermodynamic savings of 50 to 100 per cent, whereas recycling can achieve a savings of merely 10 per cent.

In fact, it is becoming increasingly clear that the close correlation between standard of living levels as measured by GNP and per capita energy consumption need to be reassessed. A.B. Makhijani and A.J. Lichtenberg contend (*Environment,* June, 1972) that although the 1964 U.S. Government study, *Energy Research and Development and National Progress,* does show such correlations between GNP and commercial energy consumption, it also shows that eight industrial countries with similar standards of living (indicated by GNPs within 10 per cent of each other), the United Kingdom, Australia, Germany, Denmark, Norway, France, Belgium and New Zealand, showed large disparities in energy consumption. Consumption for industry, commerce and transportation ranged from New Zealand only consuming 45 million BTUs per capita, while the United Kingdom at the upper level, consumed 110 million BTUs per capita. Obviously, a large portion of the differential can be accounted for by exports, but the disparity was striking enough to raise questions about relative energy-conversion efficiencies. The two electrical engineers then calculated the total energy inputs for dozens of primary extraction and manufacturing processes and the energy content of the finished consumer goods, and identified some areas where energy consumption could be minimized and overall energy conversion efficiencies improved. For example, they claim that utilizing waste heat from generation of electricity could realize thermal efficiencies of approximately 75 to 85 per cent, as opposed to the 40 percent efficiencies of current fossil-fueled and nuclear-fission plants. They also estimate that if the average weight of cars in the U.S. could be reduced by one-third and their fuel consumption reduced by one-third, and if some 30 per cent of auto mileage could be shifted to public transit, the nation's total energy consumption for ground transportation could be almost halved. By employing the best mix of energy conservation methods, it is clear that an advanced economy would be able to reduce overall energy consumption without reducing its standard of living.

Since the resources to fuel energy-wasting industrial economies generally come from less-developed countries, their stake in energy-conservation methods is doubly vital. We are much aware that the current energy squeeze in the U.S. has produced a financial bonanza in the oil-producing nations of the Mideast. These oil-rich nations are now tending to reinvest their income in the U.S., thus making the U.S. a "pollution haven" for foreign investments.

But if economics is to develop even more precise tools to assess the trade-offs in resource-allocations, it will need to incorporate much of the new data being developed by the physical sciences, concerning those actual values in the macrobiosystem of nature's chemical exchange work, which maintains global equilibrium conditions for humans. Herman Daly makes an interesting analogy between economies and ecosystems: young ecosystems tend, like young economies, to maximize production. Mature ecosystems, like mature economies are characterized by high maintenance efficiencies. From such insights came Daly's proposal for yearly depletion quotas to be auctioned off by government, which he claims are superior as a basic strategy for resource utilization efficiency than effluent taxes, which he sees as a fine-tuning tactic which only addresses itself to pollution control, rather than the primary issue of depletion.

Odum has pioneered energy modeling, a quantitative method of

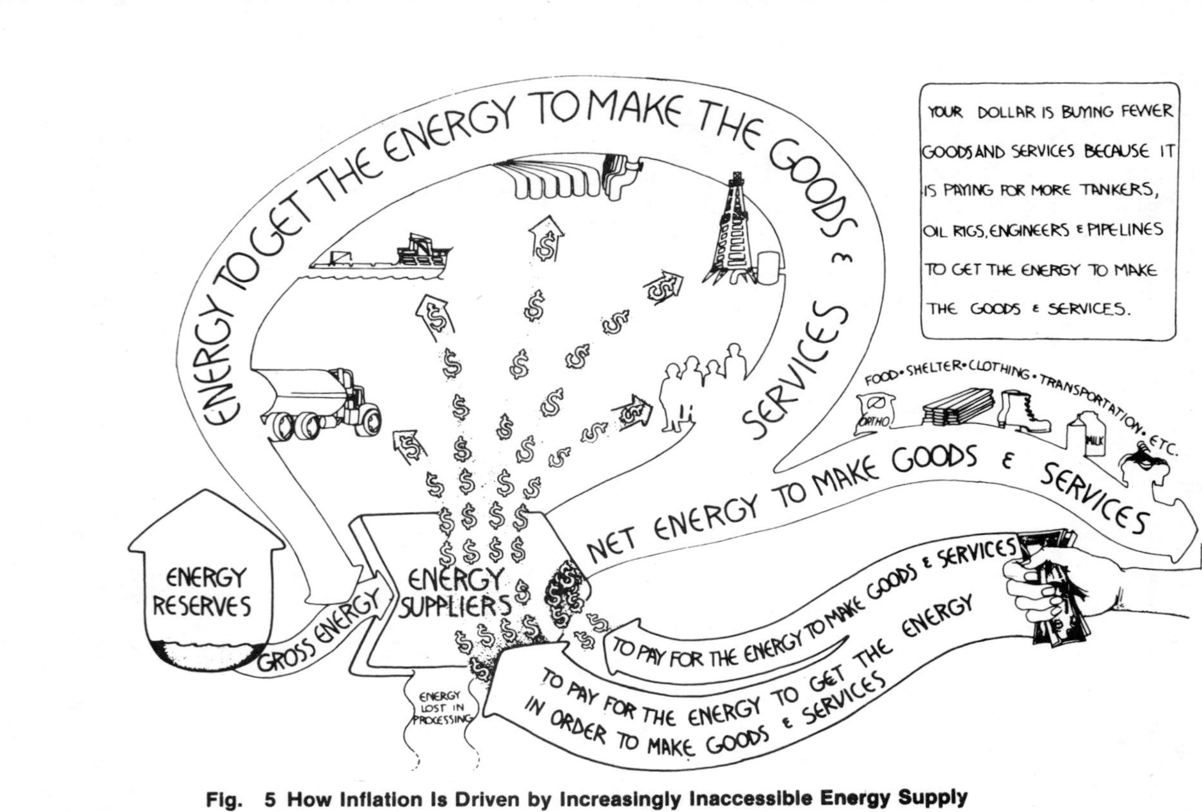

Fig. 5 How Inflation Is Driven by Increasingly Inaccessible Energy Supply

Office of the Governor, Oregon, with permission of Co-Evolution Quarterly

tracking nature's flows of energy and matter, which is fast becoming more predictive than economics. Odum's system converts kilocalories into dollars so that economists can see and account for such work performed by natural systems in their traditional cost-benefit analyses, for example, in converting carbon dioxide from combustion back into oxygen or converting industrial wastes and sewage into fuel and fertilizers. As inflation renders money an even less precise measuring rod of true efficiency, Odum's method of measuring efficiencies of production and extraction processes in the terms of "net energy" is gaining wide acceptance. Odum views inflation as the symptom of a society with a declining energy and resource base, forced to extract energy and raw materials from more inaccessible and degraded deposits. Since it takes more and more energy to extract this energy and material, more real wealth must be diverted from the purchase of goods and services. But the money supply is increased as if all this activity were productive, so the diminishing returns to all this energy-getting capital investment are expressed in the degradation of the currency, i.e. rising prices. (See Fig. 5.)

Energy-modeling is being conducted in scores of countries and by imaginative engineers, thermodynamicists and physicists, such as Stephen Berry and Thomas V. Long at the University of Chicago, Bruce Hannon at the University of Illinois and Malcolm Slesser at the University of Strathclyde, Scotland. In spite of many unresolved problems of taxonomy and differences of method, it appears to be an order of magnitude better than economics in plotting resource utilization and management processes. In 1974, the International Federation of Institutes for Advanced Study in Stockholm convened energy modelers from all over the world to map out their research agenda and agree on their terms. Other conceptual problems still faced are outlined in my letter of Oct. 30, 1974, to Odum (*Co-Evolution Quarterly,* Winter, 1974), but meanwhile, it may be the best new analytical tool at hand.

Dear Dr. Odum:

The recent energy workshop was a stimulating experience for me and I wanted to set down some thoughts for you as a result, which may be of some use as the work proceeds.

Energy accounting is an order of magnitude better than economic accounting and it should be pressed onto economists with great vigor, and I shall continue that task as best I am able.

However, energy accounting contains limitations shared by all quantitative methodologies. It cannot explain *all* phenomena and I would hope that you would not press it too far so that it becomes suspect as a new "cosmology." It is so deterministic and therefore may obscure truth in other dimensions, for example, human ethical responsibility for our decisions and actions. This is why I reacted so strongly to the unnecessary overreliance on Lotka's Principle.[4] It is so deterministic as to be almost tautological: "organisms survive because they survive." Worse, you are assuming a heavy ethical responsibility of propagating Lotka's Principle by which any and all human behavior may be justified, whether that of Hitler or our "petroleum hawks," who are dying for a good rationalization to go in and beat up the Arabs. In fact, "Social Lotkaism" could be considerably more destructive than Social Darwinism, because it would be seen as more modern, and therefore more scientific, because it would also have the blessings of ecologists. History is littered with the wreckage and fallout from powerful, simplifying ideas, from Adam Smith's damned "invisible hand," Keynes' pump-priming and macroeconomic management sophistries and genetic theories which led to German attempts to breed a "master race." And now we see the latest flirtation with the idea of triage, which hides among its grains of truth a neat way to rationalize our own greed and the extent to which we helped create the problem of overpopulation and Third World poverty.

One can even use Lotka's Principle to refute your own efforts to change our behavior and alter our national decisions by promoting the use of energy analysis: The system itself is giving us the signals, as you yourself maintain; it is telling Henry Kissinger what to say, etc., etc. So why do we need you? Inflation itself is the system's prescription for lowering lifestyles. Why do you think we can tinker with it any better, rather than relying on the corrective path of least resistance that the system has chosen, i.e., inflation? One could say that this is also Lotka at work and your efforts at intervention are as meaningless as Henry Kissinger's. Since Lotka raises all these misunderstandings, why not just soft-pedal him? The models are beautiful and don't need this as a crutch.

Now to the methodology itself: I hope that you will be able to avoid the reductionist trap inherent in the whole business of doing research under contract. I know it's unfair to hold you to a higher standard than anyone else—but I have such high hopes for energetics. There is always the danger of doing just what the "client" asks and is willing to pay for. This often leads to the fatal flaw of accepting the client's definition of the problem. In a very real sense, reality is what we pay attention to, and unless a researcher assumes the responsibility of examining the contractor's assumptions and problem definition, he may merely end up confirming the client's bias. We are now suffering from the crises of multiple suboptimization precisely because research usually follows this path of least resistance, and data is only amassed in accordance with prevailing assumptions. (The familiar story; "I wasn't paid to study that—I was paid to study this," etc.) This is why we have so little economic data on even the easily quantifiable social and environmental costs—we never *paid* anyone to collect it. I pointed out this kind of danger at the workshop in the presentation on the relative efficiency of cooling towers v. using an estuary for receiving the waste heated water. There were obviously many other options which might have been explored: e.g. other uses of waste heat, the possible cumulative effects of additional power plants in the estuarine area, and most of all, *whether* the power plant should have been built in the first place and whether reducing Florida's electrical demand curve might not have been an alternative option. In such cases (I am also trying to develop this thinking at OTA) one should go back to the client and say that other significant alternatives exist which must be explored and compared in order to arrive at a true comparison of costs, risks and benefits, and either request a larger budget to accomplish this, or refuse the contract (very hard, I know!)

Cases of Conflicting Values: for example, we might both agree that there are seldom any net energy reasons for constructing add-on pollution control equipment onto inherently inefficient, polluting processes. However, even if this is the case, say with respect to adding particulate control devices to urban power plant stacks, where at the same time large populations are at risk and the health costs and absenteeism costs can be approximated, who is to choose the trade-off between health risks and costs v. net energy efficiency? The researcher? the city voters? the doctors? Again, there is a heavy ethical responsibility for the researcher to not make assumptions that the net energy values are to be placed higher than the medical and absenteeism costs and the health values of the people involved. The best the researcher can do is to point out that there is a policy choice to be made and that while the data on the net energy efficiency may be solid, the data on medical and absenteeism costs may be sparse because we have never bothered to pay people to collect it and traditional economics has a bias against even noticing "externalities" such as these social costs. All this while agreeing with you that the whole urban structure is itself parasitic and probably unsustainable—we still have to consider the people who are trapped in this structure—no less valuable human beings than we are.

This brings me to a substantive problem in the methodology you use in energetics: the danger of too readily converting dollars (market prices, i.e., anthropocentric expectations of availability) back into kilocalories, and thus infecting the analyses with their inherent distortions. It is fine to translate kilocalories into dollars for the benefit of economists and decision-makers—

but you *cannot reverse the conversion* without incorporating all the errors of the money system. It is the kilocalories that are real, *not* the dollars. This is a new replay of Gresham's Law; bad money drives out good money! For example, when you incorporate in your models kilocalorie values for information merely based on conversions from dollar values set by the notoriously inaccurate value placed on information by the market pricing system, you incorporate the errors of a generation of economists who have *never* understood how to price information correctly. Information is undervalued because it can too often be treated as a free good—being the society's investment in its stock of knowledge, to which the entrepreneur did not contribute, and from whom society does not even exact maintenance costs! In some cases, the entrepreneur can even degrade the knowledge stock with powerfully amplified, *dis*-organizing information, i.e., advertising, which is now costing us dearly in lost adaptability and options.

You need to set up a scale, similar to your Fossil Fuel Work Equivalents (FFWE), to measure the quality of information. Also we must somehow model the rate of obsolescence/depreciation of information in relation to the speed of social change. I suspect that *mis*-information is now giving us more trouble than anything! Information can no more be valued homogeneously than can solar, coal or electrical energy. The scale runs from dis-ordering mis-information and its depressingly lengthy half-life all the way up to what has been called the "wit" (rather than the bit).

Scale of Information Quality from a Thermodynamic View

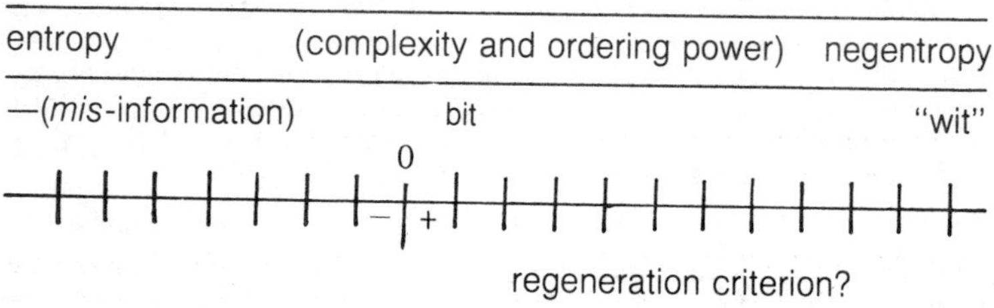

entropy (complexity and ordering power) negentropy

—(*mis*-information) bit "wit"

0

− +

regeneration criterion?

This scheme is something I have been playing with. The regeneration criterion seems to be one way of defining information quality: below a certain level of complexity it can regenerate structure; above that complexity it cannot. (The trade-off, however, of the higher ordering information is that it has lost flexibility/adaptability).

I have discussed the problem of valuing information accurately with my associate Ira Einhorn (physics), who has been thinking about it for years. He believes that the thermodynamic view of information value is a "rear-view mirror" approach and that information also exists in another dimension not yet explained by the existing laws of physics, e.g., we know that information programs and directs energy, but we do not know how to represent this process as an equation yet. Others working on this include Gregory Bateson and Stafford Beer (in Britain). I shall continue with my own amateur efforts! Again, I enjoyed the Workshop immensely and these comments are sent to you with respect and affection. Keep up the good work!

Hazel Henderson

All analytical tools and reductionist methods all suffer from the problem of narrow focus, which results in loss of the "big picture." They cannot reveal truth which exists in other dimensions. Welfare formulas for humans cannot be derived from data, but only from our own expanded perceptions of our true interdependent situation as a species marooned together on this small planet and our own striving for wisdom and ethical principles. We now turn to these issues of human welfare and interdependence and review the debate engendered by these new perceptions of resource limits.

---

[1]Nicholas Georgescu-Roegen, *The Entropy Law and the Economic Process,* Harvard Univ. Press, Cambridge, Mass., 1971

[2]Gunnar Myrdal, Distinguished Lecturer, United Nations Conference on the Human Environment, Stockholm, Sweden, 1972

[3]Howard T. Odum, *Environment, Power and Society,* Wiley Interscience, New York, 1971

[4]Lotka's Principle is named for the biologist Alfred J. Lotka, who expounded it in his book *Elements of Physical Biology,* Williams and Wilkins, Baltimore, 1925. The principle states that the maximization of power for useful purposes must be one of the criteria of natural selection in the evolution of species.

Based on "The Limits of Traditional Economics," *Financial Analysts Journal,* May-June, 1973, and a speech before the 40th North American Wildlife and Natural Resources Conference, Pittsburgh, 1975.

A.
SMITH
LTD.
Huffaker

CHAPTER FOUR

Sharing the Resource Pie

All the talk of the "energy crisis," zero population growth, the need for a new "steady-state economy," the ecological limits to growth, and the inadequacy of the gross national product (GNP) as a measure of progress signals a growing debate between ecologists and economists. Business people will need more than a superficial familiarity with this debate, because it not only raises issues that question some very fundamental economic assumptions but also generates serious criticisms of corporations as allocators of resources.

In the United States, even ordinary citizens have become aware that, by and large, economists tend to ignore social realities and frequently avoid such issues as the distribution of wealth, accepting it as a given. They perpetuate the classical concepts of the free market and ways in which these concepts are distorted by the wielding of power, and the human needs and motivations that lie beyond the marketplace.

In addition to this list of indictments, economists are embarrassed by the persistence of both unemployment and inflation that culminate in the need for wage and price controls. Arthur F. Burns, the chairman of the Federal Reserve Board, was prompted to note in July, 1971: "The rules of economics are not working quite the way they used to."[1] And Milton Friedman was even more frank in his speech given at the American Economic Association's annual meeting in January, 1972: "I believe that we economists in recent years have done vast harm—to society at large and our profession in particular—by claiming more than we can deliver." He added, "We have encouraged politicians to make extravagant promises which promote discontent with reasonably satisfactory results, because they fall short of the economists' promised land."

As these issues develop and the debate becomes more sophisticated, there will be a growing need for economists and ecologists to clarify their positions. Exaggerated, simplistic polemics between those environmentalists who cry for "halting economic growth" and those economists and businessmen who vilify environmentalists as "elitists who care nothing for the poor," previously characteristic of the debate, will no longer suffice. I shall review the issues and highlight the work of a few iconoclastic economists and other innovative thinkers so as to provide a glimpse beneath the surface of such generalities and to discuss the implications of the ecology-economics controversy for businessmen and for the economy as a whole.

Setting the Stage. Carl Madden, an economist, puts the issues in perspective in his insightful book, *Clash of Culture: Management in an Age of Changing Values.*[2] Madden finds the current argument over societal goals and priorities rooted in rapidly shifting values as people make painful mental adjustments to the twentieth century scientific revolution.

Business planners, he feels, should consider the impact on marketplace values of such scientific achievements as knowledge of the chemistry of life's origin, space exploration, medical triumphs, developments in cybernetics, and other seemingly esoteric advances. The new concern for the quality of life is based on the wide dissemination of information about these godlike new capabilities and the resulting changes in perception. This concern encompasses a powerful new sense of the nation's ability to achieve advances in human welfare and of its recognizable shortcomings that can be remedied.

Madden believes that the new value shifts are now challenging traditional concepts of what is rational. Since corporate management of economic resources derives its political and social legitimacy from the presumably rational allocation of resources, disputes over the nature of rationality itself are bound to affect corporate strategies, management, markets, and products.

This book, which Madden prepared for the National Planning Association's Business Advisory Council, outlines many possible ramifications of the new consumer values and places them in a historical, political, and economic context that is both satisfying and coherent. He believes that corporations are gradually changing from a "Cartesian view," in which products are seen as separate parts determining overall strategy, to a "holistic view," in which situations and patterns are seen as determining products. This latter perspective, for example, may help Detroit's auto companies to understand that they are in the transportation business, rather than just producers of automobiles. Such an approach should aid corporations in finding new market opportunities by fulfilling functions rather than by turning out an increasing welter of uncoordinated, ill-adapted, separate products which may not mesh with the market or the environment.

Madden also discusses the current renegotiation of the corporate mandate as citizens and stockholders confront management in the legislature and at the annual meeting. And he sees the questions that arise over the definition and goals of economic growth as embodying a very real challenge to conventional economic thinking.

Such new challenges to economics will of course concern business. Economic data determine both individual corporate decisions and the national policies of resource allocation within which companies must operate. Moreover, this attack on conventional economic practice as the basic tool for managing national resources is coming not only from an increasingly skeptical public but also from other disciplines, such as physics, the life sciences, anthropology, and psychology.

The Ecological Broadside. Strangely enough, it was the environmentalists, normally chided for their lack of realism, who began to question the premise that an economy could continue helter-skelter growth without eventually incurring severe environmental losses. The appearance of John Kenneth Galbraith's *The Affluent Society*[3] in 1958 augured much of the new debate. This book questioned why the U.S. economy seemed so well supplied with hair oil, tail-finned cars, and plastic novelties in the private sector while cities decayed, air and water became polluted, and land was despoiled in the public sector.

Elaborating on this theme, economist Kenneth E. Boulding attributed much of man's despoiling of nature to his inadequate frame of perception. In *Beyond Economics,* written in 1968,[4] he claimed that man still saw his natural resource base as a limitlessly exploitable frontier. This "cowboy economics," as Boulding termed it, did not account for environmental costs—i.e., "externalities" which are part of the true cost of production. Boulding predicted that, with the growth of air and space travel, we would finally come to realize that we live on a vast spaceship which is a closed, rather than an open, system.

I should note that the English economist, Alfred Marshall, introduced the concept of externalities as far back as 1890. He showed economists that they should be concerned with forces outside conventional economic activities. But the externalities of which Marshall wrote were mostly positive and included the rising levels of education of workers and the public services provided by government, from which the entrepreneur of that time had profited but to which he had made no contribution. His younger contemporary at Cambridge University, A.C. Pigou, became interested in the notion that there could also be negative externalities, as he watched smoke and sparks pour out of an English factory chimney.

But this concept remained a theoretical abstraction—an empty box on economists' diagrams—until K. William Kapp published his *Social Costs of Private Enterprise* in 1950.[5] Kapp documented the environmental and social effects of business activities. His basic thesis was that the maximization of net income by microeconomic units (entrepreneurs, corporations, and so on) was likely to reduce the income or utility of other economic units and of the society at large. In short, he claimed that conventional measures of the performance of an economy were misleading, since they ignored social and environmental costs.

An Environmental Price Tag. The issue of the ecological price of industrial activities took another 20 years to emerge. But, by the first Earth Day in 1970, these environmental costs had broken through the threshold of sensory awareness for millions of Americans. In 1971, systems analyst Jay W. Forrester's *World Dynamics*[6] appeared. This book attempted to develop a computer model, on a world scale, for the interactions over time of population growth, food supply, capital investment, geographical space, pollution, and resource depletion. By a different route, Forrester came to the same conclusion as had Kapp in 1950: in complex, nonlinear systems the optimization of any subsystem will generally conflict with the well-being of the larger system of which it is a part.

Most economists scoffed at *World Dynamics* (and also at the later study, *The Limits to Growth,*[7] prepared by Forrester's colleagues at the Massachusetts Institute of Technology). They argued with the very large aggregations of data and other methodological approaches used by Forrester and his colleagues. But it soon became necessary for economists to address population, resource, and distribution issues, because Forrester's work had created such an impact that these issues had entered the realm of political debate and action.

Economists React. Gradually, economists with intellectual investments in economic "growthmanship," such as Henry C. Wallich and Walter Heller, began to deal with the problems raised by Forrester and others. Their first evaluations were that these were new Malthusian scares. Since Thomas Malthus's grim predictions of overpopulation and food shortage were made 150 years ago and had not yet occurred, except in localized regions, economists argued that they would be unlikely to occur in the future (a somewhat shaky linear extrapolation). Resource shortages were discounted because, as in the past, prices would rise and encourage innovation.

The environmental view, however, was that, while technological innovation is vital, it would be foolhardy to place faith in technology as the infinite source of salvation. It might be just as likely for a technological plateau to occur, as has happened in so many other civilizations in the past, and for research investments to yield diminishing returns. The United States could, for example, be forced to extract minerals from increasingly low-grade ores at higher cost.

Moreover, although prices will undoubtedly rise to reflect specific scarcities and encourage substitution in the prescribed manner, environmentalists pointed out that prices are merely subjective expectations of availability and are not based on objective scientific

Fig. 6 The "Free Market" Equilibrium Model of Supply and Demand

(economists admit it's only a **theory**—but they often write, advise and act as if it were **real**)

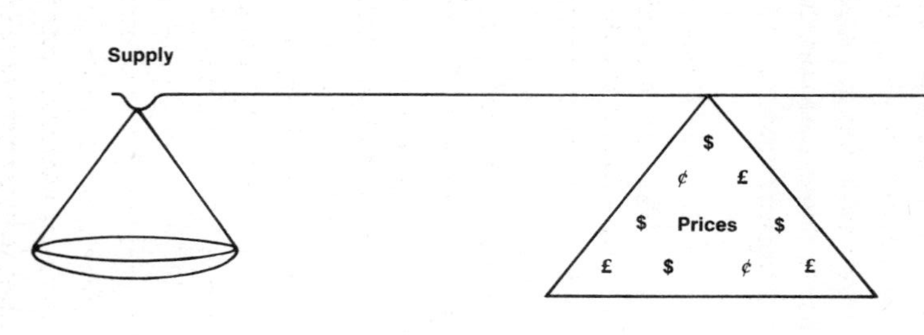

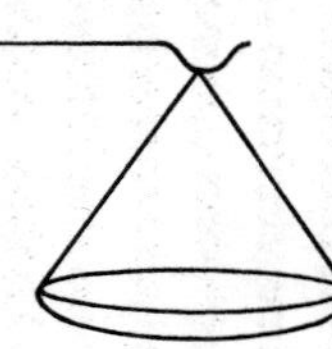

But, what about:

- **Absolute** scarcities: e.g.
  Resources running out, gas, petroleum high-grade ores and minerals, good agricultural land, etc.?
- Vital resources not counted or valued in the price, market system: e.g. pure air and water, peace and quiet, natural scenic beauty?
- Large companies with power to control supplies?
- Human services, behavior, attributes that are valuable, but not rewarded by the market system: e.g. trust, cooperation, selflessness, love, volunteer service, household production, work, nurturing children?
- "Labor" supply i.e. hiring and firing people as if they were commodities in the marketplace?

But, what about:

- Large companies spending millions on advertising to **create** demand and new wants (some irrational)?
- Need, e.g. hungry people who need food but don't have money to buy it (i.e. turn need into effective, **market** demand)?
- Role of culture in defining and determining "demand" and value. (i.e. creating high or low levels of demand for energy and material goods vis-a-vis human self-development and spiritual values)?
- Role of technology in determining energy use and materials— intensity of a society?

research. As mentioned, there are often severe time lags between the scientists' warnings of increased resource depletion or major ecosystem disturbances (e.g., accelerating water eutrophication rates) and the point at which security analysts, bankers, corporate financial officers, and economists digest this new knowledge and crank it into their forecasts. Even Henry C. Wallich, who had at first derided *World Dynamics* and *The Limits To Growth,* later wrote in *Fortune*[8] that such lags might not provide enough lead time to change pricing policies accordingly.

In addition, correct pricing must reflect information on environmental externalities. Although such information can be obtained or approximated in many cases, few social institutions support its collection or back campaigns for its dissemination. An obvious example of such pricing lags is U.S. electricity-rate policy, which still reflects earlier assumptions of continued abundance. Thus major users of electricity receive subsidies despite the power crisis.

The Debate Shifts. The controversy over the environment not only continues to expand but also has led to what is becoming the real nub of the ecology versus economics debate: the distribution of wealth. While environmentalists maintain that economic growth must be curtailed in the name of ecological sanity, most economists hold that economic growth (presumably as currently defined by GNP) is the only way to ensure that increasing shares of wealth trickle down to the poor. Here their Keynesian premises are revealed.

Such a plea for the inflation-prone, trickle-down Keynesian model of economic growth forms the basic argument of *Retreat From Riches: Affluence and Its Enemies,*[9] by Peter Passell and Leonard Ross. They are only passingly concerned with the ecological limits to growth, which they dismiss in one chapter with routine Keynesian arguments that growth is the only means of increasing the lot of the poor. Like most economists, they see environmental protection largely in terms of correct taxation of waste effluents. While environmentalists agree that effluent taxes are a useful "fine-tuning" tool for reducing pollution, they argue that such taxes do not serve to reduce resource-depletion rates or improve energy-conversion efficiencies, where the real economic and ecological payoffs lie.

The authors devote most of their effort to a well-meaning attempt to deal with intractable distributional issues. Unfortunately, they never get below the surface of the maldistribution-of-income problem, since they accept current distribution of wealth as a given. They say, in essence, that, although we know that economic growth does very little for the poor, it is still more feasible than any other form of redistribution yet devised. The bulk of the authors' argument is concerned with other influences which are "the enemies of affluence"—namely, inflation and balance-of-payments problems. Their prescription: keep on the same track, supporting economic growth policies and full employment, while cushioning the effects of inflation for those who are most burdened (e.g., people on fixed incomes).

New Ammunition. Beset by Keynesian arguments for growth and by charges from economists and corporate executives that, in their ecological zeal, they wished to stop economic growth now that they had achieved middle-class comfort, environmentalists were at first baffled about how to respond. But when Barry Commoner, in his 1971 book, *The Closing Circle,*[10] zeroed in on the distribution question, he provided the environmentalists with both new direction and new ammunition. Commoner's argument was three-pronged:

1. He noted that, if economic growth would have to be stabilized at some future point, then there could be no further moral justification for the Keynesian "trickle-down" theory of growth and distribution. The earth's resources would give out long before such uneven economic growth could ever provide for the millions still waiting in poverty. Those unlucky enough to be traveling in steerage class on "Spaceship Earth" would never even make it to the second-class deck.

2. He declared that excessive consumption by rich nations not only condemned the third world of less developed countries (LDCs) to poverty, but also caused most of the environmental devastation. These "overdeveloped" nations were disrupting the environment because of their increasingly capital- (i.e., resource-) intensive production methods. They used highly profitable but polluting technology to synthesize artificial rubber, fibers, and plastics in place of naturally produced commodities with low profit margins. Moreover, since such commodities (e.g., wool, cotton, sisal, hemp,

leather, and latex rubber) comprised a large proportion of most Third World exports, the rich nations were decimating the export markets, and thus the economies, of LDCs.

3. He maintained that the overdeveloped nations, whose populations had stabilized in step with their rising standards of living, had achieved this stability through colonial exploitation of LDC resources. Continued heavy resource consumption by the overdeveloped countries, he argued, would prevent LDCs from achieving their own "demographic transitions" to stable populations. Hence his conclusion: ecological sanity now *requires* social justice.

In addition to Commoner's analysis, which was widely accepted by environmentalists, two recent studies further question the Keynesian assumptions about growth and redistribution.

Many economists who argue for growth cite studies by Robert J. Lampman[11] that relate economic growth to increased welfare by showing a 37% gain in real per-capita consumption between 1947 and 1962. But, according to a study by Lester Thurow and Robert Lucas of M.I.T.,[12] during the years of economic growth between 1947 and 1970, the relative income shares of different groups in the economy remained essentially unchanged. Even more disturbing is a second study by Peter Henle of the U.S. Department of Labor,[13] which notes a persistent trend in the U.S. economy toward actual inequality. Henle shows, for example, that from 1958 to 1970 the share of aggregate wage and salary income earned by the lowest fifth of male workers declined from 5.1% to 4.6%, while the share earned by the highest fifth of male workers rose from 38.15% to 40.55%. Henle does not visualize a nefarious plot against the poor, but he does argue that the structure of the U.S. economy is such that it produces more high-paying, high-skill jobs while low-skill employment remains constant.

A Widening Dialogue. Commoner's analysis and the growing evidence that conventional arguments for trickle-down distribution are open to question turned the environmental movement toward a much more radical critique of economic and social arrangements. In addition, the movement had been stung by barbs from industry and confronted with corporate tactics designed to put environmentalists in conflict with labor (e.g., over a few highly publicized plant closings) and with the poor and consumers (e.g., by warning of astronomical price rises due to environmental controls).

These conflict situations further impelled environmentalists to examine the issue of distribution of wealth and income. In the process, they discovered the influence of distribution of scores of weightier factors in the U.S. economy. For example:

☐ National policies to control inflation
☐ Taxes and subsidies
☐ Discrimination
☐ Technological change
☐ Public works projects
☐ The wide prerogatives of large corporations to invest in other countries and to deploy freely their facilities and resources

It was obvious that all of these factors had infinitely larger impacts on jobs and the distribution of income than did the environmental-control measures that business leaders were fighting. Moreover, the environmental-control sector was becoming an increasingly important part of the economy and had indeed added 850,000 new jobs, even though labor markets could not always match these new jobs with the people left unemployed by the closing of an obsolete facility.

The "Stock of Wealth." A key element in the environmentalists' growing concern over the distribution of wealth and income is related to contentions by several authors that many environmental problems result from the increasing production, consumption, and waste caused by planned obsolescence and the creation of wants by advertising.

For example, Herman E. Daly asks in *Toward a Steady-State Economy,* "Why do people produce junk and cajole other people into buying it? Not out of any innate love for junk or hatred of the environment, but simply in order to *earn an income.*"[14] Daly believes that we need some principle of income distribution that is independent of and supplementary to the income-through-jobs link.

The ethic of distributing the "flow of wealth" through jobs is at the heart of the Keynesian growthmanship effort, and was institutionalized in the Employment Act of 1946. This "flow fetishism" of standard economic theory holds that everyone gets part of the flow—call it wages, interest, rent, or profit—and everything looks rather fair. But, Daly asks, what about the stock of wealth (capital)? Not everyone owns a piece of the stock; yet its distribution is usually accepted as a given—and that does not seem so fair.

If, as Robert J. Lampman has reported, the stock of wealth is held so tightly, with 76% of all corporate securities owned by 1% of the stockholders,[15] then most people must rely for survival on the flows it engenders (i.e., jobs or welfare). And, because of a growing population, the flow cycles must be continuously increased by whatever means possible. The results are built-in obsolescence, waste, creation of new wants through advertising, government pork-barrel projects, and burgeoning bureaucracy.

Environmentalists ask, "Why are we so dependent on the production of goods in the private sector to maintain employment, goods that are ill-matched with such new human needs as mass transit and clean sources of power? Why, in fact, can we not restructure our corporate and governmental institutions to meet new and future needs instead of continuing to address past conditions?" And they point to the encrustation of the federal budget with dozens of obsolete programs, such as the stream channelization projects of the Soil Conservation Service and many projects of the U.S. Army Corps of Engineers, that cost the taxpayer dearly and exact a terrible toll on the environment.

Labor and Capital. This crazy-quilt of government policies and projects is identified by Louis O. Kelso in *Two Factor Theory: The Economics of Reality*[16] as being rooted in the same problem described by Daly. He argues that ever-increasing numbers of latter-day "Works Progress Administration-type" projects will be needed if society cannot grasp the fact that, as long as capital (the stock of wealth) is concentrated in the hands of the few, the many are condemned to rely for survival on the flows from income and welfare payments.

Kelso maintains that capital produces wealth just as labor does. And in advanced, highly industrialized economies, it produces an ever-increasing portion of wealth without much human intervention (e.g., in such capital-intensive, automated industries as oil refining and petrochemicals). He claims that, to mask this disturbing and politically explosive fact, economists attribute productivity advances not to the increasing sums of capital used to increase the efficiency of each worker, but rather to the worker's own increased effort, i.e., "labor productivity."

Kelso argues that labor productivity advances are actually the result of additional capital placed at labor's disposal. And he notes that much labor strife in highly automated industries, where capital plays the major role in production, is due to the workers' understanding that they need only be present *at the scene of production* as members of a well-organized pressure group in order to derive an income from automation.

Similarly, Kelso interprets all the paraphernalia of Keynesian redistribution. Jobs have become the nation's chief device for distributing income in a thinly disguised mélange of what he calls "warfare," "workfare," and "welfare." What is archcapitalist Kelso's prescription? Spread the ownership of corporations among their workers by means of his tax-deductible employee-stock-ownership trusts, as increasing numbers of corporations are now doing.

Environmentalists find analyses such as those of Daly and Kelso supportive because they highlight the propensity of the private-sector economy to continually substitute capital for labor. This, of course, taxes the environment and increases rates of resource depletion through the massive-scale, centralized operations it allows. As Arthur Pearl of the University of Oregon put it in a recent article in *Social Policy,* "In essence, we now have a surplus of human beings and a shortage of nonrenewable resources: thus we have to reverse our historical view of efficiency." He added, "It is only in a human services society which is labor-intensive rather than capital-intensive that the resources of the earth will be conserved and human resources be expended for the benefit of human beings."[17]

The Integrated Economy. While Kelso's treatment of the capital-labor issue was important, his most useful insight for environmentalists was his underlining of the political nature of all economic distribution.

This issue was first raised by John Stuart Mill in 1848 in *Principles of Political Economy.* Mill held that, once goods or any form of wealth had been produced, society, by its laws and customs, could place this wealth at the disposal of whomever it pleased. He added that even individual wealth could not be kept without society's permission and without society's willingness to employ police to guard individuals from thieves. In this regard, the criminal justice system in the United States employs some 1% of the labor force, and in fiscal year 1969-1970 expended $8.57 billion.[18]

In *Beyond Economics,* Kenneth E. Boulding elaborates on the politics of economic distribution. He claims that there are three basic modes of human transaction:

1. The primitive threat system (e.g., "Give it to me or I'll kill you"), with its more sophisticated "blackmail" variation (e.g., "How much will you pay me to stop harming or annoying you?").
2. The exchange system of market economies.
3. The maturing, integrative system in which increasing interdependence is necessary for the viability of the whole economy.

In his book, *The Economy of Love and Fear: A Preface to Grants Economies,*[19] Boulding contends that the U.S. economy is moving (in spite of policy lags and distributional errors) toward such an integrative system, which he terms the "grants economy."

As evidence, Boulding points to ubiquitous grants and income transfers and increasing acceptance of responsibility for the disabled, unemployed, aged, and poor. He also points to the public services and amenities, which provide a viable context for market activities, and to "positive externalities." By the latter, he means commodities that benefit the society as a whole yet represent investments that are not fully recapturable. Knowledge and the flow of information, for example, play an increasing role in advanced production, innovation, technological change, and economic development but receive less than their share of attention from economists.

The Economics of Information and Knowledge,[20] edited by Donald M. Lamberton, surveys some of the crucial knowledge and information issues raised by Boulding. Contributions by social choice theorists Kenneth Arrow and Gordon Tullock, labor economist Albert Rees, and information theorist Jacob Marschak focus on such topics as uneven information availability and how it can distort the labor market, prices, and political and corporate decision making. Other papers consider optimum public investments in research that advances knowledge, the patent system, and international trade and technology transfer.

The Lamberton book is an important one for environmentalists because it underscores that, without vastly more information between buyers and sellers in this complex age, the uncontrolled aggregation of small decisions (on which market economics rests) could add up to large-scale ecological chaos.

The way such chaos evolves is described by ecologist Garrett Hardin in his now-famous treatise, "The Tragedy of the Commons."[21] In feudal England, Hardin points out, all the farmers grazed their flocks in a large communal field (the commons). Some farmers realized, however, that they could maximize their advantage by grazing more animals than their neighbors. It was only a matter of time before the idea caught on and the commons was destroyed by overgrazing. Likewise, if we arbitrarily designate a jointly shared resource—such as air, water, or even whales—as a "free good," then no individual is responsible for its overall protection. As a result, it is likely to be destroyed completely.

The anthropocentric markets of conventional economics cannot provide much information on how to cope with such free-goods problems, in which public resources can be depleted and public services jeopardized by the temptation of each individual to avoid paying their share or restraining their greed. Moreover, argue the environmentalists, economics is ill-suited to dealing with value preferences that cannot be assigned monetary weightings. Such qualitative value conflicts must be left to the political arena, where they face a key axiom of Kenneth J. Arrow's "general impossibility theorem." Arrow states flatly that, in democracies, individual preferences cannot be logically ordered into social choice.[22]

In dismay the environmentalist must ask, "If economics cannot yet provide a sufficiently rational system for public decisions, and if Arrow is right that democracies cannot order individual preferences into logical social decisions, where do we go from here?" To the rescue come many scholarly responses to Arrow's dismal prognosis for democracy, including Gordon Tullock's rebuttal, "The General Irrelevance of the General Impossibility Theorem";[23] Duncan Black's argument that reiterates the theorem's irrelevance to an understanding of how social choices are actually made in committee situations;[24] and Edwin T. Haefele's contention in his paper, "Environmental Quality as a Problem of Social Choice,"[25] that Arrow's conditions for ordering individual preferences into social choice can be met by representative governments with a two-party system.

Sharing the Pie. Once more we come to this recurring theme in environmental thinking—the distribution of wealth and income. On what basis might a new formula for such distribution be justified? Might such a new rationale lie in the growing interdependencies of advanced production processes?

As many of the papers in *The Economics of Information and Knowledge* suggest, production has now become so complex, based increasingly on such abstract commodities as knowledge, that it is no longer possible to neatly formulate the rewards due to labor and the rewards due to capital. Neither the Keynesian labor interpretation of value nor the Kelsoist capital view of value is persuasive in resolving this point. Production of wealth in advanced economies is fast becoming a social enterprise, based on a tangled web of interrelationships. The old formula—with land, labor, and capital as factors of a production process that leads to a logical distribution of wealth—is no longer adequate.

The inadequacy of this traditional distribution formula concerns environmentalists for another crucial reason. It is precisely these knowledge inputs to production which increase energy-conversion efficiencies and reduce resource-depletion rates—the only other routes to environmental sanity besides that of more equitable distribution.

The fuel cell is an example of such a knowledge-intensive, resource-conserving product, the value of which is difficult to distribute via the traditional formula. The fuel cell's energy-conversion potential is some 60% (compared to 12% for the internal combustion engine), and it is largely this advance which represents its greater value. To whom does the fuel cell belong and who shall share in its rewards? The operators? The man who invented the fuel cell? Or the taxpayers whose government grant supported the university which supported the research? The pathways through such a system of infinite interdependencies are unchartable by current economic methods, unless many arbitrary weightings and assumptions are inserted into those economic models which attempt the task.

This and other problems make it increasingly difficult to design accurate economic models and to avoid the dangers of forcing public decisions into the straightjacket of cost/benefit analysis. Paul Streeten illustrates this point by noting that cost/benefit analysis has the tendency to convert political, social, and moral choices into pseudo-technical ones; hence its psychological appeal to administrators, but also its logical flaw. "If two objectives conflict," argues Streeten, "say the requirements of industrial growth and protection of the environment, someone will have to choose. The choice may be democratic, dictatorial, or oligarchical, but choice it must be." Streeten holds that cost/benefit analysis, using the economist's often highly arbitrary weightings, conceals such value conflicts, which can only be resolved politically.[26]

In addition, the welfare economists, however well meaning, are busy trying to examine the costs and benefits of pollution control. Using economics' most trusted tool, marginal analysis, they are attempting to evaluate environmental goals in terms of the willingness to pay for some standard of environmental quality or the willingness to accept compensation for damage. As K. William Kapp notes, this economic "compensation principle" as a criterion for environmental quality leaves no doubt in anyone's mind that the common denominator is going to be money. Kapp continues, "The basically questionable point of departure consists in the fact that original physical needs for rest, clean air, nonpolluted water, and health, as well as the inviolability of the individual, are being reinterpreted in an untenable way as desires or preferences for money income."[27] He also maintains that the compensation principle does not take income distribution or information requirements into account, and does not lead to systematic research into alternative policy options.

The growing list of shortcomings in current economic concepts and methods was summed up in a witty broadside by economist Alan Coddington, who believes, with Kapp, that the main body of economic thought is ill-suited to coming to terms with ecology. "It may even be the case," Coddington wrote, "that the greatest service economists can render to posterity is to remain silent."[28]

If money is an inadequate measure for harmonizing economic activities with social needs or the ecosystem, what new criteria might be devised to evaluate the policy decisions which will face citizens in some future steady-state economy? As a few daring economists begin to respond to this question with new concepts which more accurately match new realities, we will see their discipline incorporate more hard data on resource factors and on human needs and potential.

Questions of Value. All of these new issues lead environmentalists to call for a reexamination of cultural notions of "value." The economic impact of these qualitative concepts surrounding the concern for "quality of life" is discussed by Walter A. Weisskopf in his *Alienation and Economics.*[29] He notes that economics, once based on the ethic of thrift and self-denial, now requires an ethic of "utilitarian hedonism" if it is to justify mass consumption, mass production, and advanced market economies. Yet this very hedonism, promoted by corporate advertising, is now leading to the breaking down of industrial discipline and cries of dehumanizing, boring jobs.

Weisskopf stresses that notions of value are arbitrary and culture-bound. For example, he holds that the U.S. economy overvalues material wealth while dismissing psychic wealth. Similarly, we overvalue competitive activities and undervalue cooperation and social cohesion.

The real dimensions of scarcity are not economic, claims Weisskopf, but existential. Time, life, and energy are for humans the resources that are ultimately "scarce," because of our mortality. Such psychic needs are similar to those described by psychologist Abraham Maslow: love, peace of mind, self-actualization, companionship, and time for leisure and contemplation. These needs can never be satisfied by purely economic means, although economic activity that satisfies lower-order survival needs permits them to emerge. In short, humans tend to assign values arbitrarily and then pay measurers to collect only those data which conform to prevailing assumptions of "value." The hypnotic circle is complete.

Environmentalists have intuitively come to the same conclusions as has Weisskopf. They note another example: people in the United States overvalue property rights and undervalue amenity rights, with which property rights often conflict. One can find hundreds of examples in the courts today in cases involving the conflicting interpretations of these two sets of values and rights. In such cases amenity rights are beginning to win (e.g., recent court awards for noise damage).

On this issue of value clashes, Benjamin Ward shares Weisskopf's view. Ward's book, *What's Wrong With Economics,*[30] questions the avoidance in economics of efforts to study the moving target of constantly changing human values and preferences. While acknowledging the difficulties, he takes issue with arguments that such studies are beyond the scope of rigorous, scientific methods of inquiry. Ward feels evasions of such problems have caused economists to retreat to easier, but less relevant problems. He notes that a sister discipline, the law and the judicial process, does embody an often highly satisfactory system for empirically validating the changing values of consumers. Through the continuous building up and reinterpreting of legal precedents, changing consumer values and preferences become codified in law and custom. Economists must seek to capture this type of dynamic process in their analyses.

Conclusion. Many of the environmentalists' indictments of economists may seem esoteric or even "un-American." Yet some of them are spurring more openminded economists into new efforts to create fresh concepts and paradigms, such as those discussed here. After all, economics is still the discipline concerned with scarcity, choice, and the behavior of equilibrium systems, all of which are still central concerns for the future. Although many of the young, radical economists in the United States have criticized their colleagues' overwhelming preoccupation with market economics and the elegant contours of the closed, equilibrium systems it hypothesizes, they are no happier with the bureaucracy that typifies so many centrally controlled socialist and communist economies. And five-year plans can be just as environmentally destructive as those plans of entrepreneurs or government agencies in market economies. Yet, any assault on economics, sooner or later, is an assault on the mandate of corporations, and it behooves management to understand the rational basis of the ecologists' criticisms.

---

[1]Arthur F. Burns, statement before the Joint Economic Committee, July 23, 1971

[2]Carl H. Madden, *Clash of Culture: Management in an Age of Changing Values* (Washington, D.C., National Planning Association, Report No. 3, 1972)

[3]John Kenneth Galbraith, *The Affluent Society* (Boston, Houghton Mifflin, 1958)

[4]Kenneth E. Boulding, *Beyond Economics* (Ann Arbor, University of Michigan Press, 1968)

[5]K. William Kapp, *Social Costs of Private Enterprise,* revised edition (New York, Schocken Books, 1971).

[6]Jay W. Forrester, *World Dynamics* (Cambridge, Massachusetts, Wright Allen Press, 1971)

[7]Donella H. Meadows, et al., *The Limits to Growth* (New York, Universe Books, 1972)

[8]Henry C. Wallich, *Fortune,* October 1972, p. 121

[9]Peter Passell and Leonard Ross, *Retreat From Riches: Affluence and Its Enemies* (New York, Viking Press, 1973)

[10]Barry Commoner, *The Closing Circle* (New York, Alfred A. Knopf, 1971)

[11]See, for example, Walter H. Heller, "Ecology and Economic Growth," *Economic Impact,* Spring 1973, p. 37

[12]Reported in *Business Week,* April 1, 1972, p. 56

[13]Reported in *The New York Times,* December 27, 1972

[14]Herman E. Daly, editor, *Toward a Steady-State Economy* (San Francisco, W.H. Freeman, 1973)

[15]Robert J. Lampman, *The Share of Top Wealthholders in National Wealth, 1925-1956* (Princeton, Princeton University Press, 1962)

[16]Louis O. Kelso, *Two Factor Theory: The Economics of Reality* (New York, Vintage Books, 1967)

[17]Arthur Pearl, "An Ecological Rationale for a Humane Service Society," *Social Policy,* September-October 1971, p. 41

[18]Joseph F. Coates, "Urban Violence: The Pattern of Disorder," *The Annals of the American Academy of Political and Social Science,* January 1973

[19]Kenneth E. Boulding, *The Economy of Love and Fear: A Preface to Grants Economies* (Wadsworth Publishing Co., Belmont, California, 1973)

[20]Donald M. Lamberton, editor, *The Economics of Information and Knowledge* (Middlesex, England, Penguin Books, 1971)

[21]Garrett Hardin, "The Tragedy of the Commons," *Science,* December 13, 1968, p. 1,243

[22]Kenneth J. Arrow, *Social Choice and Individual Values,* 2nd edition (New York, John Wiley, 1963)

[23]Gordon Tullock, "The General Irrelevance of the General Impossibility Theorem," *Quarterly Journal of Economics,* May 1967, p. 256

[24]Duncan Black, *The Theory of Committees and Elections* (Cambridge, Cambridge University Press, 1963)

[25]See Allen V. Kneese and Blair T. Bower, editors, *Environmental Quality Analysis* (Baltimore, Johns Hopkins Press, 1972)

[26]Paul Streeten, *Cost-Benefit and Other Problems of Method* (Paris, Mouton, 1971)

[27]K. William Kapp, *Social Costs, Neo-classical Economics, Environmental Planning: A Reply* (Paris, Mouton, 1971)

[28]Alan Coddington, "The Economics of Ecology," *New Society,* April 1970, p. 595
[29]Walter A. Weisskopf, *Alienation and Economics* (New York, E.P. Dutton, 1971)
[30]Benjamin Ward, *What's Wrong With Economics* (New York, Basic Books, 1972)

CHAPTER FIVE

The Entropy State

Many models exist of the unfolding shape of advanced industrial societies. Proposed here is yet another: that of the "entropy society."

Daniel Bell gave us the notion of a "post-industrial society" transcending via technology the ideologies of left and right, and one in which most of the labor force would be employed in service and knowledge-based industries. John Kenneth Galbraith sees a "new industrial state" of détente between business and government: a "technostructure" with power falling to cadres of bureaucrats, technicians, and managers, and with only the vestiges of a market economy.

Gunnar Myrdal describes in *Beyond the Welfare State* the future evolving from the mixed market and planned economies of which Sweden is typical. And Roger Garaudy foresees in *Crisis in Communism: Turning Point in Socialism* the shape of advancing bureaucratized communism in the U.S.S.R., as well as the more decentralized worker-managed models of communism such as that now developing in Yugoslavia. And while Karl Marx's prediction of the decline of capitalism did not count on the labor force becoming bourgeois, as it has in today's highly industrialized societies, the crystal-ball-gazing of capitalism's school of market-oriented economics has proved equally cloudy.

Another model of the unfolding pattern of industrial societies might well be that of the "entropy state." Simply put, the entropy state is a society at the stage when complexity and interdependence have reached the point where the transaction costs that are generated equal or exceed the society's productive capabilities. In a manner analogous to the phenomenon that occurs in physical systems, the society slowly winds down of its own weight and complexity, with all its forces and counterforces checked and balanced in a state of equilibrium.

We seem unwilling to come to terms with the fact that each increase in the order of magnitude of technological mastery and managerial control requires and inevitably leads to a concomitant order of magnitude of government coordination and control. Thus we see the irony of those corporate technological innovators who decry the government bureaucracies that all technological innovations call forth. Worse, as the industrial system grows more complex, specialized, and differentiated, it becomes increasingly difficult to model the labyrinth of variables in such a web of social and physical systems. Any system that cannot be modeled cannot be managed. Indeed, systems analyst Jay Forrester has noted that such complex systems tend to behave counter-intuitively and are stubbornly resistant to human manipulation.

Because advanced industrial societies develop such unmanageable complexity, they naturally generate a bewildering increase in unanticipated social costs: in human maladjustment, community disruption, and environmental depletion. All these effects of uncoordinated, unplanned activities and suboptimization are called by economists, in almost a Freudian slip, "externalities." The cost of cleaning up the mess and caring for the human casualties of unplanned technology—the dropouts, the unskilled, the addicts, or those who just cannot cope with the maze of urban life or deal with Big Brother bureaucracies—mounts ever higher. The proportion of GNP that must be spent in mediating conflicts, controlling crime, protecting consumers and the environment, providing ever-more comprehensive bureaucratic coordination, and generally trying to maintain "social homeostasis" begins to grow exponentially. New levels of expenditure to maintain this social homeostasis are augured daily, as in the recent calls for new legislation to provide government compensation for crime victims and for new agencies to counsel and assist those who succumb to chronic debt.

Another emerging fact of complex societies is the newly perceived vulnerability of their massive, centralized technologies and institutions, whether manifested in the loss of corporate flexibility, urban decline, power blackouts, skyjacking, or the many frightening scenarios of sabotage and violence now occurring daily.

Meanwhile expectations are continually inflated by business and government leaders, and it becomes more difficult to satisfy demands of private mass consumption while trying to meet demands for more and better public consumption, whether for housing, mass transit, health, education, welfare benefits, parks and beaches, or merely to keep the water potable and the air breathable. The enormous burdens of military expenditures add to this allocation problem in most industrial countries. But even without such huge arms commitments, the ever-inflating bubble of expectations is cause for concern. Denmark is a case in point. Paradoxically, a taxpayer's revolt unseated a liberal government in irritation over the costs of a highly popular social welfare program—an apparent inability of voters to understand the inevitable trade-offs between high levels of public goods and services and private consumption.

The symptoms of the entropy state are also visible in Japan. Notwithstanding military expenditures held to less than 2 percent of the GNP, the ruling Liberal Democratic Party is strained by labor unrest, soaring wage settlements, growing social dissatisfaction as inflation reaches annual levels of 16 percent, and the rising public investment costs of pollution control, sewage treatment, housing, and social security benefits. Britain too is exhibiting signs of industrialism's next stage, the entropy state. Social conflict increases as the resource base shrinks, and more equitable sharing has become the inevitable demand. Rampant inflation, soaring public investment costs and social welfare services, and the ineluctable bureaucratization follow a pattern that grows more familiar each day. There seems now to be a dawning realization on the part of the stoic British that belt-tightening is a way of life and that achievement of even a modicum of satisfaction will now require nothing less than a new frame of mind and lowered expectations.

Inflation is now so ubiquitous in advanced industrial economies that it has become one of their structural features, rather than a temporary affliction. It can no longer be described by economists as a trade-off for unemployment, since in many countries, including our own, we have both. Traditional Keynesian remedies of pumping up the whole economy in order to ameliorate areas of structural unemployment and mask the true conflict over the distribution of wealth are now beginning to be felt as too costly in that they raise rates of inflation and deplete resources. Economist Irving Friedman suggests in *Inflation: A World Wide Disaster*[1] that vastly inflated expectations for both public and private consumption are now a key factor.

Another explanation for inflation comes from thermodynamicists, who insist that economists don't yet understand the drastic multiplier effects of developing energy and resource scarcities. Simply stated, such energy researchers as the brothers Eugene and Howard Odum say that economists and federal energy officials have not yet grasped the crucial difference between *gross* energy and *net* energy. Gross energy is typified by all those theoretical barrels of oil locked in such less accessible forms as shale and tar sands. But it will take millions of barrels of oil to crush the rocks, heat and retort the shale and sands, not to mention the refining and transporting, as well as the millions of gallons of scarce water that would have to be diverted from farm use in the process. What is left over at the end of all this investment of energy and resources is net energy, only a fraction of the quantity theoretically available (gross energy). Indeed the Odums claim that so far the nation's entire nuclear power enterprise has only yielded a few percentage points of net energy, because the process is so heavily subsidized with coal and oil—for uranium extraction and enrichment and sources of other energy and capital-intensive steps that precede the final output of electricity from a nuclear plant. Inflation, in this explanation, is driven by the increasing amount of money and energy a society must keep diverting to the job of extracting and refining lower and lower grade energy and materials. Therefore, there are fewer real goods and services produced and prices soar, as the multiplier effect of additional energy-intensive processing of these resources into finished goods is felt.[2,3]

But the energy situation has merely revealed and lent impetus to what may be the unfolding "end game" of industrial societies. First there will be the frantic efforts to invest more and more capital in energy exploitation and resource extraction, despite the already visibly diminishing marginal returns to much of our capital-intensive production. Consider, for example, the case of agriculture where, according to agricultural researchers David Pimental,[4] Michael Perlman, and others, by a key measure—how much energy is used for a given output of calories—our U.S.-type highly mechanized, fossil-fuel subsidized farming is now the most inefficient in the world. In other overautomated processes as dissimilar as fishing and operating mass transit systems, the marginal returns to capital investment are falling: fish catches are now destructively overefficient, while on such transit systems as San Francisco's BART, workers are displaced by costly and erratic automated train controls.

The current stage of hurling massive quantities of capital at the increasingly fruitless endeavor of trying to produce greater supplies of energy and resources, will, in time, be played out. The learning experience will be horrendously costly because it will foreclose many other more realistic options. Capital, amassed from our previously bountiful sources of energy and materials, now represents our society's last diminishing store of low entropy (i.e., concentrated potential for useful work). As evolutionist Gregory Bateson[5] illustrates, capital is our precious stock of stored flexibility for performing an orderly social transition to adapt to new conditions, just as a chrysalis uses its stored energy to turn itself into a butterfly. Instead we see a tragic situation developing, as oil companies, electric utilities, and basic manufacturing industries all attempt to borrow larger and larger quantities of capital to squeeze new supplies from degenerating and depleted deposits of fossil fuels, materials, and minerals.

Banks, in turn, oblige their corporate borrowers, if necessary, by borrowing expensive funds themselves and issuing their own debt instruments, thus adding to the mirage. Sometimes the wasteful, disastrous capital-spending plans of corporations and utilities can be halted only by massive pressures from consumers and environmentalists. By fighting rate increases and higher prices and by forcing companies to more fully internalize social and environmental costs, such groups may deflect company plans by "upping the ante" and

Fig. 7 Gross National Product Problems

- "Gross" G.N.P. growth v. "Net" G.N.P. growth?
- G.N.P. only measures prices of those goods and services traded in the market—ignores all other valuable production services and amenities.
- G.N.P. **adds in** all social costs as if they were desirable, valuable product!

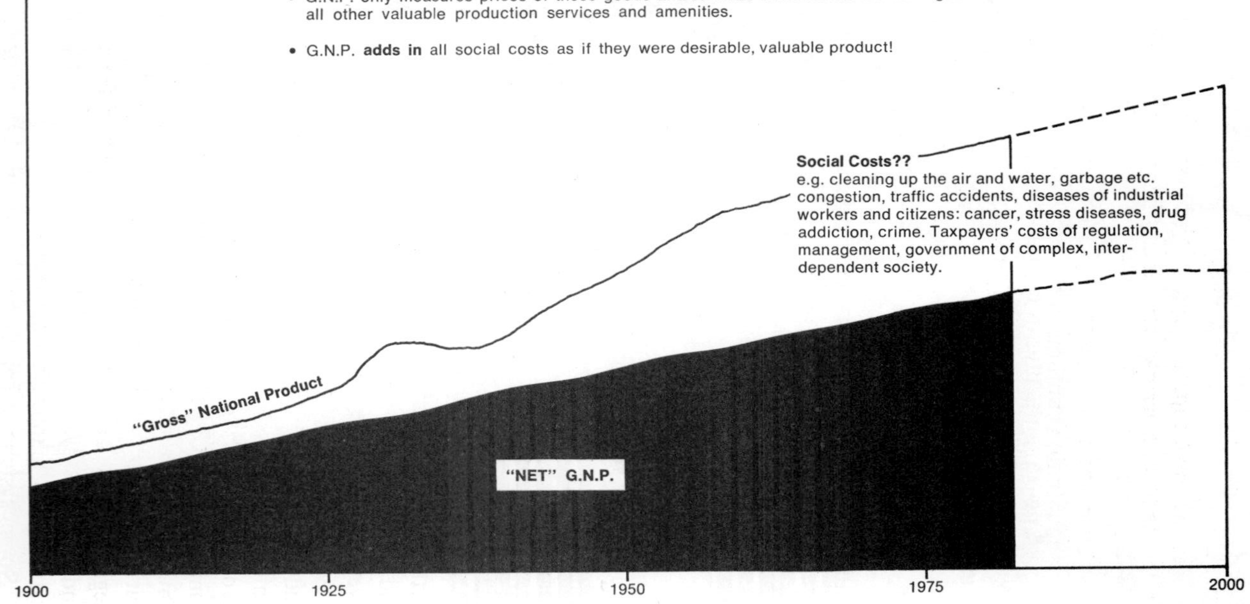

making their own capital spending and borrowing plans less viable. We can see this occurring now in many energy companies and electric utilities.

As the emerging capital shortage becomes more acute and interest rates and inflation continue to soar, we may find that debt service will become the biggest item in corporate and municipal budgets. It will then become more evident that inflation is the manifestation of a massive, futile economic wheel-spinning, where money flows faster and faster, economic activity becomes more feverish, people work harder, and the GNP appears to be climbing reassuringly. The only problem will be that fewer and fewer products, goods, and services will result from this hyperactivity, and money will simply become less and less related to real value. At some point we will recognize that investing capital to call forth diminishing supplies is a tragic misdirected effort. At that point, presumably, interest rates will fall, in spite of the increasing scarcity and value of capital, since most of the precious remaining supply will be needed to maintain existing plants, equipment, housing stocks, public buildings, and amenities, and will not be available for whatever high-yield uses may still remain. At this point, we will have drifted to a "soft landing" in the steady-state economy, while symptomatic inflation will have masked our declining condition.

The entropy state may be the future of advanced industrial societies unless as yet unimagined advances in computer science enable us to manage and control the complexity of these societies, and unless we improve our ability to devise accurate social indicators. Even then, the attempt to control such impenetrable systems will inevitably mean greater government control, further loss of freedom and individuality, and will lead us closer to the computerized state of which George Orwell warned in his book *1984*. Another path may lie in deliberately trying to reduce the interdependencies by simplifying some of the overly developed systems that have now reached some obvious diseconomies of scale. We might take, not the Luddite's axe, but the surgeon's scalpel, and with a delicacy born of desperation, begin to isolate and sever some of the interdependencies in our social and technical systems, so that the variables might once more be reduced to a manageable number.

Some systems seem to work best on a very large scale. For

example, the telephone system must be widely standardized and is, by definition, composed of interlocking elements. But other systems and institutions can be more efficiently operated on a smaller scale. Perhaps cities and many corporations fit into this category. Some have even suggested that if we convert individual homes and apartment buildings to solar, wind, and methane power generation, then the only reason that homeowners would need to be connected to central power stations would be to sell power back to utilities for resale to their own industrial customers! Certainly we are now seeing the trade-offs in building larger and costlier central power plants with longer, more expensive, energy-wasting transmission lines. But to expect existing utilities and energy corporations to develop such radically different systems would be as naive as it would have been to give the buggy whip makers the responsibility of developing the automobile.

Similarly, we might change the mix of human and machine energy that our production methods currently employ. Many of our farm families were driven off their land because of the now energy-wasteful automation and the large-scale investments that corporate farming requires. A return to smaller farms might yield benefits in human satisfaction and would save enormously on energy and transportation. This capital-to-labor equation has changed for hundreds of other production processes in our society, as the cheap energy trip comes to an end. We can communicate instead of commuting, fill in the wasted spaces in the suburban sprawl, use our existing buildings more efficiently and renovate old ones. Then we can bicycle more; perhaps one day walk the shortened distances more safely; grow our own vegetable gardens; and spend more time in family and community activities. As the Ford Foundation energy study shows, such changes in our life styles could achieve tremendous savings of energy and capital and could stretch all our resources. The entropy state might be held at bay for many generations by such new values and symbols of success.

In the last analysis, Bell's "post-industrial society," the technostructure of Galbraith's "new industrial state," the various models of socialism, communism, and welfare capitalism mentioned earlier are all too heavily dependent on increasing economic growth and technological mastery. Even service economies are wholly dependent on their primary agricultural, manufacturing, and resource bases. Buckminster Fuller's vision of technologically based abundance of the sixties seem evanescent and remote.[6] Perhaps the crumbling faith in the gods of technology will be restored and justified, and premonitions such as this, of the entropy state, can be happily banished from our minds. But at least running scared may buy us some time and retain some of our precious store of flexibility, so that we may yet transform ourselves into a new culture in harmony with the earth.

---

[1]Irving S. Friedman: *Inflation: A World Wide Disaster* (Boston: Houghton Mifflin, 1973)

[2]Eugene Odum: "Georgia Conservancy Magazine," Fourth Quarter, 1973

[3]Howard Odum: "AMBIO" (Journal of the Royal Swedish Academy of Sciences), Volume 2, Number 6

[4]David Pimental: "Science," Volume 182, Number 4111, November 2, 1973

[5]Gregory Bateson: *Steps to an Ecology of Mind* (New York: Ballantine, 1973)

[6]Buckminster Fuller, *Operating Manual for Spaceship Earth,* Simon & Schuster, 1969.

Reprinted from *Planning Review,* May, 1974.

SHELL
Coca-Cola
NESTLÉS
Ford
ITT

CHAPTER SIX

Let Them Eat Growth

It seems self-evident that wealth is based on the availability of natural resources and energy. However much human energy and ingenuity we expend, if we don't have some natural resources to manipulate and fashion for our needs, then we are not only poor, but dead. Economists seem sometimes to doubt this, particularly the technological optimists who sometimes give the impression that the human mind can engineer infinite substitution as resources become scarce. Others, such as Paul McCracken, a former chairman of the Council of Economic Advisors, seem to believe that all that is necessary to call forth greater supplies is ever-increasing levels of capital investment. Since, as we have discussed, capital itself is a store of low entropy (potential for useful work) extracted from past exploitation of resources, it would seem that McCracken's formula would only work if there are still unexploited resources present in a concentrated enough form to make their extraction and fabrication worth the effort in net energy terms. However, there is now widespread concern that capital itself is becoming a very scarce resource. For example, Senator Lloyd Bentsen of Texas has conducted hearings on the future availability of capital; it appears worrisome enough to him to prompt his interest in prioritizing investments and credit.

There are also other indications that the message of scarcity is getting through to business and national governments. Of late, *Fortune, Business Week* and other similar periodicals have reviewed the coming scarcities of resources and the spate of U.N. conferences on the environment, population and food; these, as well as the growing Third World militancy, suggest growing realization that many of our basic assumptions concerning salvation through industrialism, technology, mass production and material abundance have collided with a finite resource base, rising expectations, and exploding human population—all exacerbated by increasing global economic and political interdependence.

What all this augurs is redistribution. Not only redistribution from the affluent to the rebellious and frustrated poor within the rich countries, but redistribution from rich countries to those of the Third World. We might also expect a geographical redistribution of production, so that the raw materials of less developed countries might acquire more added value as exports while increasing their own national self-sufficiency, while rich industrial nations may divest themselves of those industries based heavily on imported raw materials as they become scarcer and more expensive.

What I am suggesting is that we may see the repeal of some of the Adam Smith concepts of international comparative advantage and the conventional economic wisdom that more world trade is always better for all concerned. Of course, world trade serves many important purposes where such exchanges distribute geographically limited, vital and nonsubstitutable resources and commodities between trading partners. Other benefits include the enhancement of global cultural exchange and communications facilities. But this should not blind us to the point at which returns to such exchanges diminish and they generate social costs. Economists who glibly justify world trade to meet the requirements of global efficiency are not convincing. Economists' models are so circumscribed and consider so many factors to be exogenous variables that it is almost arrogant to claim that they can determine what is globally efficient. Reality is what we pay attention to, and economists, like most of us, pay too much attention to short-term oscillations while overlooking larger cycles over longer time periods. Oskar Morgenstern, the brilliant mathematician, noting this propensity (*New York Times,* Sept. 23, 1974), underlines its absurdity by pointing out that the data on such apparent oscillations is itself often little better than fictitious. World markets cannot be relied upon to determine what is globally efficient because they do not account for the power wielded by large economic and political forces—whether nations or multinational corporations—and they accept the existing distribution of wealth as given. As shown in Fig. 1, such concepts as "efficiency" and "maximizing" are meaningless, particularly on a global scale, unless a time horizon and spatial boundaries are specified. In addition, we might remind ourselves that there is nothing intrinsically good about transporting goods to and fro, between or within countries. As we are now seeing in our domestic economy, our passion for transporting things and ourselves has been largely based on ideas about economies of scale in production and the historical happenstance that for much of this century we have almost treated energy as if it were a free good.

It is possible that Adam Smith also overlooked the transportation, distribution, warehousing and inventory costs that might arise with vast centralized mass production and ever-more-complex divisions of labor and failed to realize that they might eventually lead to diminishing returns from so-called economies of scale. Another factor Smith ignored was the crucial difference between exchange, or market, value and *use*-value. We tend to calculate our efficiencies in the efficiencies of exchange, which leads us to discount efficiencies associated with use. Such use-values often make small decentralized operations profitable to their owners and yet it can be difficult for such operations to exist in competition with large-scale units operating on the efficiencies dictated by markets of nationwide or global scope. Nevertheless, few economists challenge the orthodoxies concerning the unalloyed benefits of world trade based on global division of labor and comparative advantage, whatever values of use-efficiency are lost and however high the costs in transportation, warehousing and distribution. This is not to mention the increasing social costs in specific countries that such global interdependencies create; for example, the growing phenomenon of lockstep economic cycles, the disruption of domestic labor markets and the distortion of production patterns in weaker nations. In fact, world trade and the existing international monetary and financial system are based on past configurations of power and inevitably dictate inequitable participation and the exploitation of weak and poor nations. Recent U.N.-sponsored meetings on such problems as these unequal terms of trade and the power of multinational corporations attest to a growing awareness of this fact.

Even our current global inflation as economies move into synchronous inflation/recession oscillations is not seen as part of the social costs of an overcentralized, overcoherent and increasingly unstable world production and trading system. Any global or massively interlinked system whose component units retain their own suboptimizing goals and fail to develop macro-goals appropriate to the new level of organization will inevitably create a replay of Garrett Hardin's "Tragedy of the Commons" parable. Such a system breeds instability, inflexibility and vulnerability to domino-type breakdowns, as its ability to dampen oscillations is reduced. By contrast, natural systems prevent such instabilities by their characteristics of diversity, redundance and decentralization. We are reminded of the profound misunderstanding of natural systems and evolution propagated by the Social Darwinists, who saw only competition in nature and overlooked the subtle and pervasive cooperation and interdependence of all living organisms. Even as we try to shore up our crumbling world monetary system, those who question the benefits of world trade based on free-market concepts risk the label of neomercantilists advocating a regression to autarchy.

The same outdated metaphors of economics are similarly preventing creative thought on our current domestic crises. Terming these multiple crises of suboptimization "inflation" obscures the problem further. Consequently our leaders have dictated that the real debate about national-resource allocations—and, indeed, about what is "valuable"—will be conducted in the mystifying language of economics and thereby preempted by this particular priesthood, rather than being conducted in the mother tongue so that all of us might join in. Coalitions of consumer, poverty, minority and environmental groups have discovered that the current national debate is not only about inflation; indeed, the debate might be as well or better conducted by political scientists, anthropologists, social psychologists or members of any number of other disciplines as by economists. But having arbitrarily defined our problems as economic, economically trained minds, triggered by the word "inflation," lock onto Keynesian concepts and the Phillips Curve model of the trade-off between inflation and unemployment that they were taught in school, thereby short-circuiting any further useful thought, even in their own subsystem language. The debate over "inflation" masks both the real debate about sharing the economic pie and the fact that the Keynesian remedy of pumping up the whole economy has now collided with the global population/resource scarcity equation.

The drift to massive, interdependent, complex centralized systems of all kinds has now also outrun our conceptual and managerial capabilities by several orders of magnitude. Prisoners of our own semantics, we cannot even pose the right questions. Examples of our blundering attempts to model and deal with global systemic problems are legion. They are based on the notion of the technological fix, now parodied by skeptics: "Technology is the answer—but what is the question?" Take our unrealistic responses to global population increases. Based on our own experience, we conceive of the population problem as technical and educational. For example, according to Dr. Dana Raphael, publisher of the *Lactation Review,* more person-years of contraception are provided by human lactation than by all of the technical means of family-planning programs. We cannot deal either with the fact that in poor countries children are often the only social-security mechanism or with the theory of demographic transition that acknowledges the role of past colonial exploitation. Such analyses inevitably identify the U.S. and the overdeveloped countries as part of the problem and point logically to a redistribution of the world's resources.

Our approach to the world food problem is equally myopic. Former Agriculture Secretary Earl Butz, who led the U.S. delegation to the World Food Conference in Rome in 1974, placed his faith in the "free market" without inquiring too closely into how its workings are distorted by power, or into the problem of need as opposed to demand. Even our inadequate relief plans for helping rebuild reserves and setting up a World Food Bank do not acknowledge that our own appetites for meat and livestock products are a growing part of the world food problem. Meanwhile, rather than concentrating on increasing self-sufficiency, poor countries sucked into the whirlpool of world trade often act in their own worst interests by selling their protein and raising cash crops to exchange for inflation-ridden foreign currencies.

Frances Moore Lappe points out in *Diet for a Small Planet,* that for three hundred years Western colonial powers established plantation systems in their colonies to produce profitable cash crops for export in place of food for local inhabitants. The crops they selected—tobacco, rubber, tea, coffee, cocoa, cotton and other fibers—have negligible or no nutritional value. These cash crops became established in world trade as the leading exports of the Third World, so that even after emancipation their economies remained hooked on such exports rather than developing alternative strategies for self-sufficiency. According to the U.N. Food and Agricultural Organization, some 250,000 square miles of arable land is planted to nonnutritional crops and this acreage is growing faster than that for edible crops. These poor countries were also hooked into the roller coaster of world trade, and the velocity and amplitude of its cycles can cause as many grotesque distortions in the priorities of weak countries, as it can be advantageous to powerful rich countries and multinational corporations. For example, the Peruvians export fish protein to feed U.S. beef cattle while our own aid programs promote and dispense to Peru exotic protein-fortified soft drinks. Even our efforts to help such countries toward self-sufficiency via such strategies as those of the Green Revolution are arbitrary approaches to problems economically defined in our own terms. Such definitions lead to solutions based on unstable monoculture, which requires costly applications of high-energy pesticides and fertilizers and mechanical equipment—creating further dependence on uncertain outside supplies, particularly petroleum, and increasing social inequity. These convoluted policies and rationalizations permit us once again, to avoid the ever-present issue of redistribution, both between nations and within their borders.

Another example of such failures of conceptualization concerns global environmental problems. The overproduction, overconsumption and overpollution from overdeveloped countries is directly a concomitant of the underproduction, underconsumption and underpollution of the underdeveloped countries. Again the nexus is redistribution of both production and consumption in spite of what economists believe to be "efficient." In any case, efficiency claims for the U.S. economy are suspect, since while providing for such a small percentage of the world's inhabitants it requires much of the world's resources to keep it operating. And by the highly significant yardsticks of energy efficiency, researchers John Steinhart, David Pimental, Michael Perlman and others note that on an energy input/output basis the U.S.-style energy-intensive agriculture is the most inefficient in the world. And yet we are still trying to sell the world on our model of capital-intensive, consumption-oriented industrial development—or, as cynics describe such efforts, "Technology transfer is when you take money from the poor people in rich countries and give it to the rich people in poor countries."

Apart from the statistics themselves concerning the world poverty gap, the most sweeping indictments of the industrial world's destructive interactions with the Third World come from radical educator Ivan Illich. He states that underdevelopment is not an actual condition, but rather a conditioning of Third World people to demand wholly inappropriate and unattainable packages of consumer goods and services. The media power of multinational corporations propagates these consumption patterns and expectations, whether for expensive soda pop, candy, snacks or packaged infant formula; poor consumers are thus lured into spending large portions of their meager incomes on such marginal products, often with dire nutritional consequences. Illich notes that every private car sold in Latin America denies some 500 people an adequate bus service and every refrigerator forecloses on hopes for a community freezer. He castigates the current liberal-humanist plans for global crisis management, whether to provide food or population control services: "By shifting production from guns to food, they reduce their sense of guilt and increase their sense of power. Their hubris distracts them from understanding that only the renunciation of industrial expansion can bring food and population back into a balance in the so-called backward countries." Again, the nexus is redistribution.

What institutional changes might we expect in a future scenario which may require redistribution of incomes, reallocation of resources and decentralization of industrial production and human populations? None but the very optimistic can imagine that such systemic changes could be accomplished without social strife. The arguments about how to share the no-longer-growing pie that are now being conducted in mature industrial societies under the rubric of "inflation" are only auguries of sharper clashes to come. Millions of Americans and other industrial consumers are still psychologically dependent on the idea of material acquisition and are prepared to fight out another ugly round of this oldest of all zero-sum games. Even more ominous are the increasingly militant demands for oil-price rollbacks evident in U.S. foreign policy pronouncements as the search for inflation scapegoats begins in earnest. Even stalwart "free market" economist Milton Friedman has denounced such scapegoating. The suboptimizing efforts of corporations and powerful institutional forces will increasingly collide with each other and their social and ecological host systems. We often forget that such maximizing goals of sub-units stress social systems as much as ecosystems. In the economic models that still program corporate decisions, the delicate webs of familial and community relationships and social sanctions are treated as free goods in precisely the same way as are air and water. Clearly—as inflation is symptomatically showing—all this suboptimizing by corporations and other powerful interest groups has reached unsustainable levels. Americans will now have to begin the painful process of renouncing the "keeping up with the Joneses" game.

Kicking the Joneses habit will require alteration of the symbols of success and achievement, changing societal roles and rewards and reducing the emphasis on competition in favor of cooperation. Alternative systems and institutions embodying such new values are already flourishing and will grow as negative feedback from former behavior becomes more insistent and palpable.

Existing corporations and institutions will try to continue as usual until some combination of raw material scarcities, supply bottlenecks, soaring prices and capital or credit shortages strikes their particular industry or activity. Many will go out of business, especially those heavily dependent on capital- and resource-intensive production or consumption. Others in similar circumstances may have the political power to force the taxpayers to bail them out in the manner that various aerospace, airline and utility companies have used. This type of corporate parasitism is already on the increase, as banks, feedlot operators and other special interests join the line waiting for government assistance or bailouts.

Meanwhile, as the game gets tougher in the U.S., multinational corporations may step up their drives for new markets and profit opportunities in other countries and thus provoke even sharper confrontations with national governments in their bids for resources and eventually unleash more efforts to cartelize them after the style of the Organization of Petroleum Exporting Countries (OPEC). Whenever the big players preempt the game in too blatant a fashion—whether they are powerful groups in a national economy or powerful actors on the global stage—a point is often reached when the small, weak players decide that they have nothing to lose by walking away from the game. This has happened on Wall Street, for example, where the big institutional investors and gunslinging money managers sought to play the field of small investors, often referred to as "sheep." They failed to realize that their game depended on the willingness of these small investors to stay in the game and that they were the "field," or the "Commons" of Garrett Hardin. As each big investment trust or portfolio manager tried to play this "field," more small investors dropped out of the game. Today, no large investment manager can afford to play the market, because these large institutional stockholders have now themselves become the field. The small investors, no less than the OPEC nations and the producers of bauxite, copper, coffee and other commodities, lose confidence in the unequal game and either drop out, try to set up some new rules or create a new game of their own.

At the same time, at least 50% of the variables affecting the U.S. economy are now beyond the reach of national policymakers. Today's level of global interdependence, which can cause instant feedback such as rising domestic grain prices as a result of large sales to the Soviet Union, will have a lasting effect on the U.S. food industry, as well as our entire economy.

Other effects of growing interdependence have been the confrontations between the overdeveloped and the underdeveloped nations at the world conferences on population in Bucharest and on food in Rome. Increasing demands were heard at both conferences that the overdeveloped nations curb their wasteful overconsumption, which results in each American child creating fifty times the impact on resources as that created by the birth of an Indian child. At the same time, it was forcefully pointed out in Rome that meat-eating habits and wasteful use of fertilizers for lawns, golf courses and cemeteries were exacerbating the dimensions of famine in less fortunate lands.

Since the productivity of U.S. agriculture, according to David Pimentel (*Science,* Vol. 182 #4111), is declining in net energy terms, we may expect a devolution of some of its superstructure and a shift to greater labor-intensiveness in the future. The former levels of productivity were achieved with massive energy subsidies, which, in turn, permitted the proliferation of overprocessed, overcentralized, overtransported, overpackaged and overpromoted foods. This is often associated with significant nutritional depletion, such as in the case of oversugared cereals, snack foods, and TV dinners, where large proportions of the retail price are for costs of packaging and advertising. There are even depressing signs that if U.S. meat-eating habits are reduced to a more healthful level, the vegetable protein substitutes will also be overprocessed, overpackaged and similarly more costly in energy and money terms than necessary. The highly capitalized food industry can obviously make more profit selling texturized vegetable proteins and substitute foods made from them than by selling nutrition-equivalent soy grits at less than half the retail price.

The devolution of this highly capitalized structure may shorten this lengthy and costly processing, packaging and promotion chain. One can imagine that some products, for example, TV dinners, will become unprofitable to produce, as bauxite price increases eventually make their aluminum trays prohibitively costly. As transportation costs rise, regional and local food production and retailing at a smaller scale will become more efficient. And as capital becomes scarcer, many local, microunits for food production and processing requiring minimal capital for entry, such as small truck farmers, bakeries, specialty, "home style" preservers and retailers, will flourish. Similarly, there is also evidence of great interest in home gardening, city-allotment food raising and home preserving.

In addition, consumer resentment against large food companies, agribusiness operations, large retail chains, and the increasing power these companies wield, may create other pressures toward devolving such oligopolistic control over the most basic means of human survival. Inflation, as it continues to eat into purchasing power, will also cause consumers to rethink purchases of costly, convenience foods and nutritionally inferior snack foods and return to more basic, less processed foods. Current grassroots interest in sound nutrition education will no doubt grow proportionally to inflation, as well as growing fear of food adulterated with chemical additives and pesticide residues, not to mention loss of faith in large food companies and their high-powered promotion and advertising.

Lastly, I would hope that increasing nutritional awareness among U.S. consumers combined with their eroding purchasing power, would not cause U.S.-based multinational food companies to step up their already shocking exploitation of less sophisticated consumers in other countries, particularly in less developed regions of the world. We have become aware of many of the horror stories of the over-promotion of marginal snack foods, soda pop and candy bars causing poor people, lacking the cultural defense mechanisms of Americans, to overspend their meager incomes on such items, often with terrible consequences.

When I was in Singapore some time ago meeting with consumer groups, I found that much of their efforts were devoted to trying to educate young mothers not to substitute costly and inferior canned condensed milk for their own milk in the face of vast propaganda efforts of milk producers. Under a headline, "Milk and Murder," the *New Internationalist* magazine has exposed the dimensions of the problem of promotion of infant formula in less developed countries and the giveaways of feeding bottles and free samples to mothers, often employing salesgirls dressed as nurses. Often the heavy cost of such substitute formulas leads poor mothers to overdilute them, risking malnutrition of their infants. In addition, it is often difficult, the article pointed out, for such women to obtain sterile water or the means of sterilizing bottles, and so many infants are exposed to disease. The Protein Advisory Group (PAG) of the U.N. has cautioned such corporations to refrain from these promotions, to stress the importance of breast-feeding, and to improve labelling to minimize misuse of such milk products. While the PAG has no enforcement power, such problems are being exposed by *The Lactation Review.*[1]

In Kenya, while meeting with consumers groups, I encountered a similarly horrendous problem. Women in that country had gained the impression from billboards and magazine ads that Coca-Cola possessed magical, healthful properties. Many of them, unable to read, had begun substituting Coca-Cola for their own milk in feeding their babies, some of whom had died of malnutrition. Such horror stories illustrate the heavy responsibilities that multinational food companies must bear when they use high-pressure sales tactics inappropriately in vulnerable, poor countries.

In a world of food scarcity and famine, we must realize that overcentralized, capital-intensive food industries guided by market forces, tend to overprocess, overpromote and overpackage food products and then often overtransport and overdistribute them. This profit-maximizing behavior creates social and environmental costs, because their products do not bear the full share of these costs in their prices. Such companies, if permitted to continue externalizing such social costs, tend to become parasitic in terms of the larger social system.

We are entering an era where the maximizing behavior of corporations can only be pursued at the expense of suboptimizing the system as a whole. In fact, most of our social problems today can be characterized as the multiple problems of suboptimization. As systems analysts M. Mesarovic and E. Pestel point out in their book, *Mankind at the Turning Point,* the second report to the Club of Rome, when we perceive the extent of our global economic interdependence, we may learn that in a finite system competition must be balanced with cooperation, and that narrow, maximizing strategies will lead only to the worst outcomes for all global players. Indeed, we must understand that the new ballgame for the human species, whether for individuals, corporations or countries, has now become nothing less than that of maintaining the viability of the ballpark itself.

The limits-to-growth issue is a political issue. The reason is that if you have already been consuming a vast amount of the world's resources, then it behooves you to pay out a lot of money to buy "research" to justify your continuing to consume, and so you have an awful lot of studies done, and propagated, to say that there is no problem. But if you don't have your face quite so firmly buried in the sand, you tend to be a little worried about when it might all run out. And this is why the Third World countries are now talking about the New International Economic Order. I think they understand that the justification of inequality for capital formation, which is the old Keynesian "trickle-down" model of economic development, is going to leave them waiting in the back of the line forever, until all of us have our second houses and third boats. I was recently with a Third World leader, and we were talking about the inevitable subject of the limits to growth. He said: "It's like a tunnel with two lanes of traffic. You go into the tunnel with your car, and you get stuck in the lane that's not moving, and you're not allowed to change lanes. And there is the other lane going by you at a pretty good clip, and you get very frustrated." I disagree with Herman Kahn who claims that Western nations need not accept responsibility by foolishly "marrying India." The point is *we do not have the choice of whether* to marry India. We chose to marry India when we (the industrial nations) extended our global search for materials and resources to support our economies. It was not India's choice; she was a captive bride.

The real job over the next ten years is to start retooling ourselves. Herman Kahn asks, Are we worse off? Is the future going to be better? To me that's not the question. We have to redefine what's better and what's worse; we have to redefine what we mean by satisfaction. We can't talk about waste without redefining needs and greeds. There's plenty for our needs, maybe not for all of our greeds.

---

[1]The Institute for Human Lactation, founded by Dr. Dana Raphael, publishes *The Lactation Review,* is at 666 Sturges Highway, Westport, Conn. 06880.

This chapter is based mostly on a paper given at the Lake Itasca Futures Seminar in Minnesota in 1974 and on an article in *Nutrition Action,* Dec. 1975. In 1976, I debated Herman Kahn of the Hudson Institute on the occasion of our both receiving honorary doctor of science degrees. Naturally, the issue was our sharply differing views on the limits to economic growth and redistribution of wealth. Some of my remarks are incorporated in this chapter.

CHAPTER SEVEN

Japan: Industrialism's Bellwether

So awesome has been Japan's postwar boom—its economy has become, in a single generation, the world's third most powerful—that it is easy to overlook the burdens that affluence has imposed. The Japanese have almost caught up not only with American abundance but with our modern-day problems as well: pollution, inflation, labor unrest, and soaring costs of social services and public investments—in short, all the dis-economies, dis-services, and dis-amenities that we in the United States have discovered are the other side of the coin of industrial growth.

I recently toured this California-sized country of 103 million people, half of whom are crushed together on a mere 2 percent of the land, and came away feeling that American influence is on the wane. That influence has extended from mass consumption of automobiles, television sets, and other electronic gadgets to lifestyles, family relationships, and even a desire to alter one's physical appearance. (A Tokyo reporter, soliciting my opinion on whether the relatively small Japanese are "ecologically more desirable" than Americans, pointed out that since World War II many women here have fed their children protein-rich U.S.-type diets so they would grow tall.)

The warts on American society are now all too visible for it to continue serving as an unquestioned model for Japan's future. A young management consultant, Toshimasa Kurioshi, told me in Osaka: "We now see that with all their goods and good intentions, Americans are confused and unhappy, while their land is polluted and the benefits of their technology are questioned. Japan will find her own way, develop her own model of a good society." In a quiet potter's workshop in the hills of Kyushu, the southernmost of Japan's main islands, the same sentiments were voiced. Hajime Kozuru, whose skill and artistry are acclaimed as far away as Tokyo, sat sipping tea, surrounded by his drying pots, and pointed down into the valley where the town of Fukuoka lay in a Los Angeles-type haze. "We must retrieve," he said, "some of the wisdom of the past, where men did not work only for money but for their needs, their community, and their own dignity."

In fact, many Japanese seem to share the view of sociologist Yonosuke Nagai that the sum of unplanned technological developments by private enterprise in the United States has produced such an avalanche of social change that our government is floundering and our institutions disintegrating, while we await some emerging political consensus to provide new direction. Just as in the United States, affluence has brought to Japan satiety and a search for new goals and values. The critical difference between the two countries is that the Japanese "moment in the sun" has been telescoped in time and may not follow the pattern of the leisurely "golden era" of relatively untroubled affluence the United States enjoyed after World War II, culminating in the 1950s and 1960s. During that period, our resource base and land area, together with our relatively small population, enabled us to pursue the goals of industrialization and mass consumption with little concern for our social and environmental costs in the 1970s.

In Japan, however, the crunch has come almost before the mass-consumption stage has been fully realized. Anomalies abound. With freely spendable yen jingling in their pockets, the Japanese find their choice of goods, services, and amenities still limited. They cannot load up on household appliances and furniture because their houses are too cramped to provide room for such trappings of the average American home. The yearning for breathing space and *nisshoken* ("the right to sunshine") in their dense and increasingly high-rise cities finds expression in the proud ownership of *maicar* ("privately owned car") and the weekly exodus from the cities in millions of family Toyotas, Datsuns, and Hondas in search of a little spot of greenery for picnicking and outdoor exercise. As in the United States, the escape in the family auto becomes more difficult as pollution and congestion spread and parks and beaches are engulfed by crowds. One of the few other outlets for Japanese consumers is foreign travel, in which, because of language problems and their unfamiliarity with other countries, they tour in conspicuous groups, creating a new image to rival the "ugly American."

Japan's resource base and land have always been severely limited in relation to population pressures. Nature's revenge came swiftly, and the word *kogai* ("public hazard and environmental disruption") passed into the language. Land prices have soared, followed by those of food and other commodities, while the rising expectations of the good life—titillated by the barrage of consumer advertising—have contributed to general inflation and demands for higher wages. In addition, the Japanese cannot spread out spatially without the utmost care, for fear of using up their precious arable land, already intensively cultivated, and thereby becoming more dependent on undependable supplies of foreign food. Even the carefully conceived decentralization plan of Prime Minister Kakuei Tanaka was viewed by many as merely "spreading pollution around." And underlying all these problems is the recurrent headache of having to import almost all of the oil supplies—as well as many of the basic resources—needed to keep Japanese industries operating.

Japan has several advantages in dealing with its new problems and developing a unique social model of its own. Pollution and congestion, contaminated fish and rice, outbreaks of mercury- and cadmium-related poisonings (such as those at Minimata Bay), and respiratory ailments from air pollution in Yokkaichi are producing a sort of "instant feedback" for the public, as well as intellectual ferment in academic circles. A young professor of law in Kyoto, while still afraid of professional disapproval, told me earnestly about the need for "public-interest law" in Japan and the application of the Ralph Nader model of social action. Other academics believe that the rising citizens' movements for consumer and environmental protection will provide the momentum needed to galvanize the scientific disciplines. Scientific "whistle blowing"—such as the recent warning by Dr. Hiroshi Hirata, of Tokyo University, that government-set limits on the cadmium content of rice are so high they could cause kidney ailments—is now becoming respectable. Development of pollution-control technologies is accelerating. Examples include Honda's stratified-charge automobile engine, which reduces emissions, and a method developed by the Tohoku Technological Research Institute in Sendai for extracting poisonous metals from water.

At the grass roots, the Japanese are mobilizing to defend their own interests. Mrs. Shoko Mukai, an official of the Japanese equivalent of the League of Women Voters in Tokyo, proudly showed me the booklet her group had produced for housewives on the dangers of additives in foods and pesticides. She said that some business and government leaders suspect that the fuss about pollution is a Communist plot (the Communist party has picked up many seats in the Japanese Diet in the past few years). "But if there is cadmium in our rice, how can that be a Communist plot?" asks Mrs. Mukai. Back in the mid-Sixties, citizens' groups waged a successful campaign to prevent the building of a petrochemical plant in a scenic area near Mount Fuji. Marshaling scientific expertise in chemistry, biology, and meteorology, they researched and produced their own report on the social effects they believed were not adequately considered by the company and the local prefectural government. They organized tours to nearby Yokkaichi to show their Mount Fuji neighbors the air pollution caused by similar petrochemical operations. And, long before such strategies were contemplated by U.S. environmentalists, residents of Yokohama effectively intervened in the siting of a new power plant, extracting a memorandum of understanding from the Tokyo Electric Company and the city government of Yokohama; the agreement, posted in the municipal building, specifies the permissible sulphur content of the plant's fuel and the environmental safeguards to be met.

On the consumer front, newly sophisticated housewives' groups, such as Shufuren and Chifuren, suspicious of the oligopolistic practices of some of the large trading companies, are planning boycotts of fabrics. Two years ago they successfully boycotted TV sets, forcing prices down some 15 percent. Like their counterparts in the United States, these citizens' organizations have learned the value of the dissemination of information—and the problem of its distortion by advertising-supported commercial mass media. The result has been the swift growth of an underground press (called *mini-commi* by the Japanese) that reflects the grass-roots concerns of activists.

Japanese businessmen, too, are realizing the complex interdependencies of industrial urban societies. On the whole, the businessman here seems more secure in his social position than does his American counterpart. The interlocking relationships between Japanese big business and government are openly acknowledged, and they have some advantages over the more covert relationships in the United States. Such a cozy, almost feudal system engenders an accompanying spirit of *noblesse oblige.* Whereas U.S. corporate managers are still agonizing over what constitutes "corporate responsibility," their Japanese counterparts effortlessly assume womb-to-tomb responsibilities for their workers' job security, health, and general welfare. This may make them philosophically readier to deal with unanticipated side effects of their operations and may put their honor and social prestige more firmly on the line when it comes to making amends.

The Chubu Economic Federation in Nagoya, which includes among its members the chief officers of such giant corporations as Toyota Motors and NKG Insulators, is planning to build an environmental research center in central Japan. The Japan Center for Area Development Research, one of Tokyo's most noted "think tanks," recently conducted an international symposium calling for intellectual initiatives on the part of business, government, and the academic world. The work of nine competing planning teams and their "alternative futures" scenarios for Japan were compared and then integrated with the insights of experts from many other countries invited to critique the plans. Similarly, while it has taken the pressure of outsiders and militant stockholders to interest U.S. companies in conducting "social audits" of their activities, Japanese companies and trade groups are undertaking their own audits, even sending teams to interview U.S. scholars working in this new field.

There is significant activity at the government level also. Japanese control standards for pollution are in some cases twice as high as those in the United States. Dr. Moshio Hashimoto, of the National

Environmental Protection Agency in Tokyo, points out: "This is necessary because we have such tremendous population densities in our most heavily industrialized region, from Osaka to Tokyo." Dr. Hashimoto is now drafting a bill that will compensate Japanese citizens for injuries and ill health brought on by pollution.

In Japan today there is skepticism about the role of technology—and a willingness to learn from the American experience. For example, many Japanese feel embarrassed about their paucity of flush toilets and chemical sewage-treatment facilities. But before they commit a huge capital investment to water treatment, they can evaluate the new debate in the United States over the advisability of one-dimensional chemical treatment systems. This may enable the Japanese to build on the best insights of their own past: that waste nutrients are needed and belong in the soil—not in the water, where they are nothing but detrimental. In like manner, they can now capitalize on our research in biological, rather than chemical, pest controls or maybe even retain the streetcars of Kyoto, Fukuoka, and other cities to stem the proliferation of automobiles—already some eight times more numerous per hectare of level land than in the United States.

Most encouraging of all is the move among Japanese economists to reassess economic theory in light of our new understanding of environmental constraints. By contrast, American economists have felt too inhibited to step outside the politically safe bounds of market economics, for fear of being branded as radicals or Marxists; they have thereby been denied firm intellectual terrain from which to examine critically the U.S. market economy.

One of Japan's leading economists, Professor Shigeto Tsuru, president of Tokyo's Hitotsubashi University, explained to me the newly devised macroeconomic indicators that will soon replace the standard gross national product (GNP) formula in Japan. Called net national welfare (NNW), the new indicators will deduct from the GNP figures several categories of social costs—auto accidents, traffic congestion, pollution, run-down public facilities, and the like—so that Japan will have a clearer picture of whether its economy is improving the quality of Japanese life. In the United States such new economic concepts are still germinating in the halls of academia and have not yet entered the realm of serious political debate.

Lastly, Japanese cultural traditions, with their concepts of responsibility and familial respect, will continue to generate social innovations. Their traditional *ringi* method of reaching decisions—which proscribes action until all affected groups have been consulted—could well be made to include increased participation of wider circles of citizens. This might provide government and industry with more up-to-date indicators of the rapidly changing goals and values of the Japanese people, and might result, in turn, in more efficient allocation of capital and scarce resources. In addition, Japan will continue to have an important pragmatic advantage in its posture of nonmilitarism, which has kept military expenditures at about 1 percent of its GNP. This will continue to release resources for social investment and innovation—while the U.S. economy is still shackled with massive military budgets. Another trait that may have considerable survival value is the Japanese appreciation of restraint, subtlety, and miniaturization. As Buckminster Fuller has been telling us for years, "Less is more," and it is in the technology of doing more with fewer resources and less energy that the path to environmental and social harmony lies. Maybe someday, if Japan solves its problems of rapid growth, the Japanese will be teaching Americans a vital lesson—that more and bigger are not necessarily brighter and better.

This chapter reprints an article originally published in the *Saturday Review/World* of Dec. 18, 1973.

CHAPTER EIGHT

The Great Economic Transition

Let us now zero in on the U.S. as an example of a mature industrial economy facing the inevitable transition to more sustainable modes of production and consumption and review some strategies for readjustment.

If the present "slumpflation" is over, can the next one be far behind? Or will our latest turn of the Keynesian crank lead us back to the good times? I believe that it will not and instead, that it will clearly reveal the inadequacy of traditional Keynesian policies and indeed, the bankruptcy of macroeconomic management itself. In spite of an unemployment rate hovering between 7 and 8% and prospects of federal budget deficits on the order of $60 billion, officials pronounce that the light is now visible at the end of the tunnel. Meanwhile, traditional economic advice is rigidifying the responses of both Congress and the Administration and continuing to generate policies that are counterproductive and often lead to exacerbation of our problems. The new malaise of stagflation or even "slumpflation" is inexplicable by traditional economic theory[1] and has now been experienced by virtually every Western economy as well as Japan. Inflation, no longer understandable by the Phillips Curve trade-off with unemployment, has now become a structural feature of many industrial economies rather than an aberration, and recession hits all industries dependent on heavy use of increasingly scarce and expensive energy and resources. And although resource prices have fallen of late due to worldwide recession, as soon as the latest shot of Keynesian adrenalin takes hold, albeit briefly, resources prices are likely to surge again and the now familiar shortages will reappear.

It is no longer unthinkable to speculate whether industrial societies are approaching the sort of evolutionary cul-de-sac that I have described as the "entropy state." Such societies may have already drifted to a soft landing in a steady state, with inflation masking their declining condition.

We must face the fact that business cycles in these mature industrial economies are now created by economists and governments rather than by market forces and therefore market forces can no longer be relied on to right things. In their frantic efforts to deal with what they perceive as the three-headed monster of inflation, recession and energy-supply problems, government policy makers, caught in a conflicting chorus of advice from economists, heroically man the money pumps and fiscal machinery and alternatively inflate and deflate their respective economies. Sadly, after each one of these artificially created business cycles, the economy undergoing such treatment is left in a feverish, flabbier condition with residually more unsatisfactory levels of both unemployment and inflation. Obviously the problems are structural, and aggregate policies of pumping up the whole system to ameliorate structural pockets of unemployment and mask distributional inequities are now too costly in increasing rates of both inflation and resource consumption. Conversely, trying to deflate the whole economy will not touch structural causes of inflation, such as monopolistic pricing, government protectionism and other excesses in the wielding of institutional power, or deal with rising global competition for scarce resources of the newly perceived social costs involved in expanding world trade: the phenomenon of excessive interdependence and synchronously oscillating economies.

The short-term, artificial oscillations brought on by applications of conventional economic wisdom obscure longer term cycles and inexorable realities, such as the declining global resource base, changing climatic trends, together with rising expectations and unabated population growth. The real prospect of a massive world food shortage and widespread famine suggest that Malthusian predictions cannot yet be wished away by the technological optimists. It is now self-evident that there will be no way out of our economic discontents without jettisoning some of our most cherished assumptions, particularly the elegant, free market equilibrium model of supply and demand which still exerts such hypnotic power over our minds, and which has permitted economists to discount possibilities of absolute scarcity, of both resources and capital, on the supply side. Such unrealistic models of how our economy works are now shortcircuiting new formulations of our dilemma. As any citizen who watched President Ford's 1975 inflation summit on TV now knows, economics has become a substitute for thought.

Fortunately the current policy confusion and official resignation to dangerously high unemployment levels is at last opening up the debate to expression of new economic thought, formerly considered heretical and often suppressed. Those jumping into the vacuum range from well-known advocates of national planning, including Wassily Leontief; Keynesian purists, including Paul Davidson of Rutgers and Hyman Minsky of Washington University in St. Louis; to innovators such as Kenneth Boulding and E.F. Schumacher; and finally to the almost underground views of the young radicals of the Union for Radical Political Economics and Paul Sweezy, the Marxist editor of the *Monthly Review*. The extent of the disarray among the economics establishment can be gauged by the gloom and self-doubt aired at the recent annual meetings of the American Economics Association and an unusual manifesto signed by seven Nobel Prize winners at a Democratic Socialist Organizing Committee meeting in New York on January 25, 1975. The signers included Harvard's Kenneth Arrow, Sweden's Gunnar Myrdal, Jan Tinbergen of Holland, Heinrich Boll of Cologne and Mel Delbruck of Caltech. They noted, "In the advanced industrial democracies [economic crises] raise serious questions about the very nature of the economic systems in these societies," and called for the exploration of alternative economic systems.[2]

The advocates of national economic planning, including Leontief, Galbraith and others, recognize the structural imbalances and interdependencies in the economy and would try to correct these failures of the invisible hand with computerized input-output models and the largely voluntary, indicative planning, such as that used in France. They also advocate government policies aimed specifically at goals of full employment, economic growth and stabilizing inflation, for example, credit allocation, wage-price guidelines as well as fiscal and monetary tools. But sympathy for their desire to halt the current drift and home in on more rational and humane social goals cannot obscure problems associated with their approach. Such advocates often fail to point out that the underlying assumptions programed deeply within many of the economic models and statistical data, on which economic planning and government policies rely, are either erroneous or outdated. As we have noted, in spite of widespread recognition of its inadequacies of design and assumptions, no attempt has yet been made outside academe (except in Japan) to overhaul the Gross National Product, which overstates advances in national welfare by including social costs as "product," treats education as expenditures rather than investment in society's knowledge stock and ignores the value of leisure time and unremunerated housework and volunteer activities. In addition many models are still flawed by underlying Smithian assumptions of the free market which ignore institutional and political power wielders; which assume that prices adequately reflect external costs (or that such social and environmental costs are exceptions); which inadequately model the behavior of individuals, oil sheiks and ecosystems and assume that adequate information is available to all in the marketplace. Planning, using economic models in which such flawed assumptions are so inaccessibly buried, is a hazardous enterprise. In fact, planning with such inadequate conceptual and statistical tools would probably tend to reinforce counterproductive patterns and structural problems and lead to further inefficiencies in allocating resources. Indeed, the composition of Leontief's Initiative Committee for Economic Planning confirms such fears, being composed of leaders of big business, big labor and established liberal centrists, including Henry Ford, Leonard Woodcock, J. Irwin Miller and Robert Roosa, as well as academicians and others.

Meanwhile, the Keynesian "purists,' including Davidson and Minsky, have pinpointed similar conceptual weaknesses in the mainstream Keynesians, such as Paul Samuelson, Walter Heller and Leon Keyserling, which have caused them to misunderstand inflation and therefore downplay its effects in an effort to widen distribution via growth. Davidson, Minsky and other "purists" point out that

Keynes developed a basically disequilibrium view of the economy and although his policy recommendations have been adopted in most industrial economies in what has been termed the Keynesian-neoclassical synthesis, that in reality no synthesis occurred on the conceptual level. Instead of a true integration, Keynes' policies were simply overlaid on the basic equilibrium model of the free marketplace developed by Adam Smith and later rendered more elegant, but alas, less accessible to scrutiny, by Leon Walras in France some one hundred years later. Minsky has zeroed in on the fragile financial structure of today's U.S. economy resting on trillions of dollars in debt, and believes that the choice is no longer between inflation and recession, but between inflation and debt deflation, with resulting deep depression.[3] He argues for achieving full employment in the context of a low investment economy, thus joining forces with those, including the writer, who call for a shift to more labor-intensive production for reasons to be outlined further in this chapter.[4]

The radical view, espoused by many members of the Union for Radical Political Economics, including David Gordon,[5] as well as Paul Sweezy of the *Monthly Review*, sees business cycles as inevitable in capitalist systems. Inflations, where labor's position is improved relative to capital, will always be countered by recessions, whose functions are to shake out and discipline the labor force, while providing the pause in capital investment and economic growth that refreshes and restores the capitalists' profit.

However, all of these explanations (necessarily sketchy and much-abbreviated) are inadequate because they still lie within the confines of the discipline of economics, and incorporate historical lags which do not capture the changes in context which have occurred since Keynes wrote his *General Theory of Employment, Interest and Money* in the '30s. This radically changed context includes the aforementioned worsening population/resource ratio the planet faces, the limits of human adaptation to rationalistic, massively scaled organization and production systems, mounting environmental disruption and the new global interdependence and rising militance of the less industrialized world. Thus today we notice that policies hailed as good economics, such as placing reliance on the free market, or vetoing legislation for public service employment, are becoming more obviously nonviable politics, poor sociology, inadequate systems theory and almost completely ignorant of psychology, ecology and the basic laws of physics.

Let us now examine the structural problems of our economy which are rendering our traditional macroeconomic medicines ever less potent. Ours is now a highly institutionalized, interdependent society characterized by large economic, social and governmental enterprises, whether we designate them as public or private. We have not yet recognized that each order of magnitude of technological mastery and managerial control inevitably calls forth a concomitant level of government coordination effort of varying effectiveness. In such an economy, the cybernetic operation of the free market described by Adam Smith, where small buyers and sellers met each other with equal power and information, only exists in residual areas. Institutions and interest groups and their relative political and economic power dominate the resource allocation system and produce the characteristic "viscosity" of mature industrial economies described by Adolph Lowe in *On Economic Knowledge*. In short, we must now admit that we already plan and we already allocate credit, as for example, when Federal Reserve policies encouraged the banks to bail out the real estate speculators and their Real Estate Investment Trusts (REITs). Thus facing the facts of extensive existing corporate/government planning, we can address ourselves to the need to plan more openly, democratically and in decentralized, counter-cyclical ways, so as to prevent those instabilities caused by excessive synchronization and scale.

An economy is a continually changing, evolving system, with new enterprises growing at its advancing edge while older corporations and institutions die off and dissolve. In this decay process they release their components of capital, labor, land and management to be reabsorbed into the fledgling companies in the leading sector for their future growth. Since an economy is also a living system, composed of live biological components, i.e., human beings in dynamic interaction with the energy and resources around them, it obeys the basic laws of physics and conforms to the same entropy/syntropy cycles of decay and regeneration as do all biological systems. Since the First Law of Thermodynamics tells us that matter-energy can neither be created or destroyed, it follows that for some systems or components of systems to grow, others must die and decay, so that their elements can be reutilized. The Second Law of Thermodynamics, the Entropy Law, states that all these cycles of building up and breaking down involve the use of energy, some of which is lost as waste heat. Since this waste heat cannot be recycled, the system very slowly evolves qualitatively and irreversibly toward greater entropy and disorder.[6] We should note that this trend is countered by the evolutionary drive to higher complexity and order in living organisms via knowledge or information, i.e., negentropy.

This basic model of the entropy/syntropy cycle and the irreversible evolution of all natural and biological systems is crucial to our understanding the particular subsystem we call our economy. Today, policy makers are trying to arrest this inevitable process of evolution and freeze the economy arbitrarily in its current institutional pattern. By their efforts at aggregate demand management, nonprioritized investment tax credits and granting tax relief or even bailing out feedlot operators, banks, and retailers, they are trying to preserve some of the large, obsolescent corporations in our economy.

Such efforts are understandable because these corporations have grown so big and employ so many people that we believe that we cannot do without them, and secondly, because such large companies and interest groups have the political power to persuade government to bail them out with taxpayers' money. A typical example is the current pressure being generated by large financial and corporate interests, spearheaded by investment banker Felix Rohatyn, to resuscitate a new version of the Reconstruction Finance Corporation to channel capital into the expansion of business and as a lender of the last resort to financially troubled corporations.[7] Such a distortion in already pinched capital markets would assure that allocations would flow to the older, obsolescent corporations while further starving innovative small ones. Another case in point is the auto industry, which has so long dominated the economy that it is traumatic for it and the whole society to adjust to the possibility that fundamental changes in energy and resource availability may mean an irreversible shift toward mass transit and smaller, more efficient, durable automobiles. Detroit, however, accounting for one out of every six jobs in the economy, is geared up for eleven million-car years and expensive annual style changes, and has found its products all but priced out of the market, in spite of expanding credit.

Similarly, our nation's utilities are geared toward trying to meet electricity demand projections that may never materialize. In the unfamiliar new world of capital shortages, to be discussed later, they or the auto industry cannot get the capital they feel they require without starving some other sector of the economy. With all their management reward systems predicated on corporate growth, they cannot envision a stabilization at their current size, let alone a devolution to a lower level of operations. One notable exception has been Chrysler Corporation's Lynn Townsend, who sees a future where the six-million car year is the norm, and has announced that Chrysler will cut overhead and middle-management ranks and gear itself down to this projected level of demand.[8] Many utilities are also rethinking or cancelling their overblown capital spending plans; as their new troubles: rising costs, angry consumers, the disappointing performance of nuclear plants and environmental and safety issues, have reduced their attractiveness to investors. However utility managements cannot conceive that alternative power sources will emerge based on systems for which they are completely unsuited to develop; for example, decentralized rooftop solar collectors on individual houses, apartment buildings and commercial facilities, or the production of methane gas from our nation's sewage plants. And as to current demand projections, utilities must heed those such as Leo Daly, chairman of the Energy Committee of the American Institute of Architects, who notes, "An effective national program of energy conservation in buildings alone, could within the next twenty years conserve approximately as much energy as any present supply system is expected to produce"[9]—and, one must add, at an enormous saving of scarce and vital capital for the development of new technology.

It is still hard for us to grasp that these mature companies, so long on center stage, are now taking their place at the obsolescent end of our evolving economy. If they cannot retool themselves and their products to meet new needs, they must be allowed to decline or pass from the scene, as did the buggy-whip makers before them. Only in this way will their capital, human and resource elements be available to be recycled into many new areas now being starved, such as solar energy, resource recovery and methane conversion. For many corporations which are overdependent on production of energy and resource-intensive goods, the writing is on the wall, whether they must purchase bauxite, chemical feedstocks or other scarce resources to produce an array of marginally necessary products, from aluminum foil, throwaway polyethylene and paper packaging to plastic toys and novelties. Such companies will soon find themselves unable to produce such items profitably and may have to launch demarketing campaigns following the lead of oil, gas and utility companies, as described in "The Decline of Jonesism" (Chapter 20).

If the normal growth/decay cycle of our economy is disrupted further, capital, management and resources will remain wastefully impounded within obsolescent corporations and the many government agencies created in the past to address long forgotten needs. In addition, calls by businessmen for even more favored tax treatment for investment and income from capital will worsen the social inefficiencies permitted by management's ability to set high levels of retained corporate earnings, above what is needed for working capital and replacement. The ability of large oligopolistic corporations to retain earnings, rather than pass them along to stockholders, not only weakens our nation's capital markets, but permits often arbitrary, or over-investment by management for its own aggrandizement, without such decisions being submitted to the discipline of the outside capital markets. Sumner Rosen of Brooklyn College has examined this problem in his paper, *The Inflationary Bias of Corporate Investment Control.*[10] Even more unfortunate, when capital is short, the current policies of banks favoring corporate borrowers with prime interest rates and the bias of government research and development assistance help shore up politically powerful companies with declining performance. Small companies, entrepreneurs and inventors are elbowed to the back of the line for capital, credit and government contracts by the wounded, but politically well-connected giants. Ironically, the giants exhibit poor performance in fostering innovation themselves, as many investments with exciting potential do not fit with their structure or are not able to deliver large enough profits to be significant or contribute sufficiently to their massive overhead costs.

Most of such political efforts to obtain special dispensations are loudly and often solely predicated on maintaining jobs, as if this had replaced their primary function of production. But raising the issue of jobs at all costs, like a religious icon in the face of the devil of economic difficulties, whether due to technological obsolescence, saturated markets, competition for resources or capital, management ineptitude, the need for energy conservation or to preserve public health and environmental values, can no longer go unchallenged. The inevitable question such claims invite is "jobs producing *what,* and at what *cost* to the taxpayer and at the displacement of what *other* public priorities in spending?" Most pleas for federal bailouts, subsidies or tax credits raise such awkward questions and therefore lead to much closer public scrutiny of the corporation, its management, its products and the social costs it may incur, such as health hazards to workers and consumers, or excessive resource consumption and concomitant pollution. In addition, questions will be raised as to whether its products are vital or necessary and deserving of subsidy, or marginally useful, frivolous, wasteful, or even detrimental, such as tobacco and amphetamines.

When corporations thus make jobs the key issue they also force us to reappraise our prevailing economic assumptions that the dominant means of survival and entitlement to an income is to be via holding a job, and whether private sector companies can absorb a major share of the workforce in an industralized advanced technological society. This refocuses the debate over the public and private sectors of our economy and their relative roles in providing employment, goods and services, and why we can "afford" a multimillion dollar aerosol-can industry, elephantine cars, energy-wasting gadgets and several hundred different brands of analgesic of questionable therapeutic value, and why we cannot "afford" adequate police, fire protection and sanitation and other civic services taken for granted in other less affluent countries, mass transit, health care, parks and clean air and water. This issue, raised by Galbraith in his *The Affluent Society* in 1958, has never been adequately addressed by mainstream economic advisers to either the legislative or the executive branches of government. More importantly, our current economic dilemma and the overriding issue of jobs (since the debate over guaranteed incomes has foundered on the Puritan ethic) has drawn our attention back to the ancient conflict between capital and labor and how the fruits of production shall be shared between them. Earlier, I pointed out the fallacies of neat neoclassical formulas for sharing these rewards, since production itself in an advanced industrial society has become a social process. But the erroneous assumptions that income shares to capital and labor can be rationally determined still lingers, even though these shares are most often determined by the relative power between labor and capital in each situation. In many cases there is the assumption that capital's share is immutable and only labor's is susceptible to negotiation. For example, this assumption underlies the Phillips Curve and together with the belief that cities can be run as businesses, both are evident in the maneuvering to balance New York City's budget, where job cuts are portrayed as the only option, and defaulting on interest payments to bondholders as a last resort. The extent to which economists have erred in defining the relative efficiencies between these two factors of production is now clear and bears crucially on our newly perceived but as yet little understood capital shortage.

A brief historical digression may be helpful. As our productive enterprises grew in technological and managerial scale and complexity, people were lured off the land and into the growing factory system. Land was gradually fenced and redefined as a commodity to be privately held and exchanged, and working people were forced off the land or traded their former self-sufficiency as farmers, craftsmen and small producers for jobs, cash payments and the greater mobility and excitement of the town and cities. But as technological efficiency and organizational size increased the workers' dependence on the new industrial system for survival became almost complete. As we are now seeing, there is a price to be paid for this dependence—in vulnerability to the vicissitudes of technological change, macroeconomic mismanagement, corporate planning errors and energy and raw materials scarcities.

For most workers, when such large-scale economic misfortunes occur, there is almost no conceivable alternative to sitting tight, collecting unemployment checks and waiting passively for conditions beyond their personal control to improve. The alternative of regaining personal self-sufficiency by moving out of cities, changing occupations to those perhaps servicing needs in a small town, for example, in carpentry, home repairing, roof shingling, plumbing, window-glazing, tailoring, locksmithing or gardening and the thousands of similar services for which there are huge backlogs of orders in most small communities, has become unthinkable. Such simple skills have rusted, are socially devalued even though they are as vital as ever, and such moves would be seen by most workers as a retrogression. And yet there is considerable evidence during recessions, including the recent ones, that such small local businesses ride out the storm more easily than the giants, due to their greater flexibility, reliance on family workers and modest overheads.[11] Moreover, there is a very real question as to whether centralized, industrialized societies, with their unmanageable complexity, can ever be operated smoothly enough to provide dependable income streams to their blue or white collar workers without frequent and painful dislocations such as we are now experiencing. Worse, many consumers and ordinary citizens now have lost faith that the economy is being managed for the good of the little people and their families, but believe it has been captured by the powerful and is being manipulated for their own economic ends.

As inflation continues to erode their real income, many Americans are now becoming increasingly angry at the monopolistic behavior of large corporations and the consumer movement will become less interested in details and push for more vigorous antitrust enforcement. Recent surveys by Opinion Research Corporation, Yankelovich and others confirm that approval of the performance of business has fallen to an alltime low of less than 20% of U.S. citizens. Reasons cited for this disapproval of business include the health and safety issues, Watergate and a widespread feeling that U.S. companies no longer play fair but put their own interests above those of the consumer.[12] Citizens now wonder if they can be relied upon to deliver uninterrupted electricity and all the consumer durables on which we have become hooked, with adequate safety and reliability, at prices we can afford and with tolerable levels of pollution and disruption of other community values. And as global competition for resources increases and they become further cartelized and multinational corporations increase their global search for resources, cheap labor and less burdened environments to pollute, the social tensions engendered by their normal profit maximizing behavior will increase. Other mature or obsolescent corporations may be forced to devolve their managerial superstructure and reduce costly overheads. This devolution is already evident in some industries, such as retailing, and conglomerates are spinning off divisions that they were busy acquiring in the past decade, while mammoth unmanageable cities such as New York face the same need to reduce unwieldy infrastructure. All these large institutions and power centers are now suffering from dis-economies of scale, centralization and the newly perceived vulnerabilities of complex, interlinked technologies. As transportation, which in the past has almost been treated as a free good, becomes more realistically priced, the trend to decentralization will continue. Regional and local efficiencies may again be able to compete with national market efficiencies, in the same way that we are rediscovering use-value as opposed to market-value, in growing more of our own food and doing more of our own house, car and appliance repairs.

If such a devolution of institutional superstructure is inevitable and, given new conditions, desirable, economic concepts will have to be expanded to embrace more external variables. Since this probably cannot be achieved without sacrificing some of the pretensions to rigor that have seemingly separated economics from its humbler sister disciplines in the social sciences, economists will have to moderate their claims to value-free scientific method if they are not to further confuse government policy-makers. More uncertainties will have to be acknowledged, and the paradoxical fact will have to be faced that much of today's research into policy-related questions, using environmental impact and technology assessment methods, actually increases *un*certainty, only revealing more of what we do not know. For example, the cluster of related problems economists define as inflation, recession and a shortage of capital require explanations from beyond traditional economics. While many economists have reached beyond the Phillips Curve explanation and recognized the role of the Vietnam war, global interdependence and rising expectations and, of course, the oil producers' cartel in increasing rates of inflation, few have explored the following underlying problems:

The declining productivity of capital investments. This phenomenon is rooted in our declining resource/energy base as we are forced to use ever more capital to extract resources that are more inaccessible and degraded. This decline in the productivity of capital investment is visible in the food system, where yields for many crops can no longer be increased by the massive increase in fertilizer inputs, as well as in destructively overmechanized fishing boats which destroy fingerlings and diminish the available catch. Ecologists Howard and Eugene Odum drew our attention to the same problem in the energy-extraction process, where increasing quantities of capital investment yield less and less net energy.[13] These new conditions, in which we are now "gnawing on the bone" in many energy and resource extraction processes, means that more of society's activities, wealth and income must be diverted into getting the energy to get the energy. The GNP continues to climb and we work harder, but our money simply becomes worth less in real terms, and the multiplier effect is felt in manufacturing and throughout the economy as inflation.

The mounting social and environmental costs of production and consumption, not factored into prices and often unquantifiable. Economic activities, especially when defined in free market terms, not only treat air, water and the absorptive capacities of the environment as free goods, but also the delicate web of the social system: the human relationships of the family, community cohesiveness and the network of social sanctions that enables societies to maintain order without constant recourse to police and courts. The maximizing of profits stresses the social fabric in many costly ways, for example, excessive mobility demanded of corporate employees can result in less stable communities, less committed voters and citizens, alcoholic wives and disturbed children and schools.

Technical and managerial scale and complexity exacts its toll in human casualities, dropouts, drug addicts and criminal behavior and

alienation. In addition such dis-economies of scale, complexity and interdependence mandate soaring transaction costs, as described in *The Entropy State* (Chapter 5), which begin to exceed the society's production. In addition, there has been little examination of the structural tendencies of our economy to overinvest capital, for example, the favored tax treatment of capital investment and income; the ability of large corporations to retain earnings and thus generate their own internal capital; and confusions in analyses of the factors of production which have led to the definitions of productivity and efficiency skewed toward the substitution of capital for labor and normative assumptions equating technological innovation with "progress," as illuminated by François Hetman in *Society and the Assessment of Technology*. It becomes clear that both market-oriented and Marxist economists share this confusion. Marxists still believe in the labor theory of value, even though in the intervening period since Marx's analysis, the population/resource ratio has shifted, and so capital/labor ratios shift to labor in varying degrees, all over the planet. On the other hand, the confusion of market-oriented economists is based on their belief in the equilibrium model of supply and demand, which leads them to view capital and resource inputs as unlimited. Thus, all that appears to be needed to call forth more supplies of raw materials is to hurl more capital into the extraction process, even when it becomes little more than a sink. This accounts for the conventional expressions of optimism, rather than more realistic worries, when aggregate capital expenditures rise, no matter how frivolously or wastefully allocated. Similarly, the constant demand of business and economists for increased tax favoritism to capital investments are based on a Keynesian, trickle-down theory of job creation, rather than alternatives via credit allocation and social investments, and on increasing the capital available on the supply side by progressively larger interest rates skewed to affluent, large savers and owners of existing capital.

Lastly, concomitant confusion exists over the correlation between capital investment and employment. In fact, capital is often invested to *reduce* employment, as in the case of oil-refining processes, automation of supermarket check-out lines and banking institutions' drive to install electronic funds transfer systems. The Report of the

National Commission on Technology, Automation and Economic Progress tried to lay this issue to rest in 1966, but it will not go away in an economy which officially accepts 7 to 8% of its labor force unemployed. Even a cursory critique of this report reveals its conceptual flaws, for example, accepting free-market assumptions in viewing the labor market, rather than the more persuasive case that labor markets are highly imperfect and much unemployment in technologically advanced economies is structural. All of these confusions rest on a more profound misunderstanding of "efficiency." Rarely do economists ask, "Efficient for whom and for what system or subsystem?" Corporate efficiency often results in less social efficiency if costs are externalized to taxpayers, as they frequently are. Such fuzzy definitions of efficiency in an interdependent economy lead to chronic suboptimization, since it is axiomatic in many other disciplines such as general systems theory, biology and ecology, that optimizing subsystem goals is always at the expense of the larger system. We can define "efficiency" on a case-by-case basis using the chart on p. 30 (see Fig. 1). Similarly, other value-laden words, "productivity" and "profit maximizing," are indefinable unless a system boundary and a time horizon are specified. In an even wider, longer term context, we must face the question of whether what we call "profits" and centrally planned societies call "economic growth" do not always incur matching but unrecorded debit entries in some social or environmental ledger. It seems more likely that as socioeconomic systems approach boundary conditions, such as those imposed by the laws of physics and cited by Dr. Alfred Eggers, Director of the National Science Foundation's RANN Program, in his warning to economists at the Senate Conference on Economic Planning,[14] that the concepts of "profits" and "economic growth" become little more than anthropocentric figments of human imagination. In the relatively brief decades of maneuvering time available we all hope that the technologists can repeal these basic laws of physics, but prudent economic planning suggests that we had better not count on it. Furthermore, options for technological substitution are fast eroding since it is now clear that a large range of potentially substitutable resources are becoming scarce simultaneously.

An economy which has been riding on a cornucopia of resources and enjoyed a long historical period of growth cannot be shifted drastically to a new course without dislocation and hardship, at best; at worst, widespread social unrest, depression and economic gyrations and, paradoxically for economists, continued inflation. Therefore a series of stop-gap policies are called for to ease passage and enable a rolling readjustment, while attention is turned to needed long-term structural changes in consumption patterns, lifestyles and values, and effecting the mandatory shift to sustainable forms of production and energy based on renewable resources and corresponding reduction of rates of materials throughput and maximum conservation of all nonrenewable resources. Such interim policies during the vital readjustment decade ahead should be geared toward:

1. Maintaining consumer purchasing power, even at minimal levels, by extending unemployment benefits to those in dislocated, resource-intensive industries and providing a comprehensive system of negative income taxes for those for whom jobs cannot be found. Such uncomfortably austere, but basic economic security for all citizens would act as a stabilizer and permit phasing out of old production processes and allow orderly transition without unacceptable individual hardship and the danger of widespread social unrest.

2. Vigorous conservation measures to preserve maximum options in dealing with oil-producer and other likely cartels and our own declining resource base, geared toward the concept of reducing throughput and that a BTU of energy saved is always cheaper than a BTU generated. Specific policies might include:

(a) Bringing domestic natural resources under greater democratic control: e.g., public lands, publicly owned oil and gas and coal reserves. Tightening up on leasing practices geared to rapid exploitation or speculation, e.g., separating exploration from leasing and retaining exploration in the hands of the government to determine exactly what is being leased and assure the public a fairer return. Bringing all major components of the energy industry under much tighter government surveillance and control, not excluding nationalization, anti-trust action and competing public corporations.

(b) Mandatory fuel-allocation programs, less penalizing to colder climatic regions than raising import-tariffs or rationing by price.

(c) "White rationing" of gasoline and taxing horsepower and fuel inefficiency. White rationing is, in effect, more of a "market" solution to gasoline conservation than raising prices and attempting to offset social inequities by setting up complex systems of transfer payments to the poor. Not only would raising prices set off another inflationary surge, but the need for compensatory transfers to the poor would engender higher transaction costs than a system of white rationing, while the taxes to be collected and transferred would probably prove an irresistible pork barrel and unlikely to be transferred to the poor. A white rationing system, based on the issuance of a prescribed number of ration coupons covering the total consumption rate targeted, would be made available to *all* citizens over age 18, rather than merely drivers. This removes a major inequity, since some 20% of U.S. families do not own a car, and at the same time, assures enough ration coupons in circulation to make counterfeiting and black-marketeering marginally unprofitable. Such a scheme has the additional advantage of providing not only a "stick" reducing consumption, but also a "carrot" rewarding nonconsumption, by permitting those who do not own vehicles to sell their coupons to those who do at free market prices. This would also reduce the need for rationing boards and costly administration because there would be plenty of extra coupons available to those willing to pay for them. Coupons could be issued each month at post offices or through other means, and any citizen who wished to turn them back at an established price could immediately do so, or otherwise sell them to friends, neighbors or employers. Thus a socially equitable solution to gasoline conservation might be possible without adding to inflation pressures, and improving consumer freedom of choice.

(d) Tax and credit allocation policies to reduce energy consumption and materials through-put, e.g., repealing depletion allowances and tax credits for those capital investments that are wasteful or unproductive in net energy terms. Directing and encouraging investments in the needed new industries geared to recycling and based on renewable resources, such as solar, wind and geothermal energy, methane conversion, with selected tax credits for such investments and government research and development funds targeted to meet similar criteria. Repeal taxes favoring use of virgin materials and explore amortization taxes to increase product durability. Require full disclosure of corporate R and D investments, and technology assessments to determine likely social and environmental impacts, shifting the burden of proof, in recognition of capital shortages, to the instigator of technological change.

3. Policies recognizing the need to shift to low-investment. labor-intensive production.

(a) Retraining programs for all workers dislocated from resource-intensive production, including the reorientation of highly skilled engineers trained in esoteric, high-technology fields where job opportunities will temporarily shrink, as emphasis in investment is placed on achieving economies of scale in manufacturing less complex solar, thermal and wind energy components and recycling systems, geared to replacing scarce capital with abundant labor.

(b) A federally funded program of public service employment, similar to that vetoed by President Ford, or modeled after the Humphrey-Hawkins Bill, to finance productive human service jobs, particularly to fill pressing needs in cities for restoring services cut by layoffs in police, fire, sanitation, hospital and education services, and to restore confidence of urban populations that cities will not be abandoned by federal and state governments, due to pursuit of economic policies based on free-market assumptions. Human-service jobs must be recognized for their advantages in the economic transition, as well as for their social and humanitarian benefits, for example, they are labor-intensive, capital-conserving, energy-conserving and environmentally benign.

(c) All federal programs now authorized, which create needed public facilities in labor-intensive, capital-conserving, energy-conserving, and environmentally benign ways, should be rapidly funded and initiated, e.g., the $6 billion authorized under clean water legislation for water-treatment facilities, funds provided in the Railroad Reorganization Act for repair of roadbeds, Highway Trust Funds for repair, maintenance and setting up of express bus lanes on all arterial roads. The Council on Environmental Quality[15] has prepared a summary review of all such mandated programs that can be initiated without further legislation, thus exposing the artificial conflict often fomented by special interests between labor and environmentalists. Conversely, all resource- and capital-intensive projects, such as new highways, space and other high-technology programs, should be reassessed in light of capital shortages and competing needs. Mass-transit expenditures represent a gray area, since many projects are overly capital intensive boondoggles geared to Buck Rogers schemes more suited to the capabilities of high-technology vendors than the needs of riders.

(d) Tax policies and expenditures to restore neutrality in treatment of capital income and wage income, treating dividends at the same rate as wages.[16] Repeal of across-the-board tax credits for capital investments, however wasteful or socially marginal, replacing them by credit allocation to essential industries, and tax incentives to needed new industries geared to declining resource base. Repeal of favored tax treatment permitting wasteful real estate speculation, ruining of agricultural land and all accelerated amortization which encourages unnecessary write-offs and capital replacement. Tax credits should instead be enacted to give incentives to hire labor and replace capital, for example, credits to those who are self-employed and employ others, with a cut-off aimed at encouraging small business and preventing the program from subsidizing large companies' payrolls unduly. Such a program would help reduce the disastrously high unemployment rates among minorities, teenagers, women and other special and unskilled categories, by making it more feasible for people to hire each other in child care, yard care, home repairs and small proprietary retail businesses. If capital is scarce, such a package of policies will raise its marginal cost, relative to labor inputs, thus conserving it for more optimal productive uses.

(e) Reform of banking institutions to limit bank holding companies and excessive speculation. Repeal of regulations limiting interest payments to small savers to encourage more decentralized capital formation. Rather than thus forcing small savers to underwrite cheap mortgage funds, specific subsidy of mortgages for home construction may be needed, as well as for low-cost housing.

(f) Anticipatory studies on the employment ramifications of all public and private investment over a certain size, such as the Employment Impact Statements, suggested by Jerry Brady in "Putting People To Work," in the *Washington Post*, (Nov. 16, 1974).

(g) More vigorous antitrust enforcement to prevent corporations with undue market power from exercising it for internal capital investment, and to restore market competition under the new tax constraints to prevent development of socially costly and wasteful goods and services.

(h) Explore ways of controlling the volume of product advertising on radio and television through the Federal Communications Commission and the Federal Reserve Board, to develop this as a new means of aggregate demand management. No First Amendment principles would be violated, since total advertising time available is already limited by regulations and the limits of the electromagnetic spectrum. Space and time would still be available to all commercials on a competitive basis, as today, but limiting the total time and ceilings for advertisers as a means of reducing inflationary demand would tend to prevent saturation advertising by powerful corporations to achieve rapid penetration and domination of market share, which in turn, would favor small businesses which cannot compete with such massive advertising budgets. Disallow corporate advertising expenditures as tax write-offs.

(i) Policies designed to encouraging the wider diffusion of capital ownership, such as the Employee Stock Ownership Trusts recently enacted, based on the ideas of Louis O. Kelso in *Two Factor Theory: The Economics of Reality*. Diffusing capital ownership and encouraging small savers with the same rates as large savers would tend to restore the vitality of capital markets and reduce tendencies to overinvest, as well as reducing the concentration of wealth and maldistribution of incomes.

Lastly, the *sine qua non* for the successful transition to a sustained yield economy based on renewable resources will be leadership and programs of public education to explain the basis for such apparently drastic policy shifts. The current vacillation in economic policies is heightening the atmosphere of fear and the loss of consumer confidence. An all-out program is needed to illuminate the new contexts, the need for change and to reassure people that a gentle, managed economic transition can sustain full employment, if we make the tough choices we now must, and are willing to forego our former waste of energy and materials in order to invest in our new productive base. It should even be possible to portray the advantages we will gain in less pollution and environmental disruption, not to mention the psychic relief in store for those who relinquish the destructive, exhausting game of keeping up with the Joneses.

Longer term structural readjustment to a sustained-yield economy and to restore lost flexibility will require policies of decentralization of population and industrial activities. Such moves are now under discussion in Congress, where the House Committee on Public Works last year appointed a Scientific Advisory Panel to explore how public works investments could direct and distribute growth more rationally to uncrowded, less developed areas of the U.S. Chairman John Blatnik[17] voiced concern over increasing urbanization, and the Panel's Report reviews the trend to centralization and cited the potential of performing "carrying capacity" assessments of the environments of many overcrowded regions, so as to set limits on their growth and redirect it to sparsely developed areas. Many other lawmakers have focused on more rational growth policies, including Senators Humphrey, Hartke, Mondale, Jackson and Bentsen, as well as Congressmen John Dingell and Morris Udall. The political viability of such policies is augured by the victory of Governor Richard Lamm in Colorado, who campaigned on a platform of reduced, balanced growth that has also proved popular in Oregon and California.

Lastly, it is vital that we counteract the long accepted notion that economics is a science, which is now proving destructive in that it is preventing Americans from talking to each other and debating the important subject of what is valuable. Economists must be called to account for this unnecessary mystification so that we can put an end to the concealing of value conflicts under the guise of technical or economic efficiency. Two policy suggestions for making the normative nature of economics explicit are:

(1) To amend the Employment Act of 1946 so as to expand the Council of Economic Advisors from three to seven persons, and include economists from labor, consumer, minority and environmental constituencies.[18]

(2) To similarly assure that the Federal Reserve Board members are selected from major constituency groups, instead of from banking, financial and business-oriented constituencies exclusively, as they tend to be at present.

The aforementioned proposals by no means exhaust the possibilities of steering our economy through the dangerous shoals ahead, and all of them need to be explored for unanticipated impacts in other areas and for second-order consequences which require careful further research. Recognizing the inherent difficulties of implementation, we have no choice but to accept the challenge.

---

[1]See, however, the work on energy modeling which has been more explanatory of the phenomenon of "slumpflation" (e.g., H.T. Odum, N. Georgescu-Roegen, and Malcolm Slesser).

[2]*New York Times,* January 26, 1975

[3]*New York Times,* April 30, 1975

[4]See, for example, E. F. Schumacher's Buddhist economics and G. Bruckmann's value-added-intensive production concept (*Auswege in die Zukunft* by G. Bruckmann and H. Swoboda, Molden Press, Vienna-Munich-Zurich, 1974, p. 236).

[5]See, for example, Recession is Capitalism as Usual, David Gordon, *New York Times Sunday Magazine,* 1975

[6]See, for example, Nicholas Georgescu-Roegen, *The Entropy Law and the Economic Process,* Harvard University Press, 1971

[7]*New York Times,* December 1, 1974

[8]*New York Times,* January 14, 1975

[9]American Institute of Architects, 1735 New York Ave., Washington, D.C.

[10]*Inflation, Unemployment and Social Justice,* Academy for Contemporary Problems, Columbus, Ohio, May 1975

[11]*New York Times,* January 22, 1975

[12]Interview with Dr. Arthur White, Daniel Yankelovich, January 9, 1975

[13]See, for example, Howard T. Odum, *Environment, Power, and Society,* Wiley Interscience, 1972

[14]Conference on Economic Planning sponsored by Senator John Culver of Iowa, Chaired by Martin Agronsky, May 22, 1975

[15]Council on Environmental Quality. Review requested by Senator Muskie and Rep. John Dingell, April 1975. Copies available from CEQ, 722 Jackson Place, Washington, D.C.

[16]For several of these proposals and insights into efficient factor-mixing, the author is indebted to: Dr. Mason Gaffney, Director, British Columbia Institute for Economic Analysis, Victoria, B.C., Canada; also Louis O. Kelso, author, San Francisco; Dr. Robert Edmonds, economic analyst, San Francisco; and Dr. Richard Pollack in an unpublished paper. Prof. Herman Daly, Dept. of Economics, Louisiana State University.

[17]*A National Public Works Investment Policy,* Committee on Public Works, U.S. Congress, 93rd Congress, 2nd Session, November 1974

[18]From statement of the Ad Hoc Committee for Full Economic Representation, Press Release, October 15, 1971

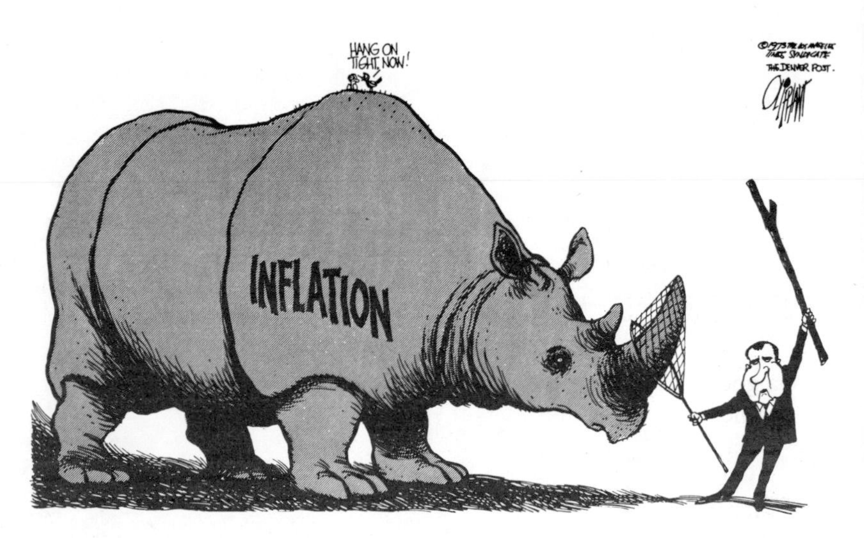

". . . then, picking up this big stick, we belt it firmly but moderately between the eyes!"

CHAPTER NINE

Inflation: The View from Beyond Economics

There still exists, after millions of words from economists on the subject, much confusion concerning the definition, causes and effects of "inflation." The term itself now mystifies more than it enlightens, since we humans still tend to assume that when we have *named* something, we have also *explained* it. The word "inflation," in fact, is a good example of a case in which taxonomy has become an enemy, rather than an aid, to thought. Consider, for instance, the dominant debate of the 1976 U.S. Presidential election campaign, as to whether inflation or unemployment is the more serious social problem. Firstly, framing our economic debate in this manner still presupposes the validity of the Phillips Curve model of the presumed trade-off between these two scourges of mature industrial societies, and the companion assumption that they are both aberrations of some imaginary equilibrium state (beloved of neoclassical economists) rather than structural features of industrial societies committed to undifferentiated, Gross National Product-defined economic growth.

The debate became ever-more circular and tautological under the Ford Administration as business-oriented economists tried a new twist to convince us that inflation is more serious than unemployment: inflation was pictured as the *cause* of recession![1] This "explanation" is cited by both former Treasury Secretary Simon and Alan Greenspan *and* Charles Schultze, President Carter's chief economic advisor! Economists still avoid taking responsibility for the fact that the entire business cycle from inflation to recession is now *created* by economists and conscious fiscal and monetary policy decisions, rather than by those mysterious forces of the "invisible hand," in which both Republicans and Democrats still believe.

Today, some economists are pointing to additional factors other than those of the traditional Phillips Curve in their attempts to explain the seemingly structural inflation of industrial economies. They include the global condition of economic interdependence, the glut of Eurodollars on world markets, the Vietnam war and the most recent cartelizing of oil prices, which, inexplicably, they discount as a one-time, transitory phenomenon, rather than acknowledging that the new higher prices are now base-line data, whose multiplier effects will remain crucial to future economic calculations.

However, as mentioned, two major sources of inflation cannot be understood in the traditional language of economics: the first involves the internal social dynamics of mature industrial societies; the second emanates from external constraints imposed by their declining resource-bases. The first, rooted in the "entropy state" syndrome, can best be understood by using the concepts of general systems theory, i.e., "inflation" can be viewed from this vantage point as a multiple crisis of suboptimization, where individuals, firms and institutions simply attempt to "externalize" costs from their own balance sheets and push them onto each other or, around the system, onto the environment or future generations. For example, as our technologies become inter-linked and more complex, they generate whole classes of new risks, which private casualty insurance companies are finding uninsurable: including massive tankers for oil and explosive, liquified natural gas, and carcinogenic or mutagenic chemicals—exposure to which may not trigger diseases for decades. Thus trade-off stages are reached between such complexity, and unanticipated risks and the necessary level of social risk-management. At some point, the costs accumulate to a point of diminishing marginal return. Such a technological society becomes inherently inflationary. It is interesting to note a similar effect in the functioning of computers: at a certain size and level of complexity, a computer system can spend more time in internal computation and transacting with itself than in generating useful output. Since, as shown in Figure 3, the social costs of all this mediation and regulation are added to the GNP as if they represented real production, such wheel-spinning economic activity expresses itself in increased rates of inflation.

The second cause of inflation as we have discussed is best modeled in thermodynamics, and concerns the sharp declines experienced in the productivity of capital investments, due to the declining resource-base of industrial societies. Thermodynamics can also pinpoint the incredible levels of waste with which traditional economics permits us to utilize these resources, and can point the way to much greater efficiencies, for example in more accurate thermodynamic matching of energy generation and distribution with end uses. Since industrial societies are resource addicts and their raw materials and energy must be extracted from ever-more-degraded and inaccessible resource-deposits, more and more capital must be cycled back into this process, with ever-lower net yields. Thus there are fewer real goods and services available and people work harder for less net return and money becomes less valuable, i.e., inflation. Energy analysts, including Odum, Slessor, Hannon, Berry, Long[2] and others, favor this interpretation, as does economist Georgescu-Roegen, author of *The Entropy Law and the Economic Process* and the severest critic of his fellow economists and their pseudo-rigorous forays into mathematical modeling.

We see that for these reasons, there is no possibility of returning to a deregulated economy without also returning to less complex, less capital-intensive, less centralized and less violent forms of technology. Such decentralized, labor-intensive technologies would be inherently less socially and environmentally costly, more benign and less needful of social control and regulation. I believe that we must now face the challenge of reconceptualizing the major policy paradigms used by most mature industrial societies. It is not merely a matter of the familiar aphorism of the computer age, "garbage in/garbage out," but a problem of "paradigm in/paradigm out." It is not lack of information; indeed, planners, policy makers, as well as the average citizens of industrial culture, are all drowning in data. But it is, on the whole, raw data, inappropriately collected, based on inadequate models and paradigms, which must now be repatterned.

These obsolete paradigms are generating statistical illusions.

Or take the unemployment figures generated by the Bureau of Labor Statistics, which, depending on how one views the situation, can vary from an "official" average figure of 7% up to over 10%. Even business publications were forced to deal with the wide disparities in these official figures, and *Business Week* has noted the confusion of economists in dealing with structural unemployment in an article entitled "Why Recovering Economies Don't Create Enough New Jobs."[3] Similarly, the *New York Times* has noted that the rate of unemployment among black teenagers has worsened steadily since 1955, when it was 15.8%; in 1965 it was 26.2% and in June 1976, it reached 40.3%.[4] Economists of the Nixon-Ford Administrations simply redefined "full employment" from the 4% unemployed level in the 1950s to 5%. By Sept. 14, 1977 former Nixon economic advisor Herbert Stein coolly claimed in the *Wall St. Journal*, that although his evidence was "partly anecdotal," we should begin accepting that 7% unemployed should now be the new definition of "full employment"! Similar manipulations of the Consumer Price Index and the rate of inflation are possible by carefully choosing the time periods used for comparative purposes. In fact, the election of Jimmy Carter undoubtedly in part was due to the revealing of such statistical deceptions and of the way in which business and political leaders and Washington bureaucracies thus altered the way issues were framed and governed by commissioning or "deep-sixing studies, deploying legions of consulting firms, and, too often, using "intellectual mercenaries" and deliberate mystification.

In Europe, popular revolts against bureaucratic, industrial capitalism have taken the more traditional form of left-wing attempts to assert greater control over the bureaucracies, as well as the campaign which kept Norway out of the Common Market and the various separatist movements and alternative, decentralized-technology groups based on traditional community control. What is now clear is that conflicts in all these industrial countries, formerly contained by Keynesian-fuelled economic growth, are becoming sharper as these forms of growth become unsustainable. Yet another way of stating the "inflation" problem is that inflation is a clear indication, as noted by James Robertson in *Profit or People?*, that economic measurement and criteria—prices and money itself—no longer provide an adequate tracking or scoring system. As this British operations researcher puts it, "An honest money system will only be restored in a society which is seen by all its members as being just and fair." This formula is not primarily idealism, but rather a correct axiom derived from general systems concepts.[5]

Let us briefly reexamine these mature industrial economies of Europe, the U.S., Canada and Japan. Since World War Two (and earlier in the cases of some) they have adopted the macroeconomic management tools associated with John Maynard Keynes without updating the earlier neoclassical equilibrium theory with his essentially disequilibrium theory. Keynes would have been shocked by today's rates of inflation, which he always inveighed against, and even more by the economic policies that bear his name. At first, these tools of aggregate demand management by fiscal and monetary means were conceived to ameliorate depressions and recessions and iron out the fluctations of the business cycle. However, it was but a short leap of economic policy makers' imaginations to begin applying regular doses to stimulate continual growth, and such economic activism has long since obliterated "naturally occurring" business cycles. Growth, of course, was politically alluring, since more people could be brought to the economic table to share the growing economic pie with those seated and already satiated. However, inequality of distribution of wealth and income and profits were justified as essential to capital formation and job creation, the major responsibility for which was to remain in the private sector.

Therefore, these economies rested on increasing of advertising-fueled mass consumption, and on the now infamous "trickle-down" model of capital-formation (consumption—jobs—income) and the use of Keynesian macroeconomic tools to mask social inequities and conflicts by government programs and—when competing claims could not be resolved politically—by printing money. Such macroeconomic management was only sustainable if cheap and dependable supplies of resource inputs were available from less developed countries (LDCs). But once these LDCs began to understand their own bargaining power and the natural resource addiction and consequent vulnerability of the industrial countries, they inevitably demanded a New Economic World Order, in which they seek to increase their share of the Gross World Product from the current 7% to 25% by the year 2000. Even short of outright, OPEC-style cartels and nationalization, some redistribution may prove inadvertent, i.e., the billions in shaky loans to LDCs by big private banks. Many of these LDCs realize also that their resources will increase in value in the ground, and are reluctant to trade them too quickly for inflation-ridden fiat currencies of the industrial countries.[6]

If industrial countries can no longer mask social conflicts by printing money, they will now have to deal with those conflicts politically and unravel the web of statistical illusion and unrealistic expectations that they have woven. Inflation, at an insignificant 2% or 3% level, has been condoned as a pragmatic tool of social policy—a pacifying device surreptitiously resolving conflicting claims on economic resources, while muddying up the waters concerning who are the winners and who are the losers. Similarly, by use of the Phillips Curve model, market-oriented economics has been able to submerge the conflict between labor and capital, characteristic of all industrial democracies. However, as Karl Polanyi pointed out in his prophetic 1944 book, *The Great Transformation,* the danger arises that the forces of labor and of capital become deadlocked, resulting in the political stalemate visible today in many industrial societies. It must also be noted that even Phillips did not postulate a Phillips Curve but considered his 1958 hypothesis of a trade-off between inflation and unemployment to be based on very scant empirical data and extremely tentative.[7] Yet this view of inflation has served capital owners and employers in their dealings with workers, unions and government policy makers, as well as the public, since it throws the discussion into supposedly "objective" areas of technological innovation and productivity. This finesses the issues of who owns access to the means of production and how its fruits are to be shared and has generally put labor on the defensive. However, the structural nature of unemployment today has forced economists to begin heeding the arguments of the growing band of young radicals in their own profession, such as those of the Union of Radical Political Economists. Some of these political economists point out that the current dilemma was predicted by Karl Marx, who held that advanced capitalism would not work without periodic recessions and the existence of an "industrial reserve army of the unemployed" such as we see today.[8] Indeed, this state of affairs was calmly acknowledged by economists of the former Ford Administration as a continuing fact of life, unless "productivity" was increased—the prevailing model of which we shall also reexamine.

This reliance on the Phillips Curve interpretation of inflation and economic reality also helped assure in cases of competing rights that those of capital owners and employers supersede those of workers. For example in New York City, the real power of government is no longer in the mayor's office or the elected City Council but resides with the Municipal Assistance Corporation. Municipal workers have learned this new fact of life, as their contracts are subject to "Big Mac's" ratification and the interests of bond holders are dominant while the job rights and security of employees seem to be infinitely negotiable.

Another aspect of the ancient struggle between labor and capital revealed by the collapse of Keynesianism is that now shaping the debate over how to achieve a full-employment society. Legislation to increase employment directly can no longer be wished away by calling it "inflationary." The entire federal budget is inflationary, habitually operating, it seems for the past few years, on massive deficits on the order of $60 billion annually. It is obviously irrational to single out one segment of it. All increments to the budget will be inflationary unless hard choices and substitutions are made with existing programs. All programs must now be weighed in light of the new conditions: whether military appropriations (the largest slice) or school lunches and food stamps; whether bailing out overextended, imprudent banks, ailing retail chains, railroads and aerospace companies or underwriting investments and subsidizing risks in energy development; whether draining revenue sources by giving tax credits for capital investments or special treatment for dividend income. The debate now focuses on *dis-aggregating* the budget and prioritizing it, thus revealing the truth of John Stuart Mill's contention that all economic distribution was essentially political.[9] We see now that economics is a normative, value-laden discipline, which can no longer parade as a value-free science. What is new in the current debate is that the battle between labor and capital can no longer continue, as before, in a vacuum. Not only are consumers and the politically underrepresented interested third parties, but the new constraints on environmental destruction and resources are now presenting a backlog of unpaid bills for past exploitation and a lien on future profit potential. Indeed, the entire economic debate has been excessively myopic. Ecological parameters are now forcing a new "reality principle" on the debate, equally inconvenient for capital, labor, consumers and taxpayers. For the first time, the question I have raised repeatedly must now be addressed: the extent to which economic growth and profits, as we reach boundary conditions, are any more than anthropocentric figments of the imagination. Even large corporations may be getting the message as they now commission more research than ever on scanning the global environment and declining resources, income-transfers among nations and political scenarios.

All of these issues, as well as the emerging conflicts between labor and capital, between corporations and consumers, taxpayers, environmentalists and advocates of the rights of future generations, are now visible to average voters and causing them to greet the banalities of economists, business and political leaders with increasing derision. Such incantations as that of Arthur Burns that the Federal Reserve Board intends to "stick to a course of monetary policy that will support further growth of output and employment while avoiding excesses that would aggravate inflationary pressures and thus create trouble for the future"[10] have an increasing air of unreality about them. Similarly, at the London summit meeting, leaders of industrial democracies still hoped that old-style stimulation, especially of the U.S. economy, could lead the way back to the good old days of consuming our way back to prosperity, the same formula enunciated at the previous summit meeting in Rambouillet, France. At the same time, paradoxically, the leaders wrung their hands over energy and raw materials shortages and lack of capital.

Such policy schizophrenia indicates clearly that two competing paradigms underlie the debate in all these industrial democracies: the dominant and increasingly obsolescent model of the private sector

"Golden Goose," versus the dawning view that the public sector is becoming the private sector's "Milking Cow." Let us examine these two competing paradigms, since the clash between them will likely characterize the economic debate over the next decade. The Golden Goose model implies that the private, "free-market" sector of the economy generates the wealth and that some of this wealth is transferred in the form of taxes to the public sector, where an increasing army of bureaucrats funnels these funds to the "unproductive" and makes social investments, as determined by legislation, to produce public goods and services, such as defense and education, unemployment insurance and welfare. In the simple decentralized age of our Founding Fathers, with a tiny population and a continent of unexploited resources, this model largely corresponded with reality. The "free market" underlying it also conformed fairly well to the conditions Adam Smith specified as essential for the "invisible hand" to allocate resources correctly: (1) that buyers and sellers meet each other in the marketplace with equal power and equal information, and (2) that no significant "external costs" would be visited upon third parties to these economic transactions.

Today, Smith's conditions for the operation of the "free market" are rarely met, while the Golden Goose has become, appropriately enough, little more than a child's fairy story. Indeed it has such a fairy-story quality that business leaders are trying to bolster it with other fairy stories, such as the "Little Red Hen" series of the Pennwalt Company's advertising, which serve as reinforcement to the Advertising Council's taxpayer-subsidized campaign of economic brainwashing, recently challenged by consumer groups under the Fairness Doctrine. Upon closer examination, we can see that the Golden Goose is still certainly excreting, but not necessarily laying Golden Eggs. Furthermore, this Golden Goose has been on a government life-support system ever since the Employment Act of 1946, when Keynesian macroeconomic management tools were instituted to give the Goose transfusions and pump up demand for its products, if necessary, by printing money. Today, the Golden Goose model of our economy conceals the extent to which private profits are won by incurring public costs, which are mortgaging our future. Taxpayers are not only coerced into bailing out private investors and

"PRESENT VALUE" ACCOUNTING HIDES LONG-TERM EFFECTS OF OUR ACTIONS

Diane Schatz

Office of the State Architect, Sacramento, California

corporations but into underwriting risky future investments as well, with no assurances of sharing in the profits. This completely inverts the free-market ideal, where investment risk is itself the chief justification for profits. Taxpayers are also called upon to subsidize private investments with tax credits with no concomitant representation in the decisions as to how they will be deployed. Neither is there assurance that these investments will be used to create jobs, the grounds on which they are justified. In fact, they are often used to disemploy people or constitute capital exported to other countries with cheaper labor and unused environmental potential to exploit.

Therefore, it is not surprising that the second paradigm now rapidly emerging, is that of that state as the private sector's Milking Cow,[11] where the politically and economically powerful raid the tax revenues and channel federal, state and local budget funds to enrich themselves, while the path of technological innovation follows producer priorities and capabilities rather than consumer choices and public needs. Thus, in this view, the taxpayers and the state are called upon to socialize costs and risks, while profits remain privatized.

This debate between the Golden Goose and the Milking Cow will provide a key to the politics of industrial societies for many years to come, since conditions have changed drastically in the past thirty years, and the issues continue to move from those of production and feasibility to distribution and the formulation of collective choices and goals. I have contended that we are in the throes of a great economic transition—from a society which maximizes production, consumption and materials throughput based on nonrenewable resources, to a society which minimizes production, consumption and waste, which instead recycles and reuses materials and is based on renewable resources and on energy sources such as the sun, wind and bioconversion and is managed for sustained-yield productivity over the long term.

The debate over this economic transition may be sharpened by clarifying the terminology of "efficiency" (see Figure 1), since it is the nexus concept linking past concerns with micro-productivity to current and future concerns for societal and long-term productivity. Traditional economics and its cost/benefit analyses suffers from a short-sighted emphasis on "present value" that tends to discount the future and long-term effects of our activities. First, we must know over what time frame efficiency is to be maximized. Economics' propensity to excessively discount the future is discussed by Talbot Page in *Conservation and Economic Efficiency* (1977). Even Buckminster Fuller's concept of efficiency—the idealized doing-more-with-less he terms "ephemeralization" (such as when transatlantic cables were replaced by less resource-intensive satellites)—cannot be assumed, apart from the issue as to the various *societal* efficiencies involved in the questions of ownership and access. In addition, there are crucial efficiency questions in the issue of availability of materials, where the location of production of the ephemeralized technology is to take place. When consideration of transportation, distribution and use-efficiencies versus market-efficiencies are weighed, it may well be more efficient in some cases to use a pre-industrial technology and locally or naturally occurring materials. Such questions are particularly crucial in solar energy efficiencies and in the newly rediscovered use-efficiencies of do-it-yourself household economics. Similarly, we have seen that as long as transportation remained cheap and undervalued, a condition unlikely to return, national market efficiencies could always manage to drive out regional and local efficiencies. This is also partly due to the propensity of national legislation to favor large-scale, national enterprises and to its being geared to statistical illusions based on concepts of national averaging as well as biases toward capital intensity in our tax laws (see Chapter 8).

We must also note that industrial societies have always chased the "mirage of efficiency" and its companion myth: that eventually bugs will be ironed out of new technology. We need to remember that some processes are not susceptible to improved efficiency. If the bugs were acknowledged in the original cost/benefit analysis of new technologies, many, including nuclear power, would never have been developed because the inevitable "cost overruns" would have been anticipated as part of the model. For all these reasons, we are still not very good at modeling efficiencies of scale and still entertain notions of mass production and high technology which pay too little attention to these subtler and more difficult trade-offs. For example.

the U.S. is gearing up to attempt to build an export food strategy to feed the world; this concentrates on the unrealistic blitzkrieg approach, while paying too little attention to the horrendous distribution problems involved. As discussed earlier, it might be better for the world's hungry if the U.S. would streamline its own bloated and energy-wasteful food system, with its overfertilizing, overuse of pesticides, overprocessed, overpromoted, overadvertised, overpackaged and overdistributed foods, not to mention its overweight consumers. This emphasis on the production side is natural, since it involves tangible, quantifiable processes and products, which of course, are easier to model than subtler factors, and so our research tends to follow such lines of least resistance. Where we have achieved some successes in modeling physical efficiencies of scale, we have completely overlooked metaphysical efficiencies of scale, such as the meta-level societal trade-off between division of labor and specialization and its transaction costs inherent in the "entropy state" syndrome. A crucial question recurs: when to average and when to dis-aggregate data, and how to combine the macro- and microviews in models, which entails explicating one's own vantage point in a system.

Such inexact criteria of efficiency have led to similarly distorted concepts of "productivity," usually measured as output-per-employee-hour. Such micro-measures of specific production processes, as discussed, usually correlate with the amount of *capital* at the disposal of each worker. Therefore, as such capital-intensive productivity gains are registered in various processes, capital tends to replace labor and the processes become ever more mechanized and automated. Such output-per-employee-hour productivity gains do not register the falling *average* productivity across the whole workforce, as more and more workers, whose productivity falls to below zero as they are shaken out at the bottom of the economy, join the ranks of the structurally unemployed. They then become part of the social cost component of the GNP, along with the other human casualties of mature industrial societies. Such skewing of capital/labor ratios toward overuse of capital occurs through such erroneous views of productivity. Even corporate leaders are coming to realize that the inevitable consequences of such corporate mechanization are growing political demands to make government the employer of last resort. In fact, the workforce of FORTUNE's 500 companies has only grown 1.4% since 1970 compared with 14.6% growth of the total civilian workforce.[12] Continuation of tax credits for investments, for which the business community lobbies incessantly, further subsidizes capital and its indiscriminate substitution for labor, even though it is the scarcer factor of production. In fact, as shown in Chapter 8, only if we move to a low-investment, labor-intensive economy, can we reduce structural unemployment and structural inflation.

To reemphasize, since capital, energy and materials are scarce and increasingly costly, rationality dictates that we conserve these natural resources, while fully utilizing our human resources. Only such a resource-conserving, full-employment, noninflationary economy can also be an environmentally benign economy. Capital is our precious "layer of fat," vital for the economic transition. Kenneth Boulding sees economic activity as a process of creating improbable, low-entropy structure at the expense of greater entropy and disorder of existing natural structure. Such thermodynamic views of economics see it as a subset of physics and conforming to its basic laws. This provides a more realistic framework for assessing available resources and their allocation and capital investment decisions and allows further incursions of thermodynamicists into the field of economic analysis. As noted in Chapter 3, many of these energy modelers believe that economic activities are best tracked and expressed in kilocalories, rather than prices and money units, since they represent a firm, scientific measure of resource potential and utilization efficiencies. Another obvious advantage is that governments could not inflate them, at least one hopes, without a howl from the scientific community. However, we have seen that one cannot derive welfare functions from thermodynamics and an energy theory of value would be societally suboptimal in many unforeseen ways and that we would not want to live in a society which maximized energy efficiency any more than one is comfortable in a society trying to maximize economic efficiency. Thermodynamic models would provide better base-line operating data from the real world. It is precisely these thermodynamic analyses that have illuminated the unsustainable productivity of American agriculture with its enormous fossil fuel subsidies: fertilizers, pesticides, gasoline and natural gas. Now these costly, petroleum-based inputs are not only yielding diminishing returns, but putting U.S. farmers in a financial squeeze. In addition, as Secretary of Agriculture Bob Bergland noted, in endorsing a return to organic farming with natural, biological pest controls, "We've got to start being more realistic in these energy-short times, about the use of oil and petroleum-based products."[13]

Amusingly, market-oriented economists, with their "output-per-employee-hour" theories of productivity and their waving aside of social costs as aberrant "externalities," are almost as wrong as those Marxists who still hew to the labor theory of value. I term this the "Houdini model of production"—it is like magic—no real oil, steel or copper or other tangible resources, with "capital" viewed as nothing more than "congealed labor." Tragically, market-oriented, socialist and Marxist economists all suffer from the same anthropocentric view of value. Only by exercising imagination, now our most valuable "software" resource, can we all hope to release ourselves from these old, myopic economic models, as outdated today as was the Ptolemaic view of the universe after Copernicus. Economics itself has become an inoperative category, almost, indeed, a form of brain damage.

Our resource-allocation debates must now be informed with meta-level concepts, as we reach the unprecedented current stage in the development of our species, now signaled by the meta-level choices we face. We have created with our technology conditions which now demand that we make these awesome choices consciously. These new cultural choices are analogous to the natural, evolutionary choices in the development path of species, and also involve these metatrade-offs and the economics of flexibility, i.e., "spending" flexibility now, versus "storing" future flexibility. Humans are now at this point of conscious genetic choice as evidenced by the nature of the policy choices we face. They are no longer choices between more roads, schools, sewers, or chemical factories, or between solar or nuclear or coal energy. They are meta-level choices between centralization and decentralization, between capital versus labor-intensive production, and lastly between conserving our store of flexibility and options versus "hard-programing" our investments into irreversible patterns which may destroy our adaptability for future changes. In fact, we can see that, in a large enough time/space framework, individual, group and species self-interest are identical. Our ethical leaders throughout our planet's history have not been unrealistic at all. Only their space/time frame has escaped our understanding.

---

[1] *New York Times,* "A New Theory: Inflation Triggers Recession," July 18, 1976

[2] See, for example, Proceedings of the International Federation of Institutes for Advanced Study, Stockholm, Sweden, Energy Analysis, Report #6, Aug. 1974

[3] *Business Week,* March 22, 1976, p. 114

[4] *New York Times,* July 11, p. 1

[5] James Robertson, *Profit or People?,* Calder & Boyars, London, 1974

[6] *Science,* Feb. 20, 1976, Vol. 191 No. 4228, p. 633

[7] A.W. Phillips, "The Relations Between Unemployment and the Rate of Change of Money Wages in the United Kingdom, 1861-1957," *Economica,* XXV: 100, November, 1958

[8] *Business Week,* March 22, 1976; ibid., p. 115

[9] John Stuart Mill, *Principles of Political Economy,* London, 1848

[10] *New York Times* editorial, July 5, 1976

[11] See for example, Michael Harrington, *The Twilight of Capitalism,* Simon & Schuster, New York, 1976

[12] Gallagher President's Report, Vol XIII, No. 39 Sept. 27, 1977

[13] *New York Times*, Sept. 27, 1977

Based on a paper given at the Fifth International Conference on Planning, International Affiliation of Planning Societies, Cleveland, July, 1976, and an article in *Planning Review,* March, 1977.

CHAPTER TEN

A Farewell to the Corporate State

I believe that private enterprise does have a future, both in this country and elsewhere in the world. But it is likely to be very different from what we see today. I would like to review the evolution of our current economy, and to suggest how private enterprise may change in the future and which of its traditional concepts are likely to be preserved.

In the Constitution, the concepts of private property and enterprise formed the basis of individual freedom. Today, two centuries later, these same two concepts are inextricably linked and confused with corporate property, bureaucratic enterprise, and state capitalism. At the same time, corporate/bureaucratic capitalism is under attack in our own country and in most parts of the world. "Property rights" and "private enterprise" are in danger precisely because the words have become synonymous to many Americans with the words' most visible manifestations—huge corporations. These vast new bureaucracies, both domestic and multinational in scope, as well as the government agencies that too often cater to them, have very little to do with private enterprise and have been giving it a bad name.

Private enterprise must be clarified and redefined if it is to survive. It will need to be clearly differentiated from those massive, quasipublic corporations that now dominate the world's economy. The extent to which they have scarred the image of private enterprise in the eyes of the American people is alarming. In the company magazine of the American Telephone and Telegraph Company,

"And though in 1969, as in previous years, your company had to contend with spiralling labor costs, exorbitant interest rates, and unconscionable government interference, management was able once more, through a combination of deceptive marketing practices, false advertising, and price fixing, to show a profit which, in all modesty, can only be called excessive."

executive John L. Curry wrote an article entitled "The Deteriorating Image of Business." Curry cited the now-familiar studies of Louis Harris, Roper Reports, and Opinion Research Corporation about the drastic decline of public esteem for business since Watergate. Between 1959 and 1973, the percentage of people who believe that major companies had become too powerful rose from 53 to 75, and 53 percent now feel that many major companies should be dismantled. The Harris 1974 study showed that over half of our citizens wanted more federal regulation of utilities, insurance companies, and the oil, drug, and automobile industries, and 87 percent agreed that too many businessmen are more interested in profits than in serving the public's needs.

Since the Sherman Act of 1890 we have attempted to constrain monopolistic power. Muckrakers from Sinclair Lewis and Ida Tarbell to Ralph Nader have drawn our attention to the social costs of unbridled free enterprise. Yet corporations have continued to grow and have now extended their power around the planet. At the same time, they continue to resist governmental regulation. Oddly enough, many misguided business spokespeople, still trying to gloss over the differences between economic totalitarianism and true private enterprise, may now be the ones who are hastening its demise. President Ford's Commerce Secretary Rogers C.B. Morton believed that the problem of public distrust could be solved with more "economic education" concerning the necessity for profits. Mobil Oil has waged a three-year campaign on the same theme, and on the need for more economic growth "to help the poor." William L. Wearly, chairman of Ingersoll-Rand, decries the U.S. government "hamstringing our activities" and adds, "We cannot understand this harassment." Firestone Tire and Rubber Company's president believes it may be too late to salvage "free enterprise." He advocates that more businesspeople "speak out."

All this has a familiar air of déjà vu about it. The latest round of corporate criticism of big government and creeping bureaucracy is as self-serving as ever. As Jefferson observed, "Those who have once got an ascendancy and possessed themselves of all the resources of the nations... have immense means for retaining their advantage." In fact, a broadening array of Americans, from Gerald R. Ford to Ralph

Nader, are also calling for deregulation. Public-interest and consumer groups have pointed out that many of the alphabet agencies set up to regulate industry in the 1930s have become little more than taxpayer-supported cartels which encourage waste and blatant inefficiency. Corporate leaders, using ringing phrases as "lifting the shackles off the backs of business," capitalized on such real concerns to push for a wholesale rollback of government regulation. This does not discriminate between big and small businesses, and it avoids the real choices which neither big business nor big government wants to face. They have not yet come to terms with the growing dilemma of complex, industrialized societies: the extent to which private profits are now being made at public costs, and the *need* for government coordination and control.

Take the case of nuclear energy, where since 1964 power plant construction costs have increased ten times faster than the Consumer Price Index, according to Richard Morgan, in *Nuclear Power: The Bargain We Can't Afford.* The only explanation for these skyrocketing costs comes from the problems in nuclear technology itself. Since 1957, more than a dozen major changes in federal safety regulations have become necessary. In addition, a study by Charles Komanoff of the Council on Economic Priorities on *Power Plant Performance,* revealed that the nation's existing nuclear plants were falling far short of delivering electricity at their promised capacity factors, and therefore in most of the U.S. nuclear-generated electricity was more expensive than that from older, coal-fired power plants. Contrary to conventional wisdom, Komanoff also found that the larger the nuclear plants, the less efficient they became.

Ironically, it was that great admirer of the American experiment, Alexis de Tocqueville, who pointed out that our revolution against *political* totalitarianism had installed a decentralized system, based on individual liberty, that contained internal inconsistencies. He believed that these might lead us to *economic* totalitarianism. In his famous 1835 study, *Democracy in America,* de Tocqueville drew attention to a major contradiction in our system: Equality of political condition would lead to increasing incomes, which would lead to greater demand for manufactured goods, which would require

"Our recommendation, then, is that these plans be shelved till the post-Watergate morality blows over."

greater division of labor. This specialization would increase the relative differences in income and "mental alertness" between workers and owners. It would result in a "manufacturing aristocracy." De Tocqueville feared that this new economic elite would then create a more centralized government to coordinate the complex social, economic, and political order. Clearly, his fears were justified.

The free market, where, as described by Adam Smith, buyers and sellers meet each other with equal power and equal information, now scarcely exists in the United States. Institutions and interest groups now dominate the huge American economy. Growth and centralization were accelerated by World War II, Korea, and the Vietnam conflict. Other contributing factors were: access to a cornucopia of resources, cheap and abundant capital, and a large domestic market on a thirty-year buying binge fueled by soaring consumer credit.

But the foundations of the postwar economy have now shifted. The Keynesian remedy of pumping up consumer demand to ameliorate structural unemployment has failed.

The very laws of physics have also made our capital investments less productive. Declining returns on investments are showing up in agriculture and resource extraction first. They occur largely because we now cycle more and more precious capital back into the processes of extracting raw materials and energy, from more degraded and inaccessible deposits. The net yields decline. Fifty years ago it was only necessary to stick a pipe into the ground in Texas to get oil. Now we have to build a multibillion-dollar pipeline from the North Slope of Alaska. Yet our corporate leaders are not fully aware of this, because most of their economic forecasters still assume fairly constant returns on capital investments. In the face of this capital shortage, paradoxically, businesspeople are calling for *more* tax credits for capital investments.

There is a growing debate over the question of the limits of growth. It now seems clear that we will encounter the social, psychological, and conceptual limits to growth long before we collide with absolute physical limits. These nonmaterial limits include the external factors and mounting domestic social costs I have mentioned and our general inability to model or manage the complexity we have created. The immediate growth issue we must face is that in all systems, if some parts are to *grow*, other parts must *die*.

If some corporations are to grow, others must decline and pass from the scene, releasing their capital, land, and human resources to new and growing enterprises. Current policies of managing aggregate demand, creating capital with tax credits, and bailing out obsolescent corporations will tend to arrest this natural evolution of the economy and attempts to revive outmoded wounded giants will exacerbate the capital shortage, aborting innovation and needed new growth.

Even the proponents of Project Independence are beginning to realize that if we diverted *all* of our capital to developing new domestic energy sources, it would still take many years to become self-sufficient and energy conservation would still be essential. To effect this difficult shift with little disruption, we need to conserve our precious capital. We must build a new productive system based on recycling and renewable materials and energy sources such as solar, wind, wave, and geothermal energy; methane conversion from sewage and wastes; tree farming as well as bioengineering methods of production; and recycling employing enzymes and microorganisms. The new economy of permanence will require much technological innovation. It affords tremendous opportunities for entrepreneurship and initiative. Many, although by no means all, such new technologies are most efficient if they are decentralized. They will therefore foster a vast new generation of inventors and small businesses. Unfortunately, the economy's normal growth/decay cycle can be disrupted further. Capital, management, and resources can remain wastefully impounded in obsolescent institutions, like corporations with saturated markets or like the many government agencies created in the past to address long-forgotten needs. Wounded, politically well-connected giants still elbow small companies, entrepreneurs, and inventors—the bearers of innovation—to the back of the line in the search for capital. We will always need some large-scale operations, such as steel manufacturing and our national telephone system. But we cannot count on the giants to encourage innovation themselves. More inventions come from small companies, as Commerce Dept. studies show, and they also employ more people per dollar of capital invested.

In 1780, over 80 percent of the American people were self-employed. Today, only 10 percent are their own bosses. Hundreds of thousands of small businesses have been driven from the marketplace, often by the market power of large companies. Small businesses and personal entrepreneurship are discriminated against on all sides, often by the very regulations designed to constrain large corporations. Firms with professional proposal-writing staffs tend to win federal research and development money. Small businesses often pay dues to giant trade associations controlled by large corporations, sometimes helping to finance legislative lobbying directly contrary to their own interests.

Complex technologies, not small businesses, usually receive subsidies. This tends to favor capital-intensive large corporations. For example, investment tax credits help substitute capital for labor, thereby increasing structural unemployment and structural inflation. Such policies are generally justified because they increase "efficiency." Using our Efficiency Chart (Fig. 1), let us take an example: it may be efficient for a large supermarket chain to automate its checkout counters. But this is not socially efficient, since 100,000 human checkout clerks may eventually be forced onto the tax-supported unemployment or welfare rolls. Nor is it efficient for consumers, since there is no assurance that cost savings will be passed on in lower prices, and comparison shopping will be impossible without individual price stamping. Finally, it may reduce market choice and neighborhood convenience by driving more small grocers out of business.

Such capital-intensive innovation is usually justified because of the overall need to stay competitive in foreign markets. But investment subsidies are still masking the true scarcity and rising price of capital. Projections of historical increase of labor costs vis-à-vis materials and energy are still masking the fact that labor is now becoming the more efficient factor of production in many industries, and, as discussed in Chapter 9, our notions of "productivity" are confused.

As global supplies of energy and resources become scarcer, more expensive, and further cartelized, the devolution of industrial superstructure may occur inevitably. It is already happening in major corporations and mammoth ungovernable cities like New York. Corporate and municipal threats of bankruptcy have become commonplace and even our Social Security System is underfunded. Elsewhere, there is increasing balkanization. Ethnic peoples within nations are deciding that central governments in faraway cities are corrupt, inefficient, and unresponsive. Large agglomerations around the world are suffering from dis-economies of scale and vulnerability because of their complex interlinked technologies. And as the price of transportation rises, the trend to decentralization and to regional and local economies will continue.

Technological mastery and the increasing social complexity it engenders have now created not only domestic interdependence, but global interdependence. In our confusion over how to manage a social transition at home while simultaneously dealing with overlinked, synchronously oscillating foreign economies, we turn to stopgap policies and incremental adjustments. In fact, we are trying to buy time to address long-term needs for structural changes. With our inadequate models of our social and economic systems, perhaps the best we can do is to correct errors and overshoots when they finally become apparent. For example, when we observe that we have begun overusing capital because of tax subsidies, perhaps we should change course and substitute temporary tax credits for employment. Even Arthur Burns has acknowledged that it would be cheaper to create a program of public-service employment than to continue Keynesian-style, enacting across-the-board tax cuts or pumping up the whole economy by printing money. In any case, it seems clear that the long binge is over. Like so many societies before us, we are lowering our overblown expectations. And if we do not manage the economic transition that is upon us, the system will continue to do it for us, with inflation.

The natural human impulse when faced with unmanageable complexity is to try to simplify it. There seem to be two very different ways to do this. One is to simplify by *controlling* some of the proliferating variables, such as when companies try to deal with the vagaries of the market by collusive action or saturation advertising. Other forms of simplification by control include the growing use of technological remedies for technological ill effects, so that nothing less than a "breakthrough a day is needed to keep the crisis at bay," in Schumacher's words. Politically, simplifying by control involves modeling and monitoring more and more microinteractions. This leads to the computerized Leviathan state of George Orwell in *1984*. Perhaps our computer scientists can develop hardware and programs for such a task. But I am doubtful, since such attempts always have failed in the past, mainly through loss of feedback.

The other path requires simplifying the hardware itself, i.e., the technological interlinkages which are so vulnerable to breakdowns, sabotage, and supply bottlenecks. Reordering the scale and interlinkage of our technologies and decentralizing our massive urban complexes will be very difficult. However, it may be easier than trying to increase our computer modeling skills and capabilities for control. Many of us are trying to revise our physical arrangements, technological configurations, and our relationship with the land and ecosystems that sustain us. The new practical philosophy is perhaps best summed up in such "sleeper" best-sellers as Robert Pirsig's *Zen and the Art of Motorcycle Maintenance,* and the "guru" status of perennial anarchists Scott and Helen Nearing, authors of *Living the Good Life.* Such writers are exploring decentralization and more humanly appropriate forms of technology.

For it is becoming apparent that technology itself tends to exacerbate maldistribution of income and social inequality. The higher the capital costs, the more restricted the access to the capital and the more concentrated are its controls and profits. For example, we have lost autonomy, increased our risks, and eliminated safer alternatives by stressing the nuclear-breeder-reactor program, which the General Accounting Office has determined will cost five times more than originally estimated. Nuclear power, or the automobile, stunt other competing modes of power generation or transportation. Ivan Illich calls this situation a "radical monopoly," far more pervasive than the usual forms of monopoly described by economists. Such technologies also tend to create human dependency, and passivity. We must ask whether or not the American consumer *uses* energy slaves, or has *become* an energy slave?

As land becomes scarcer and more vital for food production, people are exploring alternative land-use and ownership patterns in the belief that land can no longer be treated as a commodity for real-estate speculation. Groups, such as The Trust for Public Land, the International Independence Institute, and the Cooperative League of the U.S.A., are developing the concept of community land trusts. In these arrangements, land use is guaranteed, but ownership is held by a mutual trust society. These land trust organizations differ from those based on religious values, such as the Shakers. Another group of similar communities is based on Henry George's concept of "single-tax" enclaves designed to make real-estate speculation unprofitable. Among these are town corporations such as: Fairhope, Alabama; Arden, Delaware; Freeacres, New Jersey; and Tahanto, Massachusetts. It is not clear the extent to which these and other alternative institutions can promote the concept of "trustery" rather than property. But they do represent much needed diversity in our problem-solving approaches to creeping giantism. To serve and link these proliferating lifestyle experiments are publications including *Mother Earth News,* and *Organic Gardening* and *Prevention.*

Other innovators are developing technology along new lines. They generally favor small-scale and decentralized systems, designed to augment self-sufficiency rather than deliver power to a few at the expense of the many. It is variously referred to as "low-impact" or "soft" technology, because of its environmental compatibility, or "intermediate" technology because it raises human productivity by an order of magnitude over primitive tools without destroying traditional cultures and population patterns. It is based on Mahatma Gandhi's concept: not mass production, but production by the masses. It is also called "liberatory" or "demo-technology" because it is cheap and labor-intensive and accessible to large numbers of small entrepreneurs. It is discussed in various recent books such as *Small Is Beautiful,* Ivan Illich's *Tools for Conviviality,* and David Dickson's *Alternative Technology.* These address the new question of which technologies tend to create artificial scarcity and dependence while increasing aggregate human ignorance, and which can be devised to avoid these pitfalls.

The global capital/labor ratio is shifting toward labor intensivity. People are now plentiful, and resources are scarce. The old notions of productivity and efficiency, based on Western ideas of substituting capital for labor, are gradually being reversed. Some former promoters of "big-bang" technology, like the World Bank, have come around to the new thinking. United States Aid to International Development programs are specifically developing intermediate technology projects. In 1971, the U.N. highlighted the need for "appropriate technology" in its *World Plan of Action for the Application of Science and Technology to Development.* It pointed out that unskilled labor and the need to find greater use for the ample reserves of "developing" countries' primary materials require technologies which would substitute labor and local materials for capital and foreign exchange. This trend does not apply only to rural development. In urban areas, self-help neighborhoods are organizing for crime prevention, community gardening, basement fish farming, and cooperative carpentry. They are bartering skills and services from child care to TV and car repair, thus raising their welfare without money income.

The ultimate industrial revolution will be based on systemic insights. It will require a much greater ratio of intellectual software vis-à-vis machine hardware. A problem of production will not automatically trigger mental images of factories and hardware. For example, if nitrogen fertilizer is needed, it is now all too easy to decide to build a fertilizer-producing factory. Subtler thinking might involve scanning natural ecosystems for nitrogen-fixing microorganisms and encouraging their growth in the soil. This radically different approach to production is "bioengineering" or "biotechnics." It often entails the use of microorganisms and enzymes and other combinations of natural organisms and ecosystems. Such biological systems can produce nontoxic pesticides and fertilizers, methane gas, single-cell protein for animal feed grown on waste substances, commercially farmed fish, natural pharmaceuticals, and even oil.

Biotechnic methods are already being used to increase tree growth and reforestation. Ecologist Eugene Odum points out that human solar collection technology is not yet as efficient as photosynthesis. There is currently much interest in using renewable timber resources as fuel. According to one estimate, a 256,000-acre tract of forest can continuously supply a 400-megawatt plant generating electricity for 200,000 people. Projected capital costs are less than $1,000 per kilowatt, considerably lower than present sunlight-to-electricity conversion costs. Furthermore, houses built to take full advantage of solar energy, and methane-producing sewage conversion, can reduce fuel costs to zero.

Many large corporations are also researching enzymes and microorganisms. But their capital-intensive, artificial, laboratory operations are likely to be mismatched with ecological needs and may prove destructive or unsustainable or both. Among the more promising innovations from larger companies is General Mills's new nonpolluting process for extracting copper by chemical leaching. The technique, which operates without combustion, can make use of low-grade deposits and even old mine tailings. The Baghdad Copper Company in Arizona uses the process to make copper of 99.9 percent purity without any combustion. But whenever large corporations with huge overheads adopt such technological innovations, there is a danger that they will distort them by trying to develop them on a massive and disruptive scale. For example, some utilities in this and other countries are now trying to force the inherently decentralized technology of solar energy into a mode using enormous, centralized collectors. Some of the most sophisticated biotechnical research today is being carried out by small groups of inventors and entrepreneurs. These include: the New Alchemy Institute of Woods Hole, Mass.; Britain's Biotechnical Research and Development; Low-Impact Technology Ltd.; and Intermediate Technology Development Group; the Latin-American Institute for Economic and Social Planning; Sweden's International Foundation for Science; the U.S.-based Volunteers in International Technical Assistance, and the Brace Institute in Montreal, Canada.

Even more imaginative are the current efforts to set up "alternative world bank" mechanisms. One idea, for example, is based on the issuance of a nongovernmental, inflation-proof, commodity-backed currency. This concept was advanced at the Bretton Woods Conference of 1945, and its most active proponent in the U.S. has been the indefatigable Ralph Borsodi, author of *Flight from the City* (1933), *Inflation is Coming and What to Do About It* (1945) and *Social Pluralism* (1952). Recently Borsodi set up a successful experiment in Exeter, New Hampshire, where his commodity-backed currency, which he devised while working with fellow economist Irving Fisher at Yale, was issued and circulated. As reported in *Forbes* and *Business Week,* holders of the new currency, called "constants," saw their value rise 17% in relation to the dollar in just three years. Since such an alternative currency would be backed by the real value of a representative "basket" of some thirty vital commodities, from gold and silver to copper, iron, petroleum, wool, soybeans and peanuts, rather than the almost-fiat money issued by governments and central banks, it could, when quoted on world exchanges, become the "conscience" of all such inflation-prone currencies. The inflation-proof constants might be purchased by any citizen of the world whose national government did not interfere and exchanged for commodity contracts or any other national currency at rates quoted daily. The problems of transportation and storage of the commodities which back the constants have yet to be solved.

This "alternative world bank" concept has been studied by the International Independence Institute and others. Another, to set up a Women's World Bank to make loans available to enterprises started, owned or operated by women in all countries, to overcome the sexism of many conventional financial institutions, is also in the development stage, pioneered by former New York-based securities analyst Michaela Walsh. It is clear that countercyclical alternative trading systems would help stabilize and complement that existing today and which is so dominated by the power of the developed countries. Success of alternative systems may be spurred by the growing resentment of Third World nations at the increasing, and in their view, unjustified militancy of the U.S. and other oil- and raw-material-consuming industrial nations in opposing their efforts to cartelize their resources in order to extract fairer prices for them. This growing resentment was articulated in a full-page advertisement in the *New York Times* to rebut former President Gerald Ford's speech before the U.N. calling for a rollback in oil prices. Placed by the government of Venezuela, President

Carlos Andres Perez in this response to Ford stated that Venezuela was willing to work in an international forum to establish a newly balanced relationship between the raw materials produced by less developed countries, on the one hand, "and the manufactured goods and technology, on the other, which are possessed by the developed countries and are in essence, *the source of economic marginality and growing poverty in which over half of mankind continues to live"* [italics added].

Proponents of intermediate technology and new forms of productive organization, the followers of Illich, Schumacher, Dickson and others, see great advantages to rural development in using alternative currencies. Their goal is to raise manufactures and exchange from their limits of face-to-face barter at the village level to facilitate growth of markets embracing larger rural areas and groups of villages, without sucking such transactions into the vortex of national currencies and financial systems. Such discrete money systems already operate side by side in many so-called backward countries. Alternative currencies like the "constants" might permit such autonomous trading regions to flourish without provoking too much interference by national or world financial forces. Similarly, an almost forgotten experiment was conducted in Cincinnati, Ohio, at the turn of the century. Called the Time Store, it was successfully operated for two years and simply dealt in IOUs promising to perform specific labor, such as home repairs, lawn mowing, baby sitting, etc., which were exchangeable with other patrons' promissory notes to perform other labor and services, or for goods produced by the participants.

Indeed, as inflation continues and existing systems become more unstable, we are already seeing an increase in bartering of goods and services, as everyone from lawyers and entrepreneurs to small vendors, repairmen and craft hobbyists avoid taxation by exchanging goods, services, favors and referrals.

Luckily, one cannot stamp out human inventiveness. Private enterprise is not dead. Adaptation, risk taking, and even altruism, are normal human characteristics. Perhaps the bureaucratic planners of the public and private sectors may yet be checked and balanced. Perhaps the heroic global-crisis managers may yet learn from the more resilient and adaptable systems of nature and the organic life-force present in human beings. Perhaps equity, balance, pluralism, and equilibrium can be restored to humanly designed systems.

We must continually check the tendencies in human societies for knowledge and power and wealth to cluster. This clustering produces centralized and exploitable surpluses; these in turn excite greed and envy and create opportunities for despotism. Human history can be seen as an endless replay of the abuse of power and of surplus, and the struggle to check and balance the destabilizations thus caused. Perhaps there is no possibility of ordered human societies without such cycles of instabilities. Perhaps all we can hope for is the continual redress of imbalances by periodic redistribution. What may distinguish our current instabilities are their unprecedented size and concentration. As we Americans cross the frontier into our third century, the stakes have never been higher. We have exhausted the limits of empty technique. From our efforts to retool the world around us, we must now change our values and our ways of seeing things.

Reprinted from *Business and Society Review*, Spring, 1976.

CHAPTER ELEVEN

Constraints Affecting the Future of a Resource-Intensive Industry: Packaging

In this chapter, I shall examine systemic constraints and create a larger matrix for assessing the future of growth-dependent, capital/energy/resource intensive industries. The methodology can be adapted to assessments of the futures of many such industries. I shall attempt to demonstrate how much a more-inclusive, generalized matrix can lead to specific conclusions about a particular industry: in this case, the packaging industry.

In no area of corporate planning are longer term assessments of alternative futures more necessary than in the sociocultural sphere. However, the methods for forecasting shifts in social and cultural values are rudimentary, unreliable, and can do little more than extrapolate past trends in human behavior, for example, the rise in industrial societies over the past decade of the social movements for consumer and environmental protection and greater corporate accountability. The question for corporate planners is: "Will these trends continue: will they grow, or are they fads which will be replaced by even more unpredictable new behavior patterns?" Although these issues appear almost to dominate, I shall try to show that they are derivative, as well as how they interact with other factors and fit into a broad picture of constraints which, I believe, will govern most capital- and resource-intensive industries for the rest of this century.

The methodological approach subsumes economic forecasting which is inadequate in dealing with human motivations and behavior, since it has not yet incorporated into its concepts many of

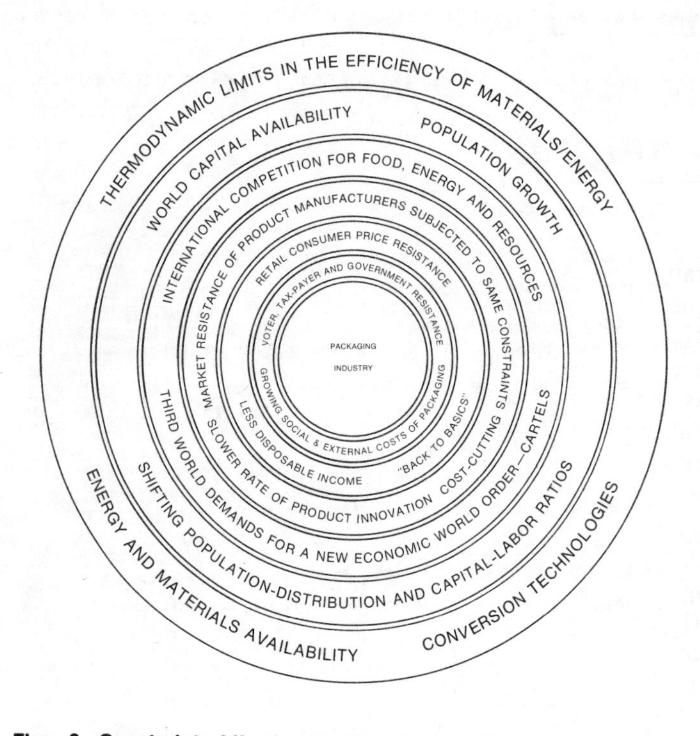

Fig. 8 Constraints Affecting the Future of the Packaging Industry

the advances in the behavioral sciences and cannot model the dynamic changes in consumer preferences except in hindsight. For example, the whole question of "consumer confidence" which is so crucial to predicting performance of industrial economies, is invoked more as a magical incantation than as a known variable. As the economist, G.L.S. Shackle pointed out, human expectations are a key to economic behavior, which in turn is a function of aggregate decisions, which can often be irrational, or based on inadequate information.[1] Prices, in fact, are nothing more than our collective, subjective expectations of availability, and often, as we have discovered recently, they are merely a measure of the time-lag in our general understanding of scientific data on actual resource availability and rates of depletion, or on unperceived environmental constraints, such as the inaccessibility of water, which is now limiting U.S. plans for coal liquification and other energy developments in our Rocky Mountain states. These basic truths about prices and resource availability were brought home to the packaging industry in the past few years by many unexpected bottlenecks and shortages in materials supplies. To forecast future economic conditions with any confidence will require constructing a larger matrix for understanding both objective and sociocultural constraints and how they interact.

I shall examine six systemic levels of contraints affecting the packaging industry, and explore options at each level for addressing them (Fig. 8). These six concentric levels range from the macroconditions to the microeffects most often felt by the packaging industry in its immediate interactions with consumers, taxpayers and government. They are as follows:

1. Global energy and materials availability, and the thermodynamic limits in the efficiencies of energy/materials transformation technologies.

2. Population growth and distribution patterns; shifts in the capital/labor ratio; and worldwide capital availability.

3. International competition for energy and resources; demands of Third World countries for a "new economic world order" and potential of more resource cartels and nationalization.

4. Market resistance of product manufacturers subjected to the

same set of constraints, leading to slower rates of innovation, cost cutting, etc.

5. Retail-consumer price resistance, due to declining disposable income, and a shift back to basic utilitarian concerns.

6. Voter, taxpaper, and government resistance to increasing externalized costs of the packaging industry.

Let us examine each level in a little more detail.

1. Global Energy and Materials Availability, and Thermodynamic Potential. These new energy and materials constraints are well summed up in the magazine *Science,* Feb. 20, 1976. The declining quality and accessibility of basic materials and energy resources is exhaustively reported. As we must move to extract our raw materials from ever more degraded and inaccessible resource deposits, there is less net yield from these processes.

Similarly, all the high-quality iron ore from the Mesabi Range is now exhausted, and we have had to turn to lower quality taconite. Philip Abelson, President of the American Association for the Advancement of Science points out that high energy prices are making prohibitively expensive the recovery of some low-grade ores. He adds that given enough time and supplies of energy, technology can provide means to gradually overcome scarcities of particular materials, but that decades must elapse before the system can adjust.[2]

It is not only the appearance on the horizon of materials shortages such as mercury, platinum, silver, tin, zinc, etc., but the simultaneous depletion of oil and gas energy sources, which together will limit our range of substitution options. Here, I find my own analyses in sharp disagreement with Herman Kahn of the Hudson Institute, who stated at the 1975 Houston, Texas, meeting of the Club of Rome that he is not at all worried about materials shortages, and offered as proof the fact that the Earth's crust is 7% alumina.[3] Since all energy/materials transformation processes, including recycling, involve the heat losses associated with the Second Law of Thermodynamics, it would seem to me that in order to extract that 7% alumina and make it available would require unheard-of levels of energy consumption and devastation of land, not to mention unacceptable buildups of atmospheric waste heat, and vast capital investments.

Our technological virtuosity on the supply side of the resource problem can carry us only so far. And even if an unlimited energy source could be brought on stream in the next thirty years, so that, in principle, all our materials problems would be overcome, we would still face the limitations of planetary heat buildup, where even a few additional degrees in temperature can trigger irreversible climatic changes. Therefore, placing blind faith in technology to increase supplies is unrealistic, and we are better advised to turn some of our attention to the demand side of our human societies.

2. Population Growth and Distribution: World Capital Availability and Shifting Capital Labor Ratios. United Nations demographic projections indicate that the world's population will grow at least until the end of the 21st Century and will reach 10,000,000,000. Many researchers, including Dr. Lester Brown, President of Worldwatch, Inc. of Washington, D.C., contend that such projections can no longer be accepted fatalistically, in light of clear dangers of the global social and ecological strains now evident in this worsening population/resource ratio.[4]

The interaction of population growth and resource depletion is not, however, a 1:1 ratio, and there is increasing worldwide recognition that per-capita consumption is the third key variable in the problem. This per-capita consumption variable, expressed in "Indian equivalents," (i.e., an American child incurs 50 times the impact on global resources that an Indian child does) highlights global inequities exacerbated by industrial economies and their wasteful, inefficient use of nonrenewable resources to sustain their relatively small populations, while their exports do not significantly redress this imbalance.

However, all is not completely bleak, since few countries still claim that population control is unimportant. Eleven countries have now achieved, or are close to achieving, population stability (East and West Germany, Luxembourg, Austria, Belgium, U.K., Finland, Sweden, Hungary, Switzerland, and the U.S.), while ten more less developed countries have achieved growth rate declines (Barbados, Taiwan, Tunisia, Mauritius, Hong Kong, Singapore, Costa Rica, South Korea, Egypt, and Chile),[5] while China is exhibiting leadership that is significant because of her huge size.

In the next few decades, it is likely that all countries will have to divert more resources and research funds to the battle for global population stabilization. Similarly, attention will need to be paid to distribution of this growing population, as to whether to allow the continued drift into urbanization or to attempt to keep people in rural areas and help them toward local food self-sufficiency. This latter approach will require assistance with intermediate technologies and renewable resources based on organic, sustained-yield agriculture. The urbanized approach will entail more fossil-fuel intensive approaches to food production, and will have to then rely on massive transportation and distribution efforts to transfer the food from producing to consuming areas. Similarly, the capital/labor ratio shifts will be dependent on the centralized or decentralized, low or high technology configurations in each region. But on a global scale, it is now clear that the capital/labor ratio has shifted back in favor of labor, since in the grossest terms, people have become more plentiful, and resources continue to get scarcer. Therefore we now have to realize, for most of the planet's inhabitants, what is needed is not labor-*saving* devices, but more ways of *employing* labor.[6]

Lastly, there is now an awareness of the worldwide shortage of capital, and that political and economic aspirations will continue to create fierce competition in international capital markets. Not only is there a shortage of capital in terms of our aspirations, but we are now experiencing sharp declines in its productivity. The declining productivity of our capital investments relates in part to the declining quality of the resource base from which we must extract our raw materials, as I have mentioned. More and more capital must be cycled back into this process of extracting resources from degraded deposits with correspondingly declining net yields. Economic activity remains high, but it becomes more of a wheel-spinning process, producing fewer real goods and services, which, in turn, is reflected in the falling value of currencies, i.e., inflation.

Another source of inflation, which we examined in Chapter 6, is the rising social and "transaction costs" of complex, interdependent, industrial economies, including the social and environmental costs of production and consumption, *the unanticipated side effects of technologies.*

I have termed this syndrome of industrial societies "The Entropy State," where they reach maturity and begin to face agonizing trade-offs in their capital investments, as between energy and materials extraction, public and private sector investments, as well as between competing industrial sectors.

For many, this came as a surprise, since the traditional equilibrium models of supply and demand used by their economic advisers tend to discount absolute scarcities on the supply side, whether of energy, materials, or capital. These traditional equilibrium assumptions still underlie too many of our economic forecasts, and because they tend to merely extrapolate historical trends and prices, they often tend to increase lag-times in responses and make policy adjustments more difficult.

Too many economies are still run on such statistical illusions, including the GNP indicators and entrenched forecasts predicting linear growth curves in demand for energy, air travel, etc., without taking into account likely damping mechanisms, such as the declining productivity of capital, social costs and other negative feedbacks I have mentioned. I have argued how easily we in the U.S. could demonstrate the extent to which our economy would encounter a capital shortage, by merely assembling in one matrix all the capital spending projections of our major industrial sectors, service sectors and our public investment plans already mandated by legislation, for such areas as health care, education, housing, public transit, defense, and space.

One could see from even a back-of-the-envelope calculating that we were all dreaming! The results are now history: soaring interest rates, followed by reckless decisions to inflate the money supply, culminating in former Vice President Rockefeller's $100 billion energy development plan, which is little more than a scheme to print more money rather than debate the hard political choices that must be made in public investments.

At least, it is now clear to economists that rising interest rates alone cannot allocate capital between all these conflicting needs and more government planning seems inevitable.

3. Rising International Competition for Energy and Resources, Demands from the Third World for a "New Economic World Order." There is still much speculation as to the extent to which resource- and energy-rich less developed countries will be able to maintain and extend cartels such as that of the Organization of Petroleum Exporting Countries (OPEC). Thus far, efforts to extract higher prices for bauxite and other materials have not been too successful, since they collided with the recessions simultaneously experienced in industrialized countries.

However, I have speculated that as soon as these countries' latest round of Keynesian stimulus takes hold, that inflation will resume its upward course and these economies will run into the same supply bottlenecks as in the 1973-74 steep rise in commodity prices. Many of the industrialized countries rely heavily on Third World or otherwise imported materials, for example, the U.S. imports more than half of its supplies of aluminum ore, tin, chromium, asbestos, and nearly two dozen other important materials. While OPEC-modeled cartels have not yet emerged, nationalization of copper, tin, bauxite, and iron mines has been accelerating, with more than 12 Third World countries taking such actions during 1974-75.[7]

These countries and many others in the less developed regions have declared their intention to take control of their own natural resources, which seems to me to be a realistic long-run policy in light of ultimate scarcities, since the trend will be for these resources to increase in value in the ground.

In addition, Third World countries are increasingly looking to China, as a model of development, using labor-intensive methods and eschewing private mass consumption in favor of communal styles of consumption, which are less materials- and energy-intensive. To the extent that China does become a cultural model for these poorer nations, they will be unlikely to provide new markets for Western-style consumer goods, as well as be hostile to Western and multinational corporate investments. Whatever the outcome, it is clear that "free markets" will no longer govern world trade as heretofore. Economics is giving way to politics as commodity agreements are discussed, as well as the idea of Third World countries pegging their currencies to the world prices of their staple resources, such as Guyana has suggested with respect to its bauxite.

Other similar proposals for "militant indexing," as advocated by Algeria, range from holding an exclusively Third World monetary conference (to develop a countercyclical trading system to the Western-dominated International Monetary Fund) to the development of a new international currency, based on a market basket of major commodities, which would enable Third World countries to participate by "capitalizing" their resources in the ground or as commodity futures and thus trade with each other, as well as gain more favorable exchange rates with Western, fiat currencies.[8]

4. Market Resistance of Product Manufacturers Subject to the Same Constraints—Slower Rates of Product Innovation—Squeeze on R & D Funds—Cost Cutting. As all of the aforementioned energy, materials and capital constraints continue to affect product manufacturers which are the customers of the packaging industry, market resistance will rise. In this area, the example of the U.S. is particularly relevant, since it is a major innovator. Dr. Jerome B. Wiesner, President of the Massachusetts Institute of Technology and former Science Advisor to President Kennedy, noted recently the decline in the pace of U.S. technological innovation. He noted that in some cases, this was inevitable and even desirable, but he cited lack of capital and government research support as reasons for this decline, and the need for much more effective management of our "man-made world," as well as for concentration on "replacement technologies," such as new energy sources, conservation techniques, and efforts to improve air quality and food production.[9] *Business Week,* in a major article in its February 16, 1976, issue, discussed the decline of innovation in such industries as steel, chemicals, paper, packaged goods, and automobiles. Capital scarcity was cited by many for dropping development of higher risk new products.

For example, Procter & Gamble took ten years and $70 million to bring its Pringle Potato Chips to market, and company spokesmen speculated that, if they had it to do all over again, Pringles might well have died in a test market.[10] Many other companies are trimming their product lines, for example, General Electric has dropped production of blenders, fans, heaters, humidifiers, and vacuum cleaners, while some food companies are dropping product lines and others are cost cutting by reducing packaging. Market researcher Professor John Howard of Columbia University's Graduate School of Business claims that products will become more functional and with fewer frills, reflecting the narrowing of product lines, substitutions and compromises and other efforts to conserve dwindling resources. He adds, "Cartons for liquor and perfume don't have to cost 50¢ each nor do match packs need to be coated with tinfoil and other flashy laminations."[11]

And according to the U.S. Office of Solid Waste Management, packaging consumption has grown in the U.S. since 1958 so that its share of consumer product costs has more than doubled for items as diverse as dairy products, produce, beverages, and candy.[12] As I pointed out in an article in *The Futurist* in 1974 (see Chapter 20), some packaging materials would probably disappear gradually, as energy became more realistically priced, for example, aluminum foil. Since then, Alcoa has dropped aluminum foil from its product line.

Packaging is a growth-dependent industry, and yet it is clear that it cannot expect that packaging demand will grow continuously any more than electrical utility companies can expect the constant growth rates they once took for granted. As F. Treu of Burmann-Tetterode Group has noted, in the sixties, growth of plastic packaging was averaging 6-8% per year, whereas now projections for 3-5% per year growth are being used, and he cites falloffs in profits and many losses being suffered, since companies have been unable to pass costs along in higher prices.[13]

5. Retail Consumer Resistance—Less Disposable Income—Trend Back to Basic Utility. The Harris Poll taken in the U.S. in December, 1975 found that 61% of the American people now feel it is morally wrong that, although they comprise only 6% of the world's population, they consume 40% of the world's energy and resources.[14] However, it is the more pragmatic factor of reduced real incomes that is forcing the U.S. consumer to change consumption habits. Even in one of the richest countries in the world, U.S. consumers are retrenching their use of energy, and switching to simpler lifestyles. Even the American dream of owning one's own house is fast evaporating under the ravages of inflation. According to an article in

Fortune, in 1950, seven out of ten American families could afford an average new house, while today, only four out of ten can.[15]

Here again, it is relevant to quote U.S. studies, since it is still the "leading indicator" of future conditions likely in many other economies aspiring to emulate its mass-consumption features.

Inflation of basic food prices has shaken the consumer's confidence, and shown how large patterns of world food trade can be instantly felt in domestic pocketbooks. With such rising food prices, millions of Americans have been forced to buy more food, and less processing and packaging for their dollar. Not only has the share of many consumer product costs represented by packaging more than doubled in the past 15 years, but in some cases, as in that of beer, packaging is the major cost factor, accounting for 56% of the retail cost while the ingredients only account for 12%.[16]

The president of Coca Cola testified recently before the U.S. Senate that "Coke sold in food stores in nonreturnable packages is priced, on the average, 30-40% higher than Coca Cola in returnable bottles. The difference lies essentially in the different costs of packaging."[17]

Overall, the consumption of food in the U.S. increased by 2.3% by weight on a per-capita basis between 1963 and 1971, while during the same period, the tonnage of food packaging increased by an estimated 38.8% per capita.[18]

Nowhere is the trend back to basic utility more visible than in the food sector, which is also a crucial segment of the packaging industry's market. This trend back to basics was thus first forced by inflation and only *secondarily* by overt consumer dissatisfactions, such as misleading packaging and advertising claims; awkward packaging, such as blister-display types which are convenient for the retailer but inconvenient for the consumer; hazardous packaging, such as aerosol cans, "puff" containers for talcum powder, which can cause inhalation of dangerous quantities of the contents; sharp-edged openings on meat cans and the beach and recreation hazards of feet cut on discarded soft-drink pull tabs. However consumer insights were sharpened by the realization of the cheaper prices available for less "convenience" and they also became aware of the costs built into products of major advertising and brand name promotion, since they could buy the same items for as much as 20% less, with a supermarket chains' private label. As people began purchasing more local fresh foods for price reasons, they have also become more aware of the extent to which packaging permits the growth of more highly processed, denatured foods and their extended transportation and shelf life, as well as the proliferation of "junk foods" sold from vending machines, now under sharp attack from consumer groups.

Gradually, such pragmatic concerns have begun to force consumers to view all of these factors and shift to a new set of trade-offs. Many are turning to the pleasures of home-food preparation, enjoying experimenting with organically grown, natural foods, growing and preserving more of their own foods and consciously rejecting overpackaged overprocessed items, as reported in a recent article in *Forbes,* March, 1976. Consumers even proved that they would return to refillable milk and egg containers, in the successful program of the Red Owl Supermarket chain in 12 Milwaukee stores. Refunds of 2¢ were given on each reused paper shopping bag, 3¢ for egg cartons and 4¢ for glass or plastic milk bottles. During the first 8 weeks of the program an average of 1,260 egg cartons were refilled each week, 1,285 shopping bags and 1,900 gallons of milk were sold in refilled containers. The consumers like the cash savings and the costs to the stores were zero, since they were equivalent to the costs of the new packaging saved. Red Owl Supermarkets have now expanded this program into the Minneapolis area.[19]

In addition, in many suburban areas of the U.S., garbage collections are made by private collecting services, whose prices rise with fuel costs, making consumers more aware of the disposal costs as well as the nuisance of dealing with growing piles of garbage.

6. Voter, Taxpayer and Government Resistance to the Social and Environmental Costs Externalized by the Packaging Industry. The very organizing of consumer, taxpayer and government resistance to proliferating packaging is a sign of a major market failure. The traditional model of the free market, as described by Adam Smith, requires as conditions that all buyers and sellers can meet each other in market places with equal power and equal information. Under such conditions, consumers were supposed to arbitrate all production decisions and control the direction of technological innovation. The extent to which such now unrealistic assumptions underlie models of technological innovation and production factor analysis, is pointed out by François Hetman in his book *Society and the Assessment of Technology.*[20] Clearly, the scale of corporate operations and the extent of their advertising budgets, as well as their ability to retain earnings without having to submit new investment decisions to the discipline of capital markets, and their ability to finance research and development out of profits, make a mockery of traditional economic models of market behavior. Since consumer choices have not always prevailed, and since the economic feedback signals cannot carry sufficient information back to producers in complex societies, consumer demand has organized and spilled over into political channels and social choices.

An additional condition must be met if free markets are to allocate resources efficiently: social and environmental "externalities" not accounted for in prices must be insignificant. The most visible and significant social costs the packaging industry has externalized to taxpayers and municipal governments are the costs of collection, waste handling, disposal and recycling. The only reason that more waste is not recycled is that the political struggle between the packaging industry and the taxpayers over who shall pay for recycling has not yet been resolved. Not surprisingly, municipal officials and taxpayers usually fight to reduce the flow of waste at the *source,* so as to reduce their costs of disposal, while on the other hand, the packaging industry depends on this very flow of materials as its source of profits and growth. Workers in the packaging industry are caught in the middle; they benefit from this waste stream from their share in incomes, while paying their share in the social costs of disposing of it as taxpayers.

Workers' jobs can be jeopardized by changing packaging methods, such as the thousands of jobs lost in the brewery and soft drink industries due to the switch to throwaway containers, and the greater capital intensity and concentration of economic activity that has accompanied the trend toward increased packaging consumption. Jobs in other sectors of packaging may increase as methods change, such as those in steel and glass manufacture for throwaways.

Consumers benefit from the essential functions of packaging for transportation and for assuring preservation and hygiene, and enjoying the convenience of its less essential fraction; but they are beginning to relate retail costs to the total social costs not reflected in prices.

Producers and investors, together with corporate managers, benefit most from the waste stream and the social subsidies it enjoys (through externalizing of full costs) in greater profits, dividends and stock options, as well as in careers. These private benefits far outweigh the costs they also bear in their roles as taxpayers and consumers. Therefore, they will resist source reduction in packaging, and instead, advocate recycling; provided they can persuade taxpayers and municipalities to pay for it.

This, then, is the complex social and political conflict now underway between the packaging industry on the one hand, and voters, taxpayers and governments on the other, who complain that decisions with heavy social impacts, such as those concerning packaging innovation, marketing, and design are made privately using normal corporate profit-maximizing criteria. Voters and taxpayers are also angered because such decisions are aggressively promoted by high-pressure advertising to gain consumer acceptance before adequate assessments of social consequences can be made. Such advertising rarely tells the full story and is contributing to the general backlash against advertising developing in many mass-consumption societies, including the U.S., Canada, Australia, Japan and others, ranging from attempts to limit the sheer volume of advertising or to regulate it as a new aggregate-demand management tool, to attempts to limit TV advertising to children, to force remedial advertising in case of misleading claims, or to force advertisers to tell the "bad news" along with the "good news," as in the case, for example, of pharmaceutical advertising to doctors which must describe contra-indications to a drug's use.

If the consumer has all pertinent information concerning the full range of trade-offs, and if the product is priced to include its social costs, then and *only* then can correct choices be made. Only under such conditions can free markets allocate resources efficiently. However, we know that in complex, high-technology societies, as systems theorist Todd LaPorte points out, markets cannot work where production and consumption consequences are indivisible, and that, therefore, government mediation and coordination are inevitable.[21]

We could only turn the clock back to free market economics if we were also willing to devolve our technologies to simpler, less centralized, and internationally linked levels. Thus, in all the macroeconomic issues of "efficiency" or "productivity," there is the issue of how the fruits of these efficiencies will be shared.

For example, in Chapter 10 we saw that current efforts to automate supermarket checkout operations may result in greater corporate efficiency, but at the expense of consumer efficiency, since prices will no longer be stamped on containers, and social efficiency, because taxpayers will have to help support the scores of thousands of checkout clerks that will become unemployed. In like manner, we saw that cost/benefit analyses conceal social conflicts by posing the issues as ones of technical or economic feasibility, if they average out costs and benefits per capita, thus masking the fact that some groups will get the benefits: (the jobs, the bond issues and contracts) and other groups will bear the costs. Thus in all these issues that involve market failures are issues of who wins and who loses, which accounts for the politicizing of all "objective" efforts to study them.

Citizens now see that there are no such things as "neutral, value-free, scientific studies" in issues involving clashes of political and economic power. Studies are commissioned and amassed as political "ammunition" and "intellectual mercenaries" hired by groups with sufficient funds, to justify their positions. In the U.S., even the august National Academy of Sciences has been accused of too readily performing research commissioned by parties with heavy stakes in its outcomes.[22]

As industrial societies reach these mature stages of complexity that I have referred to as the "entropy state," there is very little potential for any group gaining an economic advantage without incurring costs on some other group. Inflation becomes endemic and each of the sub-units tries to externalize costs from its own balance sheet in order to show "profits." These costs get pushed around the system until they can be forced into some other groups' balance sheet, or hidden in environmental degradation or pushed forward onto future generations. As human societies begin encountering these boundary conditions, profits tend to become evanescent or fictitious, and are won at the expense of an equal but unrecorded debit entry in some other part of the system, as systems analyst James Robertson,[23] economist Roefie Hueting,[24] this author, and many others have pointed out.

In fact, as we have seen in such technological societies, Western, Cartesian logic itself breaks down, since it assumes that wholes can be analyzed by examining their parts, and has led to excessively reductionist thinking, fractionated academic pursuits into a welter of compartmentalized disciplines, and led to unrealistic dichotomizing between what goods and services are to be produced in the "public" and "private" sectors, between "property rights" and "amenity rights," and assumed that the aggregating of private interests will somehow add up to the "public interest."

Possible Options for Addressing Constraints at Each Level. An important rule for planners and futures researchers is that of attempting to push all of the sociocultural and environmental conditions of any system under study to their likely boundaries. This exercise generally yields another insight: that of the law of limiting factors, i.e., a system is always constrained by the least available factor. For example, in my home state of New Jersey, the glass industry has been built on availability of cheap natural gas. Now that gas prices are rising steeply and supplies are being curtailed, these glass manufacturers' operations became unprofitable. They can only remain viable in face of this limiting factor by switching to other fuels, such as high-sulphur fuel oil, and mounting an enormous lobbying effort to have the state's air pollution regulations modified. They have won a short-term victory on variances for their high-sulphur emissions, but the voter backlash is now being mounted with aid from medical authorities, who point out that New Jersey is already the "cancer capital" of the U.S. due to its high level of environmental carcinogens, which the World Health Organization blames for over 80% of all cancers.

In looking for these kinds of limiting factors, we must first zero in on energy as the most crucial requirement. As a 1974 study by

American Can Company put it, "The increasing costs of all forms of energy means that now, more than ever, energy consumption provides a good indicator for the relative economics of all the various packaging systems."[25] The study plots energy calculations from materials acquisition to container manufacturing, processing, coating, filling, to paper outer packaging and transportation for each system. The energy used in acquiring metal can stock is as much as 85% for aluminum cans and 70% for steel cans. Therefore the biggest payoffs in efficiency of manufacture are in devising new methods of raw material production, making lighter weight containers and in recycling. Most corporations are already working in most of these areas, since they yield greater efficiencies at the corporate level. However, societal efficiency will require that materials throughput itself is reduced and that the waste stream not increase, since even if an equitable corporate/societal sharing of resource recovery costs is hammered out, and the capital can be raised, such recycling facilities could not keep up with an increasing waste stream.

It has been estimated by the Resource Recovery Division of the U.S. Office of Solid Waste Management that even if resource recovery plants, such as those now operating in Connecticut (which primarily convert the combustible waste fraction to energy) and other large-scale systems which, by 1985 are projected to recovery of 35 million tons annually, were to be *doubled*, they would not even keep pace with the projected waste stream increase![26] Clearly, a major diversion of society's investments would be required even to keep pace, and, in competition with other conflicting public and private demands for capital, would be impractical.

However, until such major societal conflicts with the packaging industry are resolved politically, as they must be, many members of the industry are working diligently in other areas of corporate efficiency, for example in reducing the current two-thirds of can-manufacturing energy accounted for in coating and lithographic processes, by ultraviolet curing and other programs. Therefore the most fruitful options open to the packaging industry at these boundary constraint levels are in reaching political accommodation on source reduction, which may mean some diversification out of the packaging business altogether, a course being pursued by American

Can Company in the U.S., as well as working out equitable formulas for sharing the societal costs of resource recovery with taxpayers. Technological innovation to increase thermodynamic efficiencies at all stages of production and distribution are obvious paths to doing more with less and insulating companies from energy and materials supply bottlenecks.

Another path is that of studying all possible methods of shifting production to renewable energy and resource bases, using chemurgic and bioengineering processes geared to sustained-yield productivity. It will also be necessary to monitor such crucial industries as electric utilities, because they are models of what may happen to excessively growth-dependent companies when caught between soaring capital costs, such as those of nuclear-power-plant construction, and unexpected elasticities of demand, which in the U.S. are now leading to a downward spiral of increased costs, leading to increased rates, leading to decreased demand, forcing even greater increased costs. In addition, as this electrical industry is forced to cancel nuclear-plant construction, due to their increasingly unfavorable economics and disappointing performance,[27] supplies of process electricity for packaging manufacturers will become less dependable, and more expensive.

At the next level of constraint (that of worldwide capital availability, population growth and distribution and the shifting capital/labor ratios these factors will dictate, from region to region) the packaging industry can best prepare itself by sharpening its planning and forecasting concepts and methods. These forecasting exercises will need to rely less heavily on traditional economic models based on outmoded concepts such as the equilibrium model of supply and demand and geared too closely to inadequate statistical indicators and national forecasts, for all the reasons I have outlined.

For example, a study by the Bechtel Corporation revealed that the capital availability assumptions inherent in Project Energy Independence were untenable, since the project called for the energy sector of the U.S. economy to gobble up 75% of all available capital between 1975 and 1980![28] Many corporations now routinely check their economic forecasts by using thermodynamic assessments of all their basic inputs and processes, while this form of analysis is fast overtaking economics as a more reliable tool for resource allocation, since all economic processes of extraction and materials conversion can be measured using accurate, scientifically exact yardsticks of British Thermal Units (BTUs) or kilocalories, which cannot be printed by governments or otherwise manipulated or inflated. As mentioned, the discipline of thermodynamic modeling has already convened two conferences under the auspices of the International Federation of Institutes for Advanced Study in Stockholm, and has developed a taxonomy and a research agenda, with researchers in over thirty-five countries. This field can be ignored by economists only at the peril of their own obsolescence.

However, too many corporations are still making crucial capital investment decisions on the advice of economists using traditional models which still predict fairly constant returns to such capital investments, rather than their more likely steep declines in productivity.

As I have mentioned, capital/labor ratios are swinging back toward labor in many regions and processes, as poor countries are beginning to favor labor-intensive technologies, for the obvious reasons of their own population growth, and the need to use Mahatma Gandhi's insight that what was needed was not mass production, but production by the masses. As capital gets tighter in industrial countries and their unemployment levels remain high, we are seeing similar trends developing. Both must be watched carefully. For example, in developing countries, markets may not expand for retail packaging, but might well expand for small-scale local packaging turnkey operations that can handle small volume efficiently, as described by E. F. Schumacher, as a result of his work with Britain's Intermediate Technology Development Group. He noted the need in Africa, filled by his organization, for a small-volume egg-carton-manufacturing facility, which could be operated by local people, where no such machinery was available at the needed small scale, off the shelf. Provision of turnkey facilities for producing basic, bulk packaging for communal distribution, rather than emphasis on retail packaging, might be a more realistic approach to developing country needs.

As noted earlier, in industrial countries, labor is fast becoming the more efficient factor of production in many processes and services and linear projections of past labor costs are still masking the fact that, relative to capital, materials, and energy, labor costs are now declining. For example in construction, where once labor was the villain in rising costs, now capital and resource inputs are the inflationary factor.[29]

Packaging companies must be aware of these errors in factor analysis and watch these shifting capital/labor ratios which will affect its markets and ultimate consumer demand. For example, one of the historical growth areas for packaging has been to increase the opportunities for self-service retailing and vending-machine sales, and in gearing up to serve automated check-out-type supermarket operations.

If industrial societies, which are the major market for such self-service-designed packaging, begin to shift back to greater labor-intensity due to higher capital, energy and material costs, such marketing innovations as automated checkout may be canceled. An added factor to watch is that at the same time in these countries, unemployment levels are politically unacceptable, and many governments are designing tax incentives which could quickly change the capital/labor ratios in hundreds of retail operations, making labor a bargain.

Other factors which will need monitoring are those concerning population distribution as between urban and rural development and how they will change production and transportation patterns. Much of the growth of packaging has been dependent on cheap transportation, and as its costs continue to increase, we will see more local and regional economies of scale in industrial countries, which will reduce the need for protective- and preservative-type packaging. In the U.S., for example, a major population shift is now underway, reversing a trend which began at the turn of the century toward the cities. Now the flows are away from the cities and their suburban sprawls and into the rural areas, as people flee the urban "rat race" and opt for simpler lifestyles and more leisure time over material possessions.[30]

Another area undergoing massive changes is the food industry and large-scale farming, processing, packaging and distribution operations (which are *currently* capital- and energy-intensive). This multinational corporate agribusiness system is under increasing attack by both nutritionists and consumer groups for its excessively profit-dominated food production decisions, and foods which nutritionists claim are concomitantly nutritionally more depleted by each increment of expensive processing. The extent to which such critics are trusted can be gauged by the explosive growth in most industrial countries of health food, or "food reform" stores and flourishing publishing enterprises based on health foods, nutrition, home cooking and preserving. At least, the packaging industry has shared this growth in the soaring sales of home-preserving jars.

It can now be seen that most of the constraint levels in closer proximity to the packaging industry are driven by the parameters of these larger constraint systems we have explored, including the prospects for addressing Third World aspirations for a new economic world order and their efforts to control their raw materials; dealing with market resistance on the part of product manufacturers, retail consumers, as well as the behavior of voters, taxpayers and government.

What I have tried to demonstrate is that all these immediate pressures in the sociocultural sphere which are impinging on the packaging industry are *effects* of much larger, systematic forces, which are now affecting all levels of our industrialized societies. These sociocultural behavior patterns, therefore are not fads, but long-term trends and value shifts, based on a new set of trade-offs at governmental production and consumer levels of decision making.

Therefore, the packaging industry can address them in two ways: either continue to commit substantial resources to legislative lobbying, public relations and advertising efforts to hold these sociocultural forces at bay, or commit a greater fraction of their resources to addressing the larger, systemic changes driving them, and attempt to realign, by some of the means I have suggested, their research and development, corporate operations, divisions, and product mixes to take advantage, where possible, of the new conditions.

Some areas will be evolutionarily blocked, such as those dependent on linear growth of the waste stream. In such cases, divestiture and diversification into new fields may be the only course. The most promising of these, in terms of utilizing existing knowledge, patents, and capital equipment, will probably be in resource recovery, and in a whole range of bioengineering and bioconversion processes, or what Dr. Ingemar Falkehag of Westvaco's Research Center calls "the regenerative resources economy, which provides its people with a minimum daily requirement (MDR) of energy and material for optimal well-being."

Such processes include manufacture of single-cell proteins from lignin or cellulose wastes for animal feed, and forest management and tree farming and other chemurgic conversion processes. New materials design can allow for easy reclamation; for example, rubbers and polymers can be reconstituted to their virgin specifications by use of prescribed solvents for each type of molecular structure, so as to avoid not only discarding such materials, or combusting them for energy, but even the need to apply energy-costly melting or other reclamation techniques.

The new scarcities will be the mother of better inventions and will force us to produce much more refined, sophisticated technology than current more primitive forms appropriate to the "meat-ax" industrial age now receding. Such careless technologies are now recognized as unsustainable, and the sooner we shift our production systems to sustained-yield, low-impact modes, the less precious capital we will waste in the necessary transition. There will always be economies of scale at all levels of production, but with better analytic models of true, long-term productivity and systemic efficiency, the better we can learn how to recognize them.

The ultimate industrial revolution will be a subtle shift from "hardware" to "software": i.e., greater knowledge inputs to production. As noted, a problem of production need not necessarily require building a factory, machinery, or "hardware" at all. Instead, we will think harder, and learn to scan natural ecosystems for signs of the capabilities we are looking for, or that we might tap into or augment.

This radically different approach to production can be used for producing everything from fertilizers and pesticides to petroleum from the lowly joruba plant, rubber from the guayule plant, and a host of pharmaceuticals and even to extraction of minerals. The roots of plants, for example, mine many vital minerals, such as potassium, calcium, manganese, nitrogen, phosphorus, sulphur, chlorine and silicon. Such "mined" minerals are concentrated by plants, fueled by photosynthesis and stored in their root systems and constitute 5% of the dry weight of the plants. These plants concentrate 5 billion metric tons of minerals each year, *five times* the production of human endeavors in mining operations![31]

Our planet is more marvelous than we yet understand. There is enormous potential to provide a stabilized population of humans with enough for all, provided we can learn its operating principles. This will require a revolution in our perception, concepts and values.

---

[1] *The Annals* of the American Academy of Political Social Science, "Decision: The Human Predicament," S. Shackle, March, 1974, Vol. 412, pp. 1-10

[2] *Science,* Feb. 20, 1976, The New World of Materials, Philip H. Abelson and Allen L. Hammond, Vol. 191, Number 4228, p. 633

[3] At the Conference Limits to Growth '75, Houston, Texas, Oct. 19-21, 1975

[4] Lester Brown, *In the Human Interest,* W. W. Norton, 1974

[5] Ibid, p. 150-153

[6] This point is perhaps best illustrated by E. F. Schumacher in *Small is Beautiful,* Harper & Row, 1973

[7] *Science,* Feb. 20, 1976 op. cit.

[8] *Co-Evolution Quarterly,* "The Gathering Challenge to Fiat Money," Carter Henderson, Fall, 1975

[9] *New York Times,* April 20, 1976, "Technology Pace Found Declining"

[10] *Business Week,* February 16, 1976, p. 59

[11] *Business Week,* January 5, 1974, p. 51

[12] *Packaging Source Reduction: Can Industry and Government Cooperate?* Eileen L. Claussen, Program Manager, Office of Solid Waste Mgmt. U.S. Environ. Prot. Agency. Paper presented at the Annual National Forum of the Packaging Institute.

[13] *Packaging 1976 Recovers Its Balance!* F. Treu, Buhrmann-Tetterode Group, Amsterdam, Macropak '76, Utrecht, The Netherlands, May 12, 1976

[14] *The Harris Survey,* Dec. 1975, *Chicago Tribune,* copyright, 1975

[15] *Fortune,* April 1976, p. 84

[16] *Bottles and Sense,* Environmental Action Foundation, p. 10. Quoted from *Beverage Industry,* Aug. 24, 1973. "Trends Indicate Top 4 Brewers Will Have 70% of Business by 1977." Sanford C. Bernstein.

[17] J. Lucian Smith, President, Coca Cola, U.S.A. Hearings Before the Sub-Committee of the Judiciary, U.S. Senate on S.3133, Aug. 8, 1972

[18] *Packaging Source Reduction,* Claussen, op. cit.

[19]Source Reduction Fact Sheet: *Red Owl Stores Program,* U.S. Environmental Protection Agency, Washington, D.C.

[20]Francois Hetman, *Society and the Assessment of Technology,* O.E.C.D., Paris, pp. 67-75

[21]Todd LaPorte, *Organized Social Complexity,* Princeton University Press, 1975

[22]Philip Boffey, *The Brain Bank of America,* McGraw-Hill, New York, 1975

[23]James Robertson, *Profit or People?* Calder & Boyars, London, 1974

[24]Roefie Hueting, Environmental Deterioration, Economic Growth and National Income, Netherlands Central Bureau of Statistics, Voorburg, 1975

[25]*Beverage Packaging Systems Energy Study,* American Can Co., Greenwich, Conn., 1974

[26]John H. Skinner, *Reduce the Incentive to Waste,* Dep. Dir. U.S. Office of Solid Waste Management. Paper before the American Inst. of Chemical Engineers, Sept. 1975

[27]*New York Times,* "Hope for Cheap Power from Atom is Fading," Nov. 16, 1975, p. 1

[28]*Energy Supply Model,* Bechtel Corp., San Francisco, Cal., July 1975

[29]*Fortune,* "The Hyperinflation in Plant Construction," E. Faltermayer, November 1975

[30]*The Futurist,* Aug. 1975, "The New Ruralism," William N. Ellis

[31]*Scientific American,* "Roots," Emanuel Epstein, Vol. 228, #5, May 1973, pp. 48-56

Reprinted from *Human Resource Management,* Spring, 1977

CHAPTER TWELVE

Autopsying the Golden Goose

The new challenges to the rights associated with capital, property and management represent a natural progression of the historical processes of human emancipation from all forms of arbitrary power: from the Magna Charta signed by Britain's King John in 1215 to our own Declaration of Independence and Bill of Rights and the United Nations Charter of Human Rights. Investors and corporate managers were naive in imagining that this drive for human emancipation and democratic participation would stop at the doors of the corporate executive suite or at the factory gate. In "Farewell to the Corporate State," I noted that after Watergate and other revelations of nonaccountable corporate power, there is a growing understanding that such corporate institutions as those typified by *Fortune*'s 500 are neither private nor enterprising and that, in fact, these large corporations are now giving private enterprise a bad name. As mentioned, the Hart Poll conducted for the Peoples Business Commission found that one out of three Americans believes that our capitalistic system is on the decline; two out of three favored basic changes in our economic system; a majority favored employee ownership and control of U.S. corporations; 74% favored consumer representation on corporate boards; and 58% believed that major corporations tend to dominate and determine the actions of our public officials in Washington.[1] If these results seem shocking, we must remember that rarely are polling firms hired to ask such radical questions, and yet even pollsters which largely serve business have

"There are plenty of jobs around. People just don't want to work."

now reached remarkably similar conclusions as to the general loss of confidence in our corporate-dominated business system.

Therefore, I shall try to show that this crisis of confidence is not merely a question of legitimacy, since it was inevitable that such highly visible non-democratic institutions in the midst of a political democracy would be so challenged. The issues are much deeper and involve the objective questions of incompetence and inability to address new societal needs, as well as the overarching concerns as to corporate enterprises' very rationality in the face of competing paradigms now mapping our evolving society and alternative images of the future. First, I shall review these issues at the macro, societal level, and, second, at the micro level, since much of our confusion now stems from our inability to categorize such problems of scale and to determine when we need to aggregate or dis-aggregate our data and models. Therefore, redefinition of the rights and responsibilities of capital must begin with an examination of this two-level struggle over defining past and present reality.

Managing the Macrosystem: Competing Models of Reality. As I noted in Chapter 9, the politics of mixed, market-oriented industrial societies in the immediate years ahead can best be understood as a battle between two major paradigms: (1) the private enterprise "Golden Goose" as the major creator of societal wealth (the dominant business image), and (2) the emerging image, captured so well by Michael Harrington,[2] of the government and the taxpayers having become little more than the private sector's "Milking Cow"—where business is seen as reaping private profits at the expense of escalating public costs, and as using its lobbying power to compel taxpayers to underwrite the risks and investments which heretofore have been the chief justification for the profit system. In addition, our macroeconomic management ineptness is a part of this paradigmatic crisis in all industrial societies, faced as they are with the new constraints of global interdependence.

At this global level, the crisis of industrialism in market-oriented, mixed and socialistic economies involves an insistent paradox: *advancing levels of technological complexity systematically destroy free market conditions and render laissez-faire policies all but impossible;* while, on the other hand, *we humans have clearly not yet learned how to plan such societies.* Facing this paradox and humbly acknowledging our conceptual confusion are the first agonizing steps to finding "the Third Way."

It is well to remember that Karl Polanyi pointed out in 1944 in *The Great Transformation* that the "free market," held in such awe by neoclassical economists, was not derived from any natural order or from any "human propensity to barter" as Adam Smith imagined.[3] Of course, markets have always existed in human societies, but we forget that until the industrial revolution in England, they were local and not interlinked, and that they had never been used as the dominant resource-allocation *system.* In fact, most of the world's cultures until that time had used two other major production/allocation systems: reciprocity and redistribution, so that free-market *systems* are rare aberrations in human history. Polanyi notes that this free-market system, far from being derived from God, began in England as a package of bitterly contested *social legislation,* over which a civil war was fought in the early 1600s. The legislative package which created the free-market system included two crucial planks: the enclosure of land, so that it could be freely bought and sold as a commodity; and the driving off the land of formerly self-reliant peasants and cottagers, so that they were forced to sell their labor as a commodity. This and other legislation breaking down regional barriers to trading and cash-based transactions set the stage for national markets and formed the basis of the industrial revolution, with its huge increases in production of goods and concomitant increase in social devastation and human misery.[4] All of this happened a hundred years before Adam Smith was born. Understandably, Smith chose to examine the fascinating cybernetic panorama that by then had arisen of small-scale economic actors: buyers and sellers meeting each other in this newly interlinked system of markets with relatively equal power and information. And he marveled that the behavior of all seemed to be regulated for the good of all the market's "invisible hand." Even then, there were two potential trouble spots: the embarrassing number of wandering, starving paupers over whom social theorists fretted (some noting that their numbers seemed to grow in proportion to each increment of productive wealth), and the appearance of occasional "nuisances" visited by these market transactions on innocent bystanders, such as the smells, smoke and fumes of manufacturers, forerunners, of course, of today's avalanche of "externalities" our own system visits upon us.

Although many deny it, this basic Adam Smith model of the free market and the elegant equilibrium of supply and demand it proposes (see Figure 6), still exercises great power over the minds of most economists and businesspeople. In fact, many propose that we can turn the clock back and return to this simpler world. Even in 1936, John Maynard Keynes showed in his *General Theory of Employment, Interest and Money* that this model was inadequate for understanding macroeconomic cycles and recessions. At last, the usually respectful press coverage accorded to economists' pronouncements is giving way to healthy iconoclasm, while an invigorating two-volume treatise, *Anti-Samuelson,* by Marc Linder, takes this conventional Keynesian apart. The allure of the Smith equilibrium model is that it lends itself so well to rewarding academic virtuosity and mathematical pseudo-rigor. In the face of new anomalies, we humans seem to rigidify and redouble our efforts rather than reconceptualize our situation. Economics is still based on reversible, Newtonian models of locomotion, while, as Nicholas Georgescu-Roegen has shown, economies are evolutionary, involve qualitative change over time, generally associated with rises in entropy and cannot be modelled by arithmetic or econometric means.[5] This confusion is now at the basis of most national economic policy debates, and even average citizens can now see that economics can be used to justify almost any set of policies, priorities or arrangements of power and wealth. Marx believed that economics arose as a set of apologies for the capitalist order he saw, while Bertrand de Jouvenal notes that in origin the term "socialist" merely described those whose form of analysis did not accept the validity of the economists' view of the world. It is in this sense that such societies may have come to what I have called the End of Economics. Now economists are widely viewed as analogous to lawyers who argue and justify their clients' actions or proposals. Their discipline remains useful for accounting purposes and keeping the books between firms, but must now forego its pretensions in macro-management and forecasting.

THE WORST THING WE CAN DO ABOUT OUR ECONOMIC PROBLEMS

IS COVER THEM UP.

American business is now spending millions of dollars each year on advertising designed to convince us that our economic system is doing just fine.

Multinational corporations like Mobil and Phillips 66 are buying "image" ads on TV and sending free films into our schools built around themes like "Free Enterprise: Sometimes We Forget How Well It Works."

The Advertising Council is getting free public service time on television for a campaign that portrays our economy as a model of perfect competition without any significant problems. Their message:

"Don't criticize, don't worry, and if we do have any problems, rest assured that American business will solve them—all we need are more tax breaks and subsidies."

Corporate America, now enjoying record profits after a planned recession, is indeed fat and happy. But working Americans are demanding equal time to reply to this all too rosy picture.

We have just lived through the worst recession since 1929. Almost 10 million Americans can't find a job, and spiraling inflation robs our paychecks. Young families are giving up the dream of ever owning their own home, while our cities and the natural environment continue to decay.

These are symptoms of a serious disease at the heart of our economic system: a system that produces corporate profits, but increasingly destroys human lives.

We don't think they can cure this disease with a public relations band-aid. Our prescription is a good, strong dose of healthy public debate on our economic problems, and how to solve them.

SPEAK UP, AMERICA!

Americans for a Working Economy is a national educational campaign designed to get our economic problems out in the open, and to get Americans talking about a healthy economy that works for all of us.

We've created TV and radio ads to talk back to the corporate advertising we've been getting lately. And we're producing educational materials—like our booklet, *A Working Economy for Americans*—for use in schools, union locals, church and community groups.

Start a campaign in your community. And write to us so that we can send you the tools.

It's time we stopped letting them cover up our economic problems. It's the first step toward a healthy economy, one that works for all of us.

American Federation of State, County, and Municipal Employees, AFL-CIO
Consumer Federation of America
Exploratory Project for Economic Alternatives
Environmental Action
Friends of the Earth
International Association of Machinists, AFL-CIO
National Education Association
Scientists' Institute for Public Information
U.S. Conference of Mayors
United Automobile Workers

An educational media campaign.
Produced by the Public Media Center.

Public Media Center, San Francisco, Ca.

Therefore, the crisis of microeconomics over the governance of corporations, and the crisis of macroeconomic management itself, are two sides of the same paradigmatic crisis in the discipline of economics. The dis-equilibrium concepts and tools provided by Keynes, which were first used hesitantly to prime the pump, soon became a bad habit and were relied on as the means to stimulate market-oriented economics to achieve wider distribution through continual growth of production. But, as noted, these tools of aggregate demand management paradoxically were assumed still to be working on an equilibrium system governed by the simple hydraulics of supply and demand. Arguments between the Keynesians and the Monetarists at a recent convention of the American Economics Association revealed that economic thought is virtually bankrupt.[6] Gleefully, systems analysts and ecologists note that it is *economists* who are hung up on steady-state, equilibrium systems; and it is ecologists and other critics of economics from the physical and biological sciences and systems theorists who understand that most human and planetary processes are governed by the *other* major kind of cybernetic systems: the dis-equilibrium systems characterized by deviation-amplifying processes which evolve structure (i.e., are morphogenetic) and are evolutionary and irreversible.[7] In other words, while economists still are fighting the Great Depression, the structure of the industrial economies they seek to understand has been irreversibly transformed.

Yet the "Goose That Lays the Golden Eggs" still exerts influence as the dominant business paradigm, i.e., that wealth is generated in the private sector and then funneled to the public sector via taxes, where it is either transferred to the "unproductive" or used to make social investments in a democratically determined menu of public goods and services. This Adam Smith model recognizes little exercise of power or infliction of social costs, and from a vantage point of 200 years ago, with a small population and an empty, resource-rich continent, this "Golden Goose" model was a fairly accurate view of reality. But today it no longer describes our economy adequately or the extent to which private sector profits are now won by incurring public costs. Indeed, I have suggested that the only portion of the Gross National Product that is growing is this social costs sector, and yet the U.S. is in the curious position among industrial economies of operating not only its private sector on laissez-faire assumptions, but maintaining a laissez-faire *public* sector as well! The best support of this hypothesis is now coming from its opponents in the corporate sector who document and protest each day these rising social costs of maintaining breathable air and adequate supplies of potable water.[8] They should also add the heavy, tax-supported burden of government monitoring and regulation and astronomical cleanup costs for such toxic effluents as mercury and polychlorinated biphenyls, pesticide and fertilizer residues and other social costs falling on the public from their enterprises. Significantly, Congress officially acknowledged the staggering social costs of private development of offshore energy facilities by passing the Coastal Energy Impact Act of 1976, which provides another $1.6 billion in *anticipated* social costs that states and localities would otherwise bear.[9] In addition we must remember that the Golden Goose has been on a life-support system ever since the Employment Act of 1946, which instituted Keynesianism. Taxpayers have, through federal fiscal and monetary policies, been obliged to give the Goose shots of adrenalin, routinely pump up consumer credit, and now are asked to underwrite its risks and investments. Even if it were still laying golden eggs, many people wonder whether we can afford to continue feeding it so rich a diet! As noted in Chapter 9, these new misgivings are leading to the newer paradigm: that of Michael Harrington's patient, much ripped-off government, as a stupid "Milking Cow" forced to socialize costs and risks and allow profits to remain private.

The fact is that economics is merely politics in disguise and that the Keynesian tool kit has been used in most mixed industrial economies as a device for masking social conflicts over distribution of the pie by printing money. This strategy was viable and probably optimal as long as natural resource inputs were cheap and dependable. Creeping inflation was considered an acceptable price to pay for industrial peace between labor and capital. Inflation was the great mystifier where open intergroup conflict could be avoided because no one could clearly affix blame.

In Chapter 9, "Inflation: The View From Beyond Economics," I have reviewed the misguided political debate in mixed, industrial economies concerning whether inflation or unemployment is the more serious social evil. But the half-life of such ideas is persistent and the Phillips Curve is still solemnly taught to economics students and appears in most of their textbooks. This Phillips Curve is another aspect of the war of images now underway between the Golden Goose and the Milking Cow forces. This interpretation of the relationship between inflation and unemployment has proved useful to the business community in its dealings with unions and other corporate critics by portraying inflation as a function of excessive wage demands over "real productivity" gains and threw labor and social reformers onto the defensive. Therefore, the current debate over the validity of the Phillips Curve sees corporations defending it while unions attack it, and consumers and taxpayers remain on the sidelines in bafflement.

Yet the major dilemma in most mature industrial societies is precisely that of structural inflation and structural unemployment, but both are related not so much to the Phillips Curve as to the excessive resource and energy dependence and capital-intensity of these economies. This substitution of capital and natural resources for labor has been justified on vague grounds of overall "efficiency," as if its benefits were to accrue to all. The issue of structural unemployment was addressed in the 1960s by proposals to guarantee incomes to victims of automation and thus distribute the fruits of this technological gain in productivity. But by the early 1970s such proposals had foundered on the Protestant Ethic. Yet the logical corollary of this work ethic was never faced squarely, i.e., if work was to be the only way to entitlement to an income (unless one was old, disabled or an owner of capital), then society should assume the moral obligation to guarantee all its ablebodied citizens the right to a job. This debate exposed the other fallacy of the free-market noted earlier: that the relative inputs of labor and capital to production processes could be accurately quantified and, therefore, provide an objective and fair formula for distribution.

While this issue languished and became politicized, private production units continued to substitute capital for labor (subsidized by tax credits and other incentives) in order to maximize that illusory variable misnamed "labor-productivity" or "output-per-

"As an equal opportunity employer, I have to tell you that you're both fired."

employee-hour." While predictable productivity increases can be recorded by microanalysis of all such processes, they ignore the inevitable macro-level result: i.e., that more and more workers are shaken out of the bottom of the economy by such processes and their productivity falls to below zero. Another inevitable result that Barry Commoner has noted is that progressive substitution of natural resources and capital for human resources also leads to increased environmental destruction and resource depletion. In recognition of this fact, the coalition, Environmentalists for Full Employment, was formed in 1975 to draw attention to the reality that an environmentally sound society *requires* full employment and the substitution of idle human resources for natural resources.

The "Limits to Growth" issue is of course, highly political. One only has to read Mobil Oil Company advertising to be overwhelmed with data supporting the notion that there is no energy shortage. Indeed, one is left with the impression that Mobil is primarily a custodian of the interests of the poor and the general public; that there is nothing to worry about and that those in the back of the queue—both the poor in rich countries and poor nations of the planet—should be patient and confident that the benefits of industrial growth will eventually trickle down to them. But the confusion of the business community over this "trickle-down" model of growth and distribution was revealed in an editorial "faux pas" in *The Wall Street Journal* of August 5, 1975, entitled "Growth and Ethics," which bluntly insisted that the U.S. now must *choose* between growth or greater equality, since the maintenance of inequality was necessary to create capital! It is not surprising that so many of the less fortunate still misguidedly hitch their fortunes to the old industrial sectors, as some large unions and uneasy civil rights groups are now doing by their support of tax subsidies to revive obsolete industries or the costly, reckless "plutonium economy." Other groups which have little hope of being dealt into the system are more ambivalent and join those raising the alarm, while the most effective response is that of "upping the ante" for the resource-addicted nations and corporations, as OPEC and many Third World countries are now doing through nationalization of their own resources and their legitimate demands for a new economic world order. Meanwhile, at recent U.S.-European economic summit

IF YOU THINK THE SYSTEM IS WORKING, ASK SOMEONE WHO ISN'T.

Even if you've got a job, chances are you know someone who's out of work. Today, more than 7 million Americans are looking for jobs, and can't find them.

These people know first-hand that our economy is not working the way it should be.

They know the terrible personal costs of unemployment—the fear, the insecurity, the bitter frustration of wanting to work and yet not being able to provide for their families.

But they're not the only ones who pay for the failure of our economic system to provide enough productive jobs. All of us pay the costs of welfare and crime and broken families. And we all lose the productive talents of millions of people who could be performing socially useful work.

High levels of unemployment seem to have become a part of business as usual, even in good times. The giant corporations that control our economy don't seem to mind having lots of people competing for a few jobs. And some of our elected leaders actually tell us that the only way to keep the lid on inflation in our economy is to keep several million people unemployed.

SPEAK UP, AMERICA!

More and more Americans are getting tired of paying the costs of business as usual. We understand that we're not going to have an economy that puts *people* to work until we make some basic changes in the way our economy works.

We're producing TV and radio ads to talk back to the corporate advertising we've been getting lately. And we've created educational materials—like our free booklet, *A Working Economy for Americans*—for use in schools, union locals, church and community groups. Start a campaign in your community. And write us so that we can send you the tools.

We want to get Americans talking about economic change. It's the first step toward a democratic economy, one that works for all of us.

Americans for a Working Economy
Washington, D.C. 20036

Public Media Center, San Francisco, Ca.

meetings, the Keynesian and business-oriented pump primers contend that they can best benefit the poorer nations and keep their own recessions at bay by injecting more of the old aggregate demand adrenalin to restart the old economic engines. Such Pollyanna remedies of traditional "economic stimulus" are still being urged by the economic shamans advising President Carter and his European, Japanese and Canadian counterparts. They will not work, and the next resort of bailing out the wounded, obsolescent corporations based on saturated markets in order to save jobs will make things worse.

The Macro-Micro Policy Interface: Paradigms in Transition. Significant competition to the Golden Goose, Phillips Curve and Keynesian "Growth by Consuming Our Way Back to Prosperity" paradigms is now coming not only from the "Limits to Growth" school but also from those arguing for an "alternative economy of permanence," based on much more sophisticated levels of technology and bioengineering and the decentralizing of *economic,* as well as political, power. This view seeks the redefinition of work in the Buddhist terms of "Right Livelihood" and a more realistic view of the limits of materialism and instrumental, Western rationality. It includes exploration of the extent to which human well-being can be enhanced by sharing, reciprocity, community cohesion, more satisfying work-roles and human interactions, and other noneconomic "software," rather than the single-minded emphasis on hardware and material accumulation. This subjective view of the need to retool ourselves and our culture has many interfaces with the more objective view of population/resource constraints and the new paradigms of the Milking Cow and the "Entropy State" I have described. I believe that we must now begin to map and deal with the economic transition that is upon us.

At the micro-level, the obsolete, resource-intensive corporations and industries on which we have relied for our jobs must be allowed to devolve and, in some cases, go out of business. One can only imagine the distortions we would now face, for instance, if we had bailed out all the buggy whip manufacturers in order to save jobs! Less costly interim strategies are clearly at our disposal to redeploy our work force into needed areas and the growing alternative economy. Even the watered-down version of the Humphrey-Hawkins Full Employment Bill mandating the government to drive unemployment below 4%, would be preferable. Yet the old corporations still have the political clout to force the taxpayers to bail them out, while they divert credit and capital which could be better used to build the new alternative, sustainable economy and develop its renewable resource base. These corporate bailouts are justified to save jobs, as if corporations had now become charitable institutions whose role was to employ people rather than to produce needed products. Thus they turn all the basic tenets of free enterprise on their head and open themselves to legitimate demands from taxpayers concerning "jobs producing what?" and the even more vital question "At what cost to create each new workplace."[10]

Further evidence of the structural nature of our economic problems is that we now have excessive liquidity in this obsolescent corporate sector, which still enjoys high credit ratings and the benefits of retained earnings but whose often stagnant markets do not warrant new capital investment, while in the innovative, alternative economy and the small business sector—where the greatest potential for growth and employment lies—capital at affordable interest rates is not available. This is yet another example of market failure, because our private capital markets only recognize demand (however structurally distorted) and cannot flow to meet the needs of higher risk innovators, co-ops, land trusts, community development corporations and all the other very important new institutions of the alternative economy of sustained-yield productivity. Merely lowering interest rates across the board will exact a heavy price of irrational over-investment in the declining industrial sector for whatever trickle-down benefits accrue to the growing innovative sector. Thus, markets can no longer be relied on at the macro-level to adequately reallocate societal resources to meet new conditions, just as their inadequacies in reprogramming corporate activities have led to the reexamination of the legitimacy of private enterprise at the micro-level. However, as I have tried to show, the new challenges also involve serious concerns over the inaccuracy of its dominant paradigms and its consequent incompetence and irrationality.

The Micro-level Issue: Corporate Governance and the Right to Manage. A sure sign of the growing conceptual ferment is the heightening rhetoric in our mass media. Business recently launched (with some commandeered dollars from the U.S. Commerce Department) a $2.5 million "public-service" campaign prepared by the tax-exempt Ad Council, to "educate" Americans about their economic system. An immediate howl of wrath was registered from a coalition of public interest and corporate accountability groups, and of advocate organizations for consumer and environmental protection, minority rights and economic alternatives. Their successful challenge of the Federal Communciations Commission's "fairness doctrine" led the way for their counter-advertising campaign, whose theme is, "If You Think the System Is Working, Ask Someone Who Isn't."[11] Predictably, economists of all stripes lined themselves up on both sides of the issue. Meanwhile, the other skirmishes continue: corporate accountability forces back legislation for federal chartering of corporations; electricity consumers win minimum, "lifeline" rates from utility commissions; some of the costly tax burden of proof of the impacts of technological innovation is beginning to shift to the innovators. On the corporate side: over the past decade the contributions of corporate income taxes to the federal budget has been whittled away by a third (it is now only 15% of the total tax revenues), thus placing a much increased burden on individuals (now 44% of the total), while the steeply progressive Social Security payments have grown from 9% to 29% of the total.[12] Business has also managed to protect its tax credits and in some cases increase them and is now lobbying for reduction of its state taxes with similar success; it has so successfully purveyed the Golden Goose paradigm to hard-pressed states and municipalities that it exacts even greater tax breaks and incentives as its price for not pulling out and seeking a better "business climate" or greener pastures either here or overseas. The seesaw battle between the rights of investors and the job security of workers continues in New York City and other municipalities. Utilities have gone to court to force customers to continue paying for their advertising even if they disagree with it and still lobby public service commissions and Congress to underwrite their capital expenditures through the infamous "construction work in progress"

levies on consumers' bills and have won further subsidizing of the risks in the extension of nuclear accident insurance in the Price-Anderson Act.

The events taking place in Britain, the world's oldest industrial society and the cradle of our U.S. political system, provide a useful microcosm of how the last act in the drama of industrialism might play out. In Britain, the "Entropy State" syndrome is far advanced and the obsolete and impacted production system, together with the legacies of class oppression and colonial exploitation, are now generating soaring social and transaction costs. The society has sunk into an accidental steady state, with all its social forces checked and balanced, and where, in lieu of clear policies, inflation is now managing the system. The British syndrome merits our study, because it may be a forerunner of difficulties we may encounter and we can learn from comparing our similarities and differences in circumstances. Historically, Britain is one more example of the old adage, "the higher they fly the harder they fall." At her zenith at the turn of this century, Britain, an island the size of Maine, controlled a worldwide empire on which the sun, literally, never set. Seventy-five years later, the pound sterling has been battered to its nadir and Britain has called on its last line of credit from the International Monetary Fund of $3.9 billion, and now the specter of bankruptcy is in sight. Typically, Britain's problems have been portrayed in the world press as stemming from the laziness and intransigence of its work force. This view does not embrace the realities of Britain's deep-seated class system and the extent to which British workers seek revenge and reparations for the past exploitation in the industrial revolution; nor does it assess the extent of nepotism and incompetence in the ranks of British management. Just as the Labor Party emerged with a majority after World War II, the high expectations of workers were dashed on finding that Britain's coffers were almost bare: depleted by war and by the dismantling of the empire, with bills coming due on a host of promises made by their forefathers to colonial peoples seeking their fortunes as British citizens. All of these social bills were over and above the more familiar backlog of social and environmental costs now coming due in all industrialized societies.

It is little wonder that Britain's political situation began to degenerate into today's acrimonious class war over the no-longer-growing pie. Britain is now polarized between labor and capital in an electoral tug of war that seems unable to produce a majority to end the stalemate. This situation is just as predicted by Joan Robinson, Karl Polanyi and others, who foresaw that the weakness of market-oriented democracies might result in the forces of labor (votes) and the forces of capital (money) becoming deadlocked, with the danger that any demagogue who promised to run the railroads on time could get elected, as occurred in Germany and Italy after the Great Depression.

The rhetoric of the current British class struggle is being conducted, on the side of business and management, in terms of the Golden Goose and the Phillips Curve and, on the workers' side, in terms of the "social contract," state socialism, worker control, and to a lesser extent the ideologies of Marx, Trotsky and Mao. Meanwhile, consumers, unorganized taxpayers, small business people and farmers look on in horror, while environmentalists, students and younger trade unionists favor the rhetoric and experimentation of decentralism, alternative technology, and themes similar to those typified by E. F. Schumacher's *Small Is Beautiful.*[13] In the British situation it is now possible to prove the invalidity of the Phillips Curve, since one can always check such a hypothesis if one of the variables can be shown pushed to the boundary. For example, it can be shown to British industrialists that even if they managed to force workers to the point of slavery Britain would still experience high rates of inflation. This illustrates clearly that wages constitute only a fraction of Britain's startling inflation rates, the rest comes from other factors: (1) the basic reality that Britain's 52 million citizens occupy a land area whose carrying capacity can support only one half its population and that even Britain's food import bill is crippling; (2) fuel and raw materials imports to keep its industrial and consumer sectors operating at current levels represent an almost monthly balance-of-payments crisis; (3) its industrial base is now inappropriate and excessively energy-materials-intensive and geared to saturated markets; (4) its infrastructure of ports, cities, housing, transportation and communications facilities is obsolescent and an excessive fraction of its resources must go to servicing this dead capital. In addition, British decision making is overcentralized, nepotistic and error-prone, still strains to make capital investments in its declining heavy industrial base, and has committed itself to a catastrophically costly nuclear energy program that British analysts have pointed out may not produce any net energy for decades. In fact, the measure of governmental confusion was illustrated in the Labor government decision to use precious British tax funds to bail out the U.S.-based Chrysler Corporation to save (at an approximate prorated cost of $13,000 each) a relatively small number of jobs.[14] Britain will continue to muddle along until the old conceptual framework changes to admit the existence of structural changes and drastically new conditions. Only then can a transition strategy emerge, out of a new consensus as to the nature of reality.

The lesson that we in the U.S. might draw from all this is that we are facing a similar economic transition and are trapped in a similar conceptual cul-de-sac. But the U.S. has several important advantages: (1) It is still an enormously rich continental mass whose carrying capacity, I would contend, has not yet been reached. We have, perhaps, two decades of maneuvering room and flexibility, e.g., it has been conservatively estimated that we could squeeze 50% of the energy waste out of our economy without affecting our GNP.[15] For Britain, the crunch is *now* and there is no room to maneuver. (2) The U.S. has a reservoir of flexibility in its political system, which is not yet polarized into the sterile Left-Right axis now compounding Britain's problems. Our pluralistic system is still best expressed as a circular spectrum (Figure 9). This pluralistic system in the U.S. still permits new ideas to emerge and allows vigorous dialogue, which will be necessary for us to achieve a consensus as to the nature of our new conditions. It is still possible for corporate critics to debate executives, for labor, environmentalists, civil rights groups and consumers to form coalitions on pragmatic concerns. In Britain, by contrast, if one comes from a working class background, one is often considered a "traitor to one's class" for aspiring to a managerial position, or cooperating with other social groups.

Out of such social desperation come interesting initiatives, one of which we will review briefly.

Fig. 9 Comparative Configurations in Britain and the U.S.

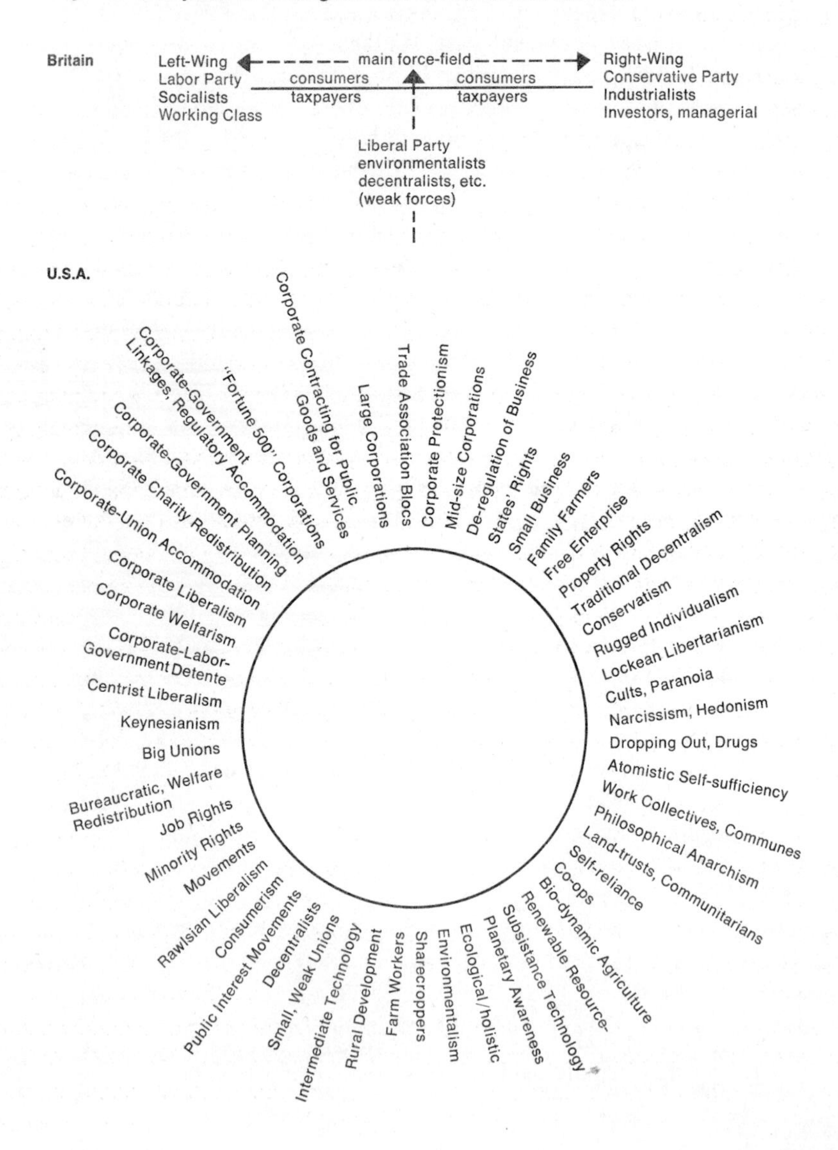
Britain
Left-Wing
Labor Party
Socialists
Working Class
main force-field
consumers
taxpayers
consumers
taxpayers
Right-Wing
Conservative Party
Industrialists
Investors, managerial
Liberal Party
environmentalists
decentralists, etc.
(weak forces)
U.S.A.
Corporate Contracting for Public Goods and Services
Large Corporations
Trade Association Blocs
Corporate Protectionism
Mid-size Corporations
De-regulation of Business
States' Rights
Small Business
Family Farmers
Free Enterprise
Property Rights
Traditional Decentralism
Conservatism
Rugged Individualism
Lockean Libertarianism
Cults, Paranoia
Narcissism, Hedonism
Dropping Out, Drugs
Atomistic Self-sufficiency
Work Collectives, Communes
Philosophical Anarchism
Land-trusts, Communitarians
Self-reliance
Co-ops
Bio-dynamic Agriculture
Renewable Resource-
Subsistance Technology
Planetary Awareness
Ecological/holistic
Environmentalism
Sharecroppers
Farm Workers
Rural Development
Intermediate Technology
Small, Weak Unions
Decentralists
Public Interest Movements
Consumerism
Rawlsian Liberalism
Movements
Minority Rights
Job Rights
Bureaucratic, Welfare Redistribution
Big Unions
Keynesianism
Centrist Liberalism
Corporate-Labor-Government Detente
Corporate Welfarism
Corporate Liberalism
Corporate-Union Accommodation
Corporate Charity Redistribution
Corporate-Government Planning
Corporate-Government Linkages, Regulatory Accommodation
"Fortune 500" Corporations

Workers employed by the Lucas Aerospace Company were faced with layoffs due to loss of military contracts and British government spending cutbacks. On their own time and with their own resources and the aid of technical advisers, Lucas shop stewards developed an alternative corporate plan to convert the company into a producer of sophisticated and intermediate technology-based products to serve civilian markets in Britain and overseas. The Lucas Plan—which would have done credit to a management consulting firm—was backed up by 800 pages of technical documentation inventorying Lucas' productive capabilities and matching them to unexploited markets which had been surveyed. The plan included a range of new products: solar electric cars and simple electric vehicles and prosthetic devices for the handicapped; cheap solar-powered irrigation pumps, cookers and plowing "snails" (as developed by E. F. Schumacher's Intermediate Technology Development Group); bicycle-operated machinery for grinding; and environmentally benign technologies for home heating and energy generation. This alternative plan for Lucas' future was made a part of the labor-management contract negotiations.[16]

As in the unfolding of a Greek tragedy, Lucas management rejected the plan out of hand. *The Engineer* in its May, 13, 1976, issue, commented, "Management based its reply on a reaffirmation of its established business strategy serving aerospace and defence. In doing so, it paid no regard to the damage to personnel morale inflicted on highly-qualified senior engineers, technicians and shop-floor engineering workers." On the other side of the issue, an editorial in a management-oriented journal opined on "What Industry Can Learn From the Lucas Affair." It stated that the plan was a forerunner of a development that would affect all of British industry and noted a surge of union recruitment of middle management, whose plant authority had been emasculated by paternalistic corporate head offices. It warned that in any atmosphere of layoffs, workers would be more likely to anticipate issues, such as the obsolescence of a company's product line, in the same way as had the Lucas workers. The editorial even conceded the "legitimacy" of the issues raised but could not conceptually deal with the plan's espousal of less capital-intensive, intermediate technologies and products, whereby the capital cost of creating and maintaining each workplace would be reduced. Parroting management's error in factor mixing and its self-destructive dedication to capital-intensive technology under new conditions of capital shortage, the editorial summed up the dying industrial paradigm: "What the Lucas employees should be pressing for is greater automation—at the same time ensuring that the displaced men are found other jobs."

For their part, the Lucas workers submitted the plan as a direct challenge to Lucas' management's competence and right to manage. Even if management had instantly accepted the plan and agreed to invest in the modest conversion, it is unlikely that the workers would have been satisfied. The bottom line for them was to demonstrate management's incompetence and publicly embarrass their bosses. What we in the U.S. would consider the pragmatic solution, i.e., worker stock options, bonuses and more participation in the production decisions and other incentives, would have been rejected by the Lucas workers as "co-option" and traitorous to their class solidarity. Thus, the atmosphere in Britain is now one of worker confrontation with the managerial class and only victory over the bosses is enough to assuage their dignity. As one British worker put it, "There isn't going to *be* any more production until it's fair. We want to end the divine right of management. Although we're a poor country and it's a luxury, let's get on with the class war and settle it."

Other equally instructive and less dismal examples of the redefinition of the rights of capital are available in Europe. Most of them involve various forms of accommodation by management of the aspirations of workers for more control over their working lives, constitutionally defined rights to job security and democratic participation in shaping industrial production and the goals it serves. In Yugoslavia, these aspirations are well articulated and constitutionally recognized. Yugoslavian concepts of property rights and responsibilities are similar in many ways to those of our own Founding Fathers. For example, Benjamin Franklin's view in 1785 was, "Superfluous property is the creature of society. Simple and mild laws were sufficient to guard the property that was merely necessary. When, by virtue of the first laws, part of the society accumulated wealth, and grew powerful, they enacted others more severe and would protect their property at the expense of humanity. This was abusing their power and commencing a tyranny."[17] In Yugoslavia there is a similar distinction between property rights to assure personal security and autonomy (a greater percentage of Yugoslavs than Americans own their own homes) and the right to endlessly accumulate property which may then be used to oppress others. In the Yugoslavian economy, private enterprises run by employers are legal up to fifteen workers, but any enterprise employing more must govern itself through a workers' council. Potential workers councils can also draw up a business plan and present it to one of the country's similarly worker-self-managed banks, and a loan committee of peers will decide whether the enterprise should be capitalized.[18]

The key tenets of Yugoslavs appear to be (1) keeping industries and thus populations decentralized and preventing the catastrophic urbanization that has plagued market-driven industrialism all over the planet; (2) maximizing worker participation in their unique brand of economic democracy and limiting the arbitrary right to manage associated with capital in market economies; (3) avoiding the equally feared pitfalls of Stalinist, bureaucratic communism and centrally planned state capitalism, which they see as characterizing much of Western Europe and the U.S.[19] However, as a small country situated between the Soviet giant and a European economy they see as dominated by U.S.-based multinationals, there are new attempts to bring in 50-50 joint ventures to build large, capital-intensive facilities, such as a planned petrochemical plant and a nuclear power station. Such dabbling with Western capital-intensive technology will alter significantly the decentralized character of modest well-being of their society and may well lead to the centralized decision making and concomitant inequality they have so assiduously sought to avoid. Unfortunately, as we are now learning here and in Europe, high technologies, such as nuclear power, are inherently totalitarian and cannot be democratically managed, politically controlled or even understood by most government leadership. In fact, nuclear scientist Dr. Alvin Weinberg admits that they must be overseen by what he calls "a technological priesthood."[20] Yugoslavians also emphasize their worries, that the media pollution of U.S. advertising, movies and TV programs will inevitably penetrate and degrade their culture, along with the multinational joint ventures. Many are in favor of censorship of the horror, perversion and violence typical of profit-motivated U.S. films and TV, echoing similar complaints by Americans.

Basically, the Yugoslav insistence on smaller, decentralized enterprises is spreading and focuses needed attention on the fact that we humans are not yet competent at national planning or at distributing surpluses. Also it highlights the intellectual arrogance of administrators, managers and bureaucrats, whether planners of multinational corporations or national governments. In this same spirit of growing humility, the 1976-80 French Regional Plan includes new "Small Is Beautiful" incentives to decentralize France's industries and limit their size.[21] European socialists and labor movements are turning away from old concepts of highly centralized socialism and nationalization of industries, and both Italian and French communists have demonstrated their freedom from Moscow. European socialism, as political scientist Norman Birnbaum points out, has deep, *pre*-capitalist, cultural roots, based on earlier notions of community, religious ideas about the value of work and human dignity.[22] The decentralist fervor led to the extraordinary secret summit meeting in Trieste in 1975 of Europe's oppressed ethnic minorities—the Basques, Catalans, Galicians, Croats, Occitanians, Bretons, Irish, Scots, Welsh, Corsicans, Sardinians, Flemings, Frisians and Piedmontese—where they aired their demands, ranging from language rights to full independence, and their shared vision of what they call "a Europe of Peoples."[23] It is ironic to remember that the very production efficiencies of the free market system helped to create the national powers and multinational corporations that dislocated such ethnic peoples and gave rise to the current "social costs" of their rebellions and terrorist demonstrations.

Human emancipation in Europe now focuses on expanding *economic* democracy as well; from the worker-management parity now achieved on German corporate boards to Sweden's much-discussed Meidner Plan, which, if enacted, would transfer annually 20% of the profits of all Swedish companies with fifty or more employees to workers' trust funds over a ten- or fifteen-year period.[24]

Other proposals favor specifically changing corporate charters so that they are managed, not for the primary benefit of their stockholders, but to balance the interests of all their "stakeholders," including stockholders, employees, customers, franchisees and the general public. Similar ideas are often rhetorically embraced by U.S. corporate executives, but if such existing informal power is thus acknowledged, such managers are surely courting its constitutional legitimation, as in Ralph Nader's proposed federal chartering of large corporations. One thing is clear: global communications are now assuring that all these European initiatives in decentralizing economic power will continue to be exported to the U.S., as our own delegations visit Norway's Industrial Democracy Project and Sweden's self-managed Volvo plants, and European delegates composed half of the attendance at the Third International Conference on Self-Management held in Washington.[25]

Self-management in the U.S. and Canada is considered in its infancy by European standards and is viewed as mostly timid efforts at "job enrichment," which, as one Canadian theorist stated, were "merely cosmetic efforts to co-opt workers into more efficient self-exploitation so as to increase corporate profits."[26] A U.S. personnel manager, who wished to remain anonymous, commented that many worker-self-managed experiments in U.S. companies had proved successful in raising both productivity and job satisfaction but were often discontinued at the inevitable stage when the workers began to question whether management was pulling its weight or was idling and featherbedding. Thus, we are forced to examine our "efficiency" criteria on yet another level. We have not yet learned that there are immense efficiency gains in fostering such seemingly intangible human qualities as trust, cooperation and honesty (not valued by the market system) and designing company charters that specifically respect and acknowledge these qualities in all their employees. For example, the charter of International Group Plans, Inc. a worker-owned and self-managed insurance corporation in Washington, D.C., states that the corporate goals include maximizing the humanness of all participants in the enterprise. It provides for open board of directors meetings and that any employee may become a member of the board. Supervisors are elected by the people they will supervise, and salary levels, work hours, vacations, etc., are all determined by the employees and managers jointly. The company sells approximately $60 million worth of insurance annually. When workers are trusted to organize the basic circumstances of their work, there is an enormous saving in the need for overseers, managers and administrative overhead costs.

Similarly, if a society is governed fairly—and this fact is perceived by all—not only is inflationary pressure reduced, as pointed out by James Robertson in *Profits Or People,* but also the cost of conflict and litigation (which in the U.S. has risen six-fold in the past decade and now threatens the entire legal system with collapse from sheer overload) is lessened.

Unions such as the United Auto Workers are taking up the cause of improving the quality of working life, as well as countering unequitably shared costs and benefits of automation by calling for work spreading. It is hardly surprising that the U.S. is considered backward in moving toward more industrial democracy, in spite of Alexis de Tocqueville's more persuasive case than that of Marx that we might end up as a "manufacturing aristocracy."[27] The U.S. resource-base and the rapid exploitation that free-market systems allow produced an avalanche of goods and increased incomes which, up to now, has kept workers reasonably well satisfied. In many cases, they have been dealt in with stock options and pension plans, bonuses and other incentives. Indeed, Peter Drucker has claimed that through the growth of employee pension funds we have now unwittingly become a socialist economy.[28] However, this case is unpersuasive, since it does not constitute socialism in the accepted sense, nor does it acknowledge that the workers do not control or vote these stock holdings, which are controlled by fiduciaries and banks, normally much more closely allied with the interests of management than labor. Indeed, the divorcing of control from such share ownership was fully explored in the 1930s by Adolph Berle and Gardiner Means.

It is ironic that the newly mandated Employee Stock Ownership Trusts (ESOTs) of Louis O. Kelso to aid workers in acquiring stock or actually in purchasing companies from their employers could result in workers being palmed off with unprofitable "bum steers." Enactment of the Kelso trusts was won at the price of further corporate tax credits, thus spreading their costs onto taxpayers rather than sharing them with stockholders and managers. It is even possible that these ESOTs might slow down the economic transition to the sustainable economy by giving workers an additional stake in the declining industries. Negotiating this economic transition will certainly require greater worker participation in order to speed the flow of innovative ideas and help create the new consensus for concerted action in all sectors of our economy. Bridging mechanisms will also be needed, such as those I have discussed here and elsewhere, and particularly those like the Canadian Local Initiatives Program (LIP), which functions on the macrolevel as a countercyclical fiscal program for creating public-service-type jobs. But whereas the U.S. job programs are centrally controlled and often abused or ineffective, LIP jobs are created by the funding of community-developed projects at the local level—where a much closer matching of the locally unemployed workers to local needs has been achieved—with a minimum of central bureaucracy. Other bridging mechanisms that might serve as models include the purchasing of local plants by residents of an area and their municipalities. A recent case is the purchase in the face of a threatened shutdown of a library-supply and furniture company from Sperry-Rand Corporation by the town of Herkimer, N.Y., and its residents and the plant's workers.

While all these processes of devolution and irrigation of the decision structures of the declining industrial sector continue, other modes of operating and innovating must go on outside the corporate system, so as to provide additional livelihoods and niches for the entrepreneurial and the self-reliant. New intermediate and evolutionary economic institutions are needed, modeled after the Scott Bader "commonwealth" enterprise described by E. F. Schumacher in *Small Is Beautiful* and such worker-owned and self-managed enterprises as International Group Plans of Washington, D.C., already discussed. New venture capital and credit-union-type institutions are needed to underwrite the emerging regenerative-resource economy. Such institutions would specifically disavow high returns on investments in order to plough them back into the fund pool so as to assist new ranks of talented young inventors and entrepreneurs. The investors might specifically agree to forego maximizing monetary returns in order to reap greater rewards in "psychic income." That such concerned, reform-minded investors exist in the rich U.S. economy is attested to by the enormous pool of funds available for all kinds of humanistic and charitable purposes, and many such new types of investment vehicles might be funded as part of foundations' program-related investments. Up to now, there have been few alternatives available to the traditional profit-maximizing-type investment vehicles and advice provided by Wall Street and other capital markets, and many idealistic young inventors, as well as investors, shun such sources for fear of becoming enmeshed in what they see as their counterproductive goals and value system. Many public-interest groups in the U.S. are working to set up such new financing vehicles, including the Exploratory Project on Economic Alternatives, the Federation for Economic Democracy, the Institute for Local Self Reliance, the Peoples Business Commission, all based in Washington, D.C. and the National Community Land Trust Center of Massachusetts, while journals such as *Self Reliance* and *Working Papers for a New Society* cover the news and develop the concepts in this field.

Urban and rural co-ops, intentional, self-supporting work collectives, land trusts, alternative technology innovators and biodynamic agriculture communities and the magazines catering to them are growing. Links are being forged between this new generation of entrepreneurs and the more traditional small business groups and rural co-operatives and family farmers. All these entrepreneurial efforts are informed by an understanding of use-value, rather than merely exchange-value and are developing theories of household economics which verify the efficiencies of small-scale operations and their success in meeting the needs and goals of their own participants. In future, it is likely that most productive enterprises will begin to define their success more broadly than in the old terms of "bottom-line" profits and move toward redefining "success" in terms of their own participants' varied criteria and goals. This will relieve the rigid, totalitarian effects of the single "profit" yardstick and restore lost pluralism to our economic life. Indeed, Professor Sol Tax of the University of Chicago advocates that we help reverse the current tax incentives favoring economic giantism by wholesale incorporation of families, so as to protect this basic institution of all societies from further breakdown. We will examine this emerging, regenerative "counter-economy" in Chapter 23.

Lastly, a prerequisite in the ongoing transformation of concepts of the rights and responsibilities of capital, property and management is the redefinition of profit. We need to invalidate legislatively the destructive and erroneous notions of speculators concerning "free lunches" and "windfall profits," which are usually won at the expense of others, of the environment or of future generations. The longer-term, more holistic view of planetary interdependence must eventually prevail because it is scientifically accurate. Our narrow concepts of time and space must be expanded so that we see in a large enough context that all our individual self-interests are identical and are coterminous with community and species self-interest. This view must modify our illusory concepts of today, where "profit" is becoming little more than "creative accounting" and externalizing costs to others or society. "Profit" is now too vague and unscientific a concept to program accurately private resource allocations, but should give way to specific system-defined efficiency criteria and the new micro-accounting model where "profit" is redefined as "success in terms of the democratically determined goals of the participants of the productive enterprise." On the macro-level, these enterprises must be governed with legislative checks and balances incorporating more accurate concepts of efficiencies of scale and quantifying of short- and long-term social costs, as reflected in a more appropriate measure of GNP (such as Japan's new net National Welfare indicator), while "wealth" must shed some of its present connotations of capital and material accumulation and give way to a redefinition as "human enrichment." Even then, a brighter future will depend on our ability to mesh our efforts at retooling our institutions and technological means with even greater efforts at retooling ourselves.

[1]Press release, Sept. 1, 1975, Peoples Bicentennial Commission, Washington, D.C.

[2]Michael Harrington, *The Twilight of Capitalism,* Simon & Schuster, New York, 1976

[3]Karl Polanyi, *The Great Transformation,* Beacon Paperback, New York, 1944, p. 44

[4]Ibid., pp. 77-110

[5]Nicholas Georgescu-Roegen, *The Entropy Law and the Economic Process,* Harvard, Univ. Press, Cambridge, 1971

[6]*Business Week,* Oct. 4, 1976, p. 48

[7]See, for example, "The Second Cybernetics," M. Maruyama, *American Scientist,* Vol. 51, No. 2, June 1963

[8]*Wall Street Journal,* Aug. 10, 1976, "Ecology's Missing Price Tag," editorial

[9]*New York Times,* July 27, 1976, p. 11

[10]See, for example, "Environmentalists for Full Employment," Washington, D.C.

[11]Public Media Center Newsletter, Spring, 1976

[12]Institute for the Future, *Corporate Associates Report,* Feb. 1976, Menlo Park, Calif.

[13]E. F. Schumacher, *Small Is Beautiful,* Harper & Row, 1973

[14]*New York Times,* Oct. 1, 1976, p. 3, "How Chrysler Steered Out of a Rut"

[15]Worldwatch Institute, *Energy: The Case for Conservation,* by Denis Hayes, Jan., 1976, p. 14

[16]*New Scientist,* "Dole Queue or Useful Projects?", July 3, 1975

[17]*Voices of the American Revolution,* Peoples Bicentennial Commission, Bantam, 1975, p. 147

[18]Personal discussions with Prof. Mirko Bunc, chairman, Dept. of Economics, University of Maribor School of Management, Yugoslavia, May, 1976

[19]Ibid.

[20]Statement during a debate with the author at the Atomic Energy Museum Bicentennial Lecture Series, Oak Ridge, Tenn., June 1, 1976

[21]*World of Work Report,* June, 1976, Scarsdale, N.Y.

[22]*New York Times,* "The New European Socialism," by Norman Birnbaum, May 15, 1976

[23]*New York Times,* July 8, 1975

[24]*New York Times,* May 26, 1976

[25]Third International Conference in Self-Management, American University, Washington, D.C., June 10-13, 1976

[26]Personal discussion with Dr. Jean-Claude Guedon, assistant professor, Institut d'Histoire et de Sociopolitique des Sciences, Université de Montréal, Canada, April, 1976

[27]*American Sociological Review,* Vol. 35, Dec. 1970, "De Tocqueville and the Morphogenesis of America," by Eberts and Whitton

[28]*New York Times* editorial, "Oh We've Been Trojan Horsed!" by Peter Drucker, June 4, 1976

Paper presented at The Minneapolis Foundation Seminar on the Future of the Free Enterprise System, Brainerd, Minn., October, 1976.

PART TWO

Creating Alternative Futures

CHAPTER THIRTEEN

Coping With Organizational Future Shock

Why do all organizations, including corporations, experience difficulties in adapting to changing conditions? To paraphrase Alvin Toffler, why do they suffer "organizational future shock" when faced with accelerating social and technological change—now a common problem of all industrialized societies? And how will organizations learn to cope with the new shock waves that may be in store for them if societies must readapt to the slower growth, less capital-intensive and energy-rich conditions that many futurists now predict?

Today, individuals are learning faster than institutions, causing many to feel enmeshed and constrained by them. As sociologist Bertram Gross notes, organizations are devices for screening out reality in order to focus attention on their own specific goals. Thus they regularly intercept, distort, impound, or amplify information, structuring it for their own needs and channeling employees' efforts toward their own goals. In the extreme, this can lead to what Joseph Coates, deputy director of the U.S. Office of Technology Assessment, calls "functional lying," a step beyond the more overzealous public-relations efforts of most organizations.

In addition, to unify their participants around a common purpose, institutions develop their own folklore and pep talk as well as more structured methods such as management information systems and management-by-objectives programs. Unfortunately, during periods of rapid societal change, these objectives themselves must be changed. At the 2nd World Future Society Assembly, systems scientist Magorah Maruyama pointed out that not only must organizational goals be changed but also the old logic behind the industrial era based on standardization, competition, and hierarchy must be replaced by a new logic for the postindustrial era based on destandardization, heterogeneity, interaction, and a new ethics in harmony with nature.

These information-structuring activities of organizations are functional only in early growth phases; as organizations grow larger, their ability to distort information and screen out feedback increases and eventually becomes dysfunctional. Commenting on how this affects management's performance in pyramidally structured organizations, futurist Robert Theobald notes, "A person with great power gets no valid information at all." Thus attempts to grow and dominate more variables in the immediate environment eventually become self-defeating, causing loss of feedback and maladaption.

The basic evolutionary law that "nothing fails like success" is the mechanism that keeps human society in balance. It eventually checks overgrowth of sub-units that have reached the dinosaur stage and prevents dis-economies of scale while encouraging diversity, experimentation, and continual learning and adaption of the whole system to change.

As we have seen, human societies and their economic subsystems operate within the basic laws of physics and conform to the evolutionary processes of growth and decay: the syntropy/entropy cycles of all natural systems. Just as the decay of last year's leaves provides humus for new growth the following spring, some institutions must decline and decay so that their components of capital, land, and human talents can be used to create new organizations.

But the idea that obsolete organizations should be allowed to die often alarms many individuals; many of them have grown so large and employ so many people that their decline can cause great social dislocations. Yet as resources, energy, and capital become more scarce and precious, less-productive organizations can wastefully divert resources and human talent, sometimes starving needed innovation and new organizations.

Often, however, many organizations can be revitalized by

restructuring themselves and changing their goals. Some can only survive by devolving to a lower level of superstructure. Some must decline and pass from the scene, such as those too rigidly programed to fill needs that are becoming saturated or those relying on resource inputs that become so scarce or expensive that profit margins in production erode to the vanishing point. These companies may have to mount "demarketing" campaigns as they phase out unprofitable products.

Other organizations become too complex and diverse and begin to spend more effort transacting with themselves than in producing their desired output. They create so many interdependent variables and interfaces that they can no longer be modeled accurately, and any system that cannot be modeled accurately cannot be managed. Corporations, government bureaucracies, and even nations are susceptible to this syndrome—"the entropy state"—in which an organization's own weight gradually winds it down into a state of equilibrium where no further useful work or output is possible.

Some familiar examples of organizations approaching the entropy state include conglomerates that have been forced to spin off some of the divisions they acquired during the 1960s and government bureaucracies (as in New York City) that suffer from dis-economies of scale and unmanageable complexity. As energy and transportation become more realistically priced, we will see a return to regional and local patterns of production and distribution—just as in our personal lives we are substituting for market value, use value, or even psychic value as we learn how to repair our own houses, appliances, and cars and grow more of our own food.

Learning Networks. How can we help our organizations to adapt and alleviate their future shock?

First, we need to recognize the different roles of "insiders" and "outsiders" and how they can mesh creatively to promote the vital adaption process. Both can function in the role that organization theorist Warren Bennis calls the "change agent."

Most people are insiders, employed or otherwise enmeshed within organizations, both the primary ones in which they earn their livelihoods and the secondary ones such as clubs, churches, and voluntary associations. Most people also have a pretty good sense of how much these organizations lag in adapting to new conditions and absorbing new perceptions, values, and goals.

When many individuals begin to notice that these institutional lags and discrepancies exist, they coalesce into informal groups and networks within their organizations, sharing their perceptions with each other at the water cooler, in the cafeteria, or elsewhere. Such interactions are frequent in openly administered organizations, which are based on Douglas McGregor's now famous Theory Y principles of participatory, cooperative, integrative functioning (as opposed to the more hierarchical, competitive, authoritarian style of Theory X). But in more rigidly structured organizations, new insights sometimes are repressed, either by superiors or through fear of "sticking one's neck out." And if these insights are valid, they will be shared by large numbers of other employees and will find expression in discontent, faster turnover, reduced productivity, or increased apathy and alienation.

In some cases, alienation has even prompted workers to take over corporate production facilities and lock out management (at a French plant and a British shipyard, for example). Or sometimes insiders feel compelled by their own consciences to "blow the whistle," as did cost accountant Ernest Fitzgerald, who complained of cost overruns on Department of Defense contracts and was fired for his trouble. Many organization theorists endorse the concept of whistle blowing, either to a professional society (perhaps when its safety standards are being violated) or to public interest researchers and the press. Such seemingly drastic action is sometimes the only recourse if the upward flow of vital information is impeded.

At the 1975 World Future Society Assembly, panelists Donald Schon of the Massachusetts Institute of Technology, Donald Michael of the University of Michigan, and Daniel Gray of Arthur D. Little, Inc., argued that our very survival now depends on the renewal of our institutions. These experts in organizational adaption discussed the need for supporting people in their efforts to change rules and redefine problems, permitting employees to learn from failure. In this way people may overcome their fear of making decisions and learn to develop group skills and greater self-knowledge.

Skirting the Issue. These brave sentiments, however, neatly avoid one organizational taboo—the taboo against questioning whether an organization itself has outlived its usefulness or whether its purposes or products have become irrelevant or even counterproductive. In such cases, outsiders in the form of consultants and other corporate critics are the only ones likely to raise this issue, even though their message will be screened out or rejected. But often consulting firms themselves are large and carry heavy overhead costs, and consequently they become overanxious to please their clients. Thus they may confirm, rather than buck, prevailing corporate values.

If management cannot always hear its own well-motivated insiders or rely on outside consulting firms, how can it develop less orthodox ways of scanning society and picking up the often faint signals that may, if correctly interpreted, portend awesome change?

Quantitative Forecasting Ad Absurdum. Many of the methods used for mapping change rely on the traditional tools of technological forecasting: cost/benefit analysis, demand forecasting, and planning, programing, and budgeting systems. However, these linear, extrapolative methods are only of limited usefulness in situations where many variables in the social, political, and economic environment are interacting and changing simultaneously.

Systems and input-output analyses, technology assessments, environmental-impact statements, Delphi techniques, cross-impact analyses, and relevance trees are some of the developing methods for trying to predict how these multiple, shifting variables may interact with each other over time. Most of these methods are still highly experimental and try to design nonlinear, dynamic models to map such complexities. But ironically, each order of magnitude of technological mastery and managerial virtuosity inevitably requires greater orders of magnitude of sophistication in modeling techniques and greater coordination and control. Eventually, the law of diminishing returns sets in as it becomes increasingly more difficult to define the boundaries of the system under study. Ultimately, nothing less than an "all-by-all matrix" may be required.

Economists are lagging behind in their attempts to expand predictive models and include more variables. Indeed, economists already have lost their former preeminence in planning and forecasting, both because of their glaring failures in this field and in macroeconomic management and because of the rise of new, more inclusive methods such as technology assessment. Similarly, microeconomic tools, such as cost/benefit analysis, are also in trouble. Often these methods are used to mask difficult political choices as if they were pseudo-technical questions of feasibility or efficiency rather than matters of conflicting values and social equity. For example, cost/benefit analysis often obscures who will pay the costs and who will receive the benefits. Not surprisingly, these conflicts later erupt into public opposition to projects justified by mere intellectual sleight-of-hand.

Intuition—the Missing Link? The purpose of performing research is to provide information that significantly reduces uncertainty for decision makers. Paradoxically, the more likely result of the information gathered through broader research methods such as technology assessments and environmental-impact statements actually *increases* uncertainty for the poor decision makers, only telling them more about *what is not known.*

As new quantitative methods develop, it becomes clearer to their practitioners that all are based on the world view inherited from Aristotle and the French philosopher Descartes. But the development of human knowledge and, hopefully, wisdom requires both imaginative hypothesis and careful validation by logical, quantitative methods. Currently, organization theorists are showing much interest in reintegrating intuitive processes into management sciences, which have grown excessively quantitative and reductionist and consequently miss important, nonquantifiable variables, particularly in human behavior.

Society's dynamically changing values and goals, far from being peripheral, are the dominant, driving variables in all human systems. Since these values change consumer preferences and create and destroy markets for corporations, managers must learn to use some of the more informal methods of tapping imagination, intuition, and

Fig. 10 Build-up of Social Energy in Response to an Emerging Public Issue

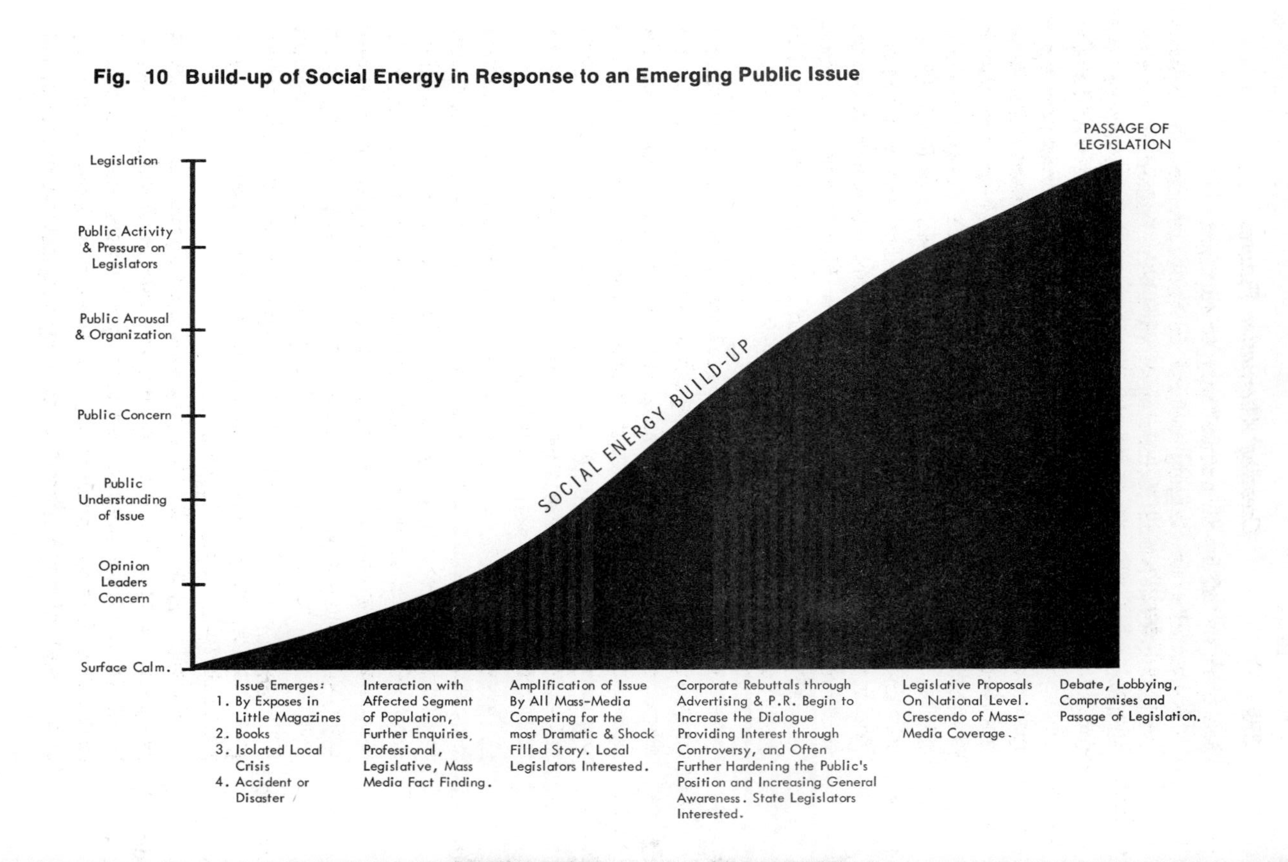

artistic expression to monitor the social scene.

Todd A. Britsch of Brigham Young University in Utah believes that most forecasting techniques, with their emphasis on quantification, have ignored the uses of art and literature as social indicators. Britsch points out that the insights of artists and writers are often predictive of perceptual shifts before they become obvious in actual societal value changes or innovations. Perhaps the best example of this is the famous novel by British writer George Orwell, *1984*, which still qualifies as one of the best futuristic scenarios of the darker possibilities of the computer age—even though it was written in 1948. But until now, only science fiction has been widely utilized by futurists. Even in science fiction, instrumental, machismo, "yang" thinking predominates over the intuitive, female, "yin" modes. Yet in many stories when warring cultures are overtaken by superior galactic races, such beings are often described as intuitive, peaceful and cooperative—all idealized "feminine" values! Britsch suggests that social forecasters monitor novels, plays, and poems for key words and attitudes on topics such as the work ethic, divorce, and child bearing and then compare this data to sociological material for the same period and for subsequent years.

Similarly, Britsch notes that satire is generally a social indicator, telling us that new iconoclasts are emerging to challenge the old order. Indeed, corporations perhaps should monitor "little" magazines and other obscure journals for signs of value shifts and use them to plot possible consumer preference changes. Such monitoring methods have been adopted by the Institute for Life Insurance's Trend Analysis Program, the Urban Research Institute, Weiner Edrich Brown Inc. and other organizations.

Counterculture Consultants. Citizen movements for social change are obviously social indicators of high visibility, whether their causes are consumer and environmental protection, peace, social justice, or economic opportunity. But too often, corporate and government executives prefer not to confront these groups directly; instead, they prefer to purchase very expensive, secondhand information from public relations firms, further contributing to the general confusion.

Social scientists are constantly constructing more easily quantifiable social indicators such as data on park acreage per citizen, educational levels, health statistics, and other objective indicators of

Fig. 11 Typical Curve of Corporate Response to Social Issue

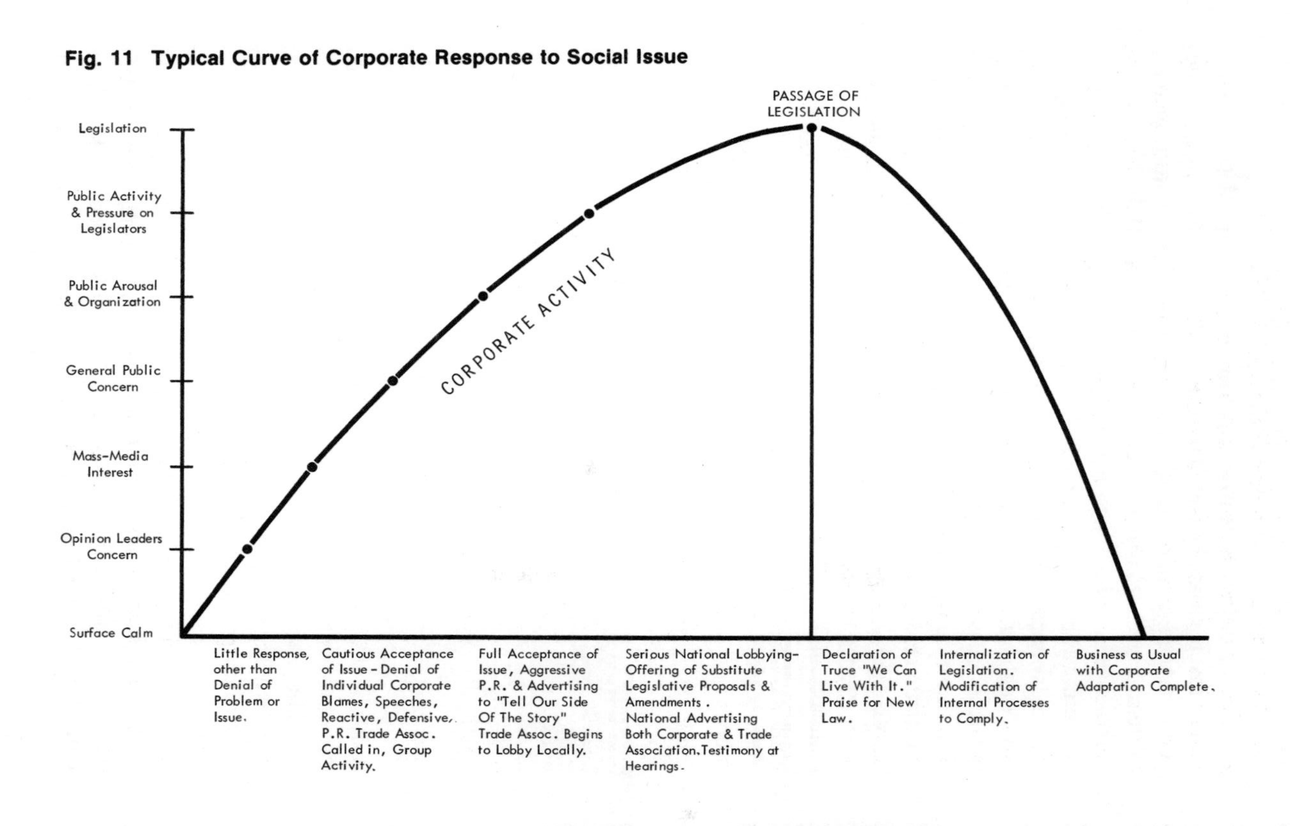

the quality of life. However, much more work on subjective social indicators is needed—for example, polling citizens on how they perceive the changing quality of their own lives or implementing exercises in "anticipatory democracy" such as Hawaii 2000 and Alternatives for Washington, where whole communities are linked by communications media and feedback devices in order to explore regional values and goals, which will be reviewed in Chapter 14.

Consulting unorthodox, imaginative dropouts from our highly institutionalized society provides still another innovative approach to monitoring value changes. The president of one large corporation, for instance, maintains regular communication with Ira Einhorn, a young poet and prominent counterculture figure since the 1960s. Since money is of little interest to this unorthodox "consultant," the company underwrites the purchase of the books on his huge reading list and gives him copying and mailing facilities to help him stay in contact with his far-flung network of equally brilliant, iconoclastic friends.

Networks—the "Un-Organizations" of Tomorrow. Out of all our current social ferment, organizations are slowly learning that if they and our society are to survive, they will need to reformulate their goals and restructure themselves along less pyramidal, hierarchical lines. Such participatory, flexible, organic, and cybernetic design is now mandatory in the face of cataclysmic changes. As Bennis points out, "Democracy becomes a functional necessity whenever a social system is competing for survival under conditions of chronic change." Adds Richard Cornuelle in *De-Managing America,* "People will no longer be used." He believes that their self-discovery and self-expression are now creating a final revolution and transformation of authority.

In fact, the ultimate organizational design is already visible to those whose perception is attuned. These new organizations already exist, although they are metaphysical. They are most often referred to as networks, and their participants describe themselves as "networkers." They have no headquarters, no leaders, and no chains of command. They are free-form and self-organizing, composed of hundreds of autonomous, self-actualizing individuals who share a similar world view and similar values.

Few organizational theorists have yet studied networks because they are evanescent, ebbing and flowing around issues, ideas, and knowledge. It is impossible to guess how many of these networks exist today in the United States and other industrialized countries. The numbers certainly run into the many thousands in this country, and many hundred exist transnationally. Their chief product is information processing, pattern recognition, and societal learning.

Networks are a combination of invisible college and a modern version of the Committees of Correspondence that our revolutionary forefathers used as vehicles for political change. Luckily, networks are linked by the mimeograph machine, the postal system, and the telephone—all decentralized technologies accessible to individual users with constitutional guarantees of privacy. Networks can now create a recognizable, media-reportable, national event expressing grass-roots interest in a political issue in a matter of hours. Even a nation as communications-rich as our own now requires this kind of instant political signaling system to its decision centers in order to overcome bureaucratic inertia and hardening of our political arteries.

Networking cross-hatches all existing structured institutions and links diverse participants who are in metaphysical harmony. Sociologists are beginning to evidence interest in studying networking. However, these spontaneous organic forms will elude such outside observers since they create "static" on the lines, and this is instantly picked up by networkers who then regroup, using alternative channels.

To try to analyze this new organizational form with traditional, reductionist social science approaches would be unfortunate; networking is the most vital, intelligent, integrative organizational mode on our turbulent social scene. Perhaps this self-organizing mode represents a new maturing of human intelligence. It may even augur the next evolutionary step in developing human consciousness, which is now necessary if we are to survive on spaceship Earth.

CHAPTER FOURTEEN

Creative Social Conflict: New Approaches to Social Mediation

Feedback from a succession of ecological crises signals our economic system to the nonnegotiable demands of nature and the changing values of our citizens. The decade ahead will be similarly marked by manifestations of these growing discordances in two major categories: (1) the objective environmental crises related to the misuse of actual physical human resources and ecosystems, such as fish kills, and oil spills, and cancer epidemics among workers; (2) the wave of subjective, value-based changes in the behavior and attitudes of consumers, employees, stockholders, concerned individuals, and social groups that all these social costs will produce, thus increasing conflicts concerning environmental and social issues.

Referring to the wholesale revolt against the internal combustion engine, Henry Ford II said, "Never before has American business been under such great pressure to change. It seems clear that neither business in general nor the auto business in particular will survive in its present form."[1] The problem facing business people is how to adapt to a growing public sensitivity to the broader social effects of traditional economic theories and practices, and how to cope with the resulting conflicts when diverging interests collide.

Just as the Sherman and Clayton Acts and child-labor laws clamped down on yesterday's exploitation, so will today's environmental, social, and consumer legislation further narrow the possibilities for making profits by externalizing costs—i.e., shifting them to others outside the enterprise. It seems apparent that the only paths to future profits will be found as managers pick their way through increasingly complex trade-offs and learn to wring tomorrow's slimmer earnings from better resource management and utilization and from more efficient energy-conversion ratios.

The traditional predisposition of manufacturers toward the practice of externalizing costs is not just a "capitalist" phenomenon; it is equally evident in socialist and communist economic systems. The problem faces all advanced industrialized economies, irrespective of whether the goal is making profit or amassing capital for state-directed economic expansion. Even more disturbing is our growing suspicion that what we call "profit," or what state-controlled economies might term "economic expansion," has probably incurred an unrecorded debit in some previously invisible social and environmental ledger. In theory, socialized enterprises, by their very definition, should not be able to externalize *any* costs, but in practice, these industries create social and environmental costs similar to those in market-oriented societies.

Need for Broader Views. The current clamor over the misuse of global resources cannot be expected to subside of its own accord. As these ecological disparities are felt, we can expect increasing conflicts between concerned social groups and institutional interests. Moreover, with the new environmental awareness comes the potential for conflict over a wide range of social issues, for "environment" has come to mean the social as well as physical surroundings in which we live.

But this group of conflicts need not be regarded as dysfunctional. In fact, it is not conflict that is alarming, but rather its mismanagement.[2] The expanding environmental awareness produces its own relentless logic that illuminates the inadequacies of two-dimensional economic planning and decision making; it underscores the need to develop more realistic three-dimensional models of the economy that include social and physical resource factors.

In the global aspects of these new conflicts we can begin to perceive patterns of energy use, of population flows, and of available resources and the demands on them. This in turn could produce a broader awareness of the structural problems that transcend narrow political considerations and recognition of the need for:

☐ Political, economic, and social systems capable of managing and applying technology more rationally.
☐ More flexible, humanly-scaled institutions that can be responsive to those whose lives they shape while taking advantage of true economies of scale.
☐ Additional social structures to mediate society's intergroup conflicts.

To achieve these goals we must manage conflict so that parochial views are expanded to a higher and more balanced plane. It is in the interest of business and other institutions, as well as society generally, that we develop new "societal bargaining tables" that can turn conflicting views of these new crises into a truly integrated awareness of the wider implications of debated issues.

How, then, can enlightened U.S. business people join with government officials and citizen leaders of good will in a serious effort to minimize the advertising and promotional grandstanding by all parties, and get down to the vital job of mediating this mounting backlog of conflicts that bid fair to characterize the coming decade?

Many of the methods that seem most promising have already been applied and widely tested by corporations, and I shall attempt to show how these and other methods of rationalizing conflicts can be extended and adapted to facilitate the management of new social and environmental confrontations. First, I shall examine the potential of such conflicts, if properly managed, to expand parochial viewpoints to a higher level of awareness. Then I shall discuss the methods of conflict management for realizing this potential—there are two broad categories: (a) "software," which constitutes new forms of societal collective bargaining and other social and quasipolitical mechanisms, and (b) "hardware," which constitutes the actual and potential technology of social choice.

The Edict of Conflict. A persistent condition underlying social conflict is the differing set of subjective assumptions and levels of awareness by which groups perceive the same objective set of circumstances. Often the only time that such underlying perceptions can be made explicit, then explored and mediated, is when they clash in an open confrontation.

The urgent need to prevent the repression of groups by the powerful as well as seeing that clashes do not ignite into lawlessness or violence provides both an incentive to examine these subjective assumptions and an opportunity to expand the differing groups' understanding of each other and the wider circumstances of their dispute. If we can stretch the minds of quarreling institutions, individuals or groups so they can see how their parochial viewpoints fit into a much larger whole system, a higher plane of understanding can be developed that integrates seemingly diverse community goals.

From Confrontation to New Understanding. One example of local values in conflict involved the United States Steel Corporation's plant in Duluth, Minnesota, which was ordered to comply with minimum air-quality standards. Company officials announced that the plant would have to close if the standards were enforced; 2,000 workers' jobs were at stake. The divisive potential of this kind of conflict is truly explosive, especially when a large proportion of workers are black or comprised of other minority group members. The issue could be unscrupulously exploited as a crude confrontation between minority employment and middle-class environmental sensitivity, or even black stomachs versus white lungs.

It is possible, for example, for various economic interest groups—whether business, labor, or political subdivisions—to manipulate the conflict in its basest form to achieve selfish ends. These groups can deliberately narrow the focus of the issue in order to heat up the political climate and force a swift resolution in their favor. Sometimes, simplistic mass-media coverage of the issue can make matters even worse. Any narrow-interest group that plays this game of chicken to bluff its opponents must eventually do so at the community's expense, for the issues are merely submerged to fester in the ill will thus created, ready to flare again at any provocation.

The management and eventual resolution of a problem like the steel plant's threat to close in Duluth requires *expanding*, not narrowing, the focus of the issue. For example, an enlightened leadership might have used the crisis to explain why some national system of welfare payments is needed, so that if a particular plant had to close for valid economic reasons, the workers could be protected from hardship, without bankrupting the city, until new jobs were developed.

Local leaders might also point out that such a plant is a marginal operation and an unreliable source of employment anyway; that management, by asking the people to put up with the plant's pollution, could well foreclose on the town's opportunities to attract new, cleaner, more profitable industries serving future growth markets (i.e., computers, electronics, communications, mass transit, tourism, and pollution control).

Many industries need a clean environment for their operations and want to locate in a community that can attract and keep a highly qualified labor force. Thus a well-informed community, with active leadership organizations in health, conservation, and community services, might forge a coalition with local businesses to attract cleaner new investments and alternative forms of economic development.

At the same time, town meetings and the mass media could help educate the community concerning various national welfare and full-employment proposals and other methods of cushioning individual hardships during dislocations caused by recession, corporate relocation, environmental hazards or the general effects of technological change. Such a wide-angle, conciliatory approach would, in turn, help citizens appreciate the problems of businesses based on obsolescent technology, like their need for new capital to diversify into growth areas or to finance pollution control and recycling systems.

Similarly, dozens of local conflicts concerning the location of new power plants are at last forcing a national discussion of the full range of options, from stabilizing consumer demand for electricity by clamping down on advertising that promotes load growth, to a lively exploration of new technological possibilities for improving energy-conversion ratios. Less polluting, more efficient energy sources (i.e., magneto-hydro-dynamics, solar energy) are receiving increasing attention as substitutes for thermal and nuclear-fission electricity generators.[3]

Many of these possibilities would have remained dormant unless a crisis had forced the issue, capturing public attention via brownouts,

Drawing by Reilly; © 1973
The New Yorker Magazine, Inc.

personal inconvenience, and subsequent media coverage. Through crisis we learn of the trade-offs inherent in growth and begin to consider what might be the optimum size for a city, or what priorities should be set for the competing uses of power. As a result of battles over new generating stations, the Federal Power Commission has developed a "uniform and reasoned set of priorities as to who gets cut off first" during a power shortage.[4]

Of course, there is no easy solution to the coming power crunch in any case, because it was brought on by fragmented, inadequate resource management and the lack of an overall national energy policy. So we learn too, that overcommitment to past investment (in this case, nuclear fission) can become a drag on technological innovation by diverting research funds from more advanced systems, such as nuclear fusion, sustainable solar energy and hydrogen.

Six Steps to Chaos. While crisis, as previously described, can serve to enhance awareness, there still remains the difficult task of mediating the conflicting interests that grow out of it. But our existing channels for mediating social conflicts are already seriously overloaded, and just as the labor movement remained in the streets until the infrastructures of collective bargaining were developed, so the coming social conflicts will find even more unorthodox and disruptive modes of expression until we devise the infrastructure needed to contain and resolve them with civility. Without it, social conflicts will progress through the following six stages with increasing rapidity:

1. There is the traditional, now faltering competitive market mechanism.

2. If this fails, the problem spills into the governmental or political arena.

3. If this arena fails, the conflict overflows into the judicial process, perhaps as a class action or a consumer lawsuit aimed at a malfunctioning regulatory agency.

4. If this avenue is blocked, the grievance may be directed at an appropriate corporation or industry, through attempts to mobilize stockholders and politicize the annual meeting as an ad hoc forum for discussion of the issue.

5. If still unresolved, the conflict will erupt into the mass media, its

protagonists using the effective "guerrilla theater" mode of attracting the attention of drama-seeking news editors by such means as picketing, emptying truckloads of cans on the doorsteps of container companies, or burying internal combustion engines.

6. When all else fails, the conflict descends into the streets in open confrontation, civil disobedience, and even sabotage.

A confrontation that reached stage two of this conflict progression resulted from the decision of Western Electric to move part of its corporate headquarters from New York City to the nearby rural town of Bedminster, New Jersey.

In an unprecedented attempt to influence the zoning decision of another municipality, New York City's Economic Development Administrator, Kenneth Patton, personally appealed to a meeting of the Bedminster town planning board that they deny Western Electric's application for a zoning change to permit construction of a $10-million office complex on 300 acres of land.

"Communities can no longer view their decisions in terms of narrowly defined advantages; neither can companies," said Patton. Citing the obvious economic advantages for a company that leaves the city and locates "where the poor are not housed . . . where sewage is not treated," he added, "In their own interest, and in our nation's interest, business must begin to compete on the level of performance rather than continue to practice evasion or escapism." He further argued that since the new facility would be surrounded by a "moat" of five-acre zoning, the poorer workers would have to travel long distances to the site, creating more social "apartheid," not to mention the problems of additional auto traffic and pollution.[5]

This conflict of interests was prevented from progressing through further disruptive stages by Western Electric's decision to cancel the project. More important, however, is the fact that the dispute reflects an ad hoc attempt to compensate for the lack of a suitable mediating infrastructure.

Software of Social Conflict. In the absence of needed new mechanisms, social decisions that incrementally change the face of our nation are being made daily, as in the previous example, with not much more than rhetoric, uninformed opinion, and pressure-group politics to illuminate them.

The first step toward filling this gap is to develop the software of conflict resolution to the point that it functions as a legitimate mechanism of social mediation. The most obvious group of techniques in this category is already familiar to corporate managers: role playing, encounter groups, sensitivity training, and the host of organizational methods for enriching jobs, disalienating workers, and democratizing corporate structures. It is almost certain that these forms of mediating human affairs and helping people understand differing positions will be further adapted by communities, government, and private institutions.

More promising, however, are the social and quasipolitical mechanisms of collective bargaining and advocacy planning, the potential of new governmental structures, and the development of social change agents as representatives of the public interest. It is to these that we now turn.

Collective Bargaining. In mediating labor-management disputes, collective bargaining is more effective than anything else yet devised. Its essence is the finding of an area, however small, of common agreement between parties and expanding it. Its assumptions are that the issues to be resolved are objective and substantive—and do not merely represent the irrational, or aggressive behavior of parties in the dispute. It also assumes that the participants have been duly mandated by the conflicting parties and are empowered to negotiate and settle the issues under formalized and agreed-on rules. All these components of the collective-bargaining process are well suited to wider use, both for mediating the growing conflicts between corporations and the public at large, and for arbitrating other community conflicts.

As exasperating as collective bargaining negotiations often are, they can represent one of the most important functions of social conflict management: the recognition and acceptance by various interest groups of the legitimacy of each other's competing claims on resources, within an orderly framework of law and custom. Indeed, the style of community organizing favored by the most alienated of society's outgroups deliberately militates *against* collective bargaining, probably in conscious recognition of its legitimizing function. These alienated groups tend to shun existing structure and leadership so as to better resist what they see as "co-option into the system." This problem could still be overcome if sufficient trust were developed and if the new bargaining structures were able to demonstrate their ability to arrive at equitable settlements.

New Power Blocs. If "wildcat" types of activities against corporations by alienated groups increase, we may soon see the day when companies welcome organized power blocs of consumers, citizens, and stockholders. These groups would have duly constituted leaders with whom management could deal. By this measure, the truly conservative role of the "Campaign to Make General Motors Responsible" can be appreciated—the campaign's young lawyers meticulously used only existing legal machinery and Securities and Exchange Commission rulings in their attempt to seat three "public-interest" candidates on GM's Board.

Just as the "vice president of labor relations" function grew in symbiosis with the power of the labor movement, so the position of "vice president of public affairs" may grow in relation to the power of organized citizen, stockholder, and consumer-interest blocs. The expanding corporate responsibility movement that seeks to organize these new power blocs is a sophisticated attempt to politicize the corporation, irrigate its decision-making processes, and force it to develop new internal negotiating functions.

Even in their relations with their own employees, corporations can anticipate and reduce bargaining conflicts in the future by, in effect, transforming employees into citizens of the enterprise. This can be achieved by widening employee stock ownership, and democratizing the structure of the company itself. This method of conflict resolution, already successfully operating in a number of corporations, is based on the age-old technique of giving every member a similar stake in the goals of the business.

Community Mediation Boards. Another evolutionary trend may stem from the growing practice of permanent collective bargaining, in lieu of permitting contract deadlines to dictate the start of negotiations. This practice would be even more useful in the widespread social adaptation of collective bargaining. There might, for example, be permanent panels of highly trained citizens to serve on "community mediation boards," appointed as quasipublic bodies and available at all times for any disputes that would otherwise fester, clutter court calendars, or be resolved by power plays.

A typical example of a dispute which could have been settled by a community mediation board centers around the plan of a major New York City retailer to build a department store in an old, less fashionable, ethnically mixed neighborhood. The site had been selected and there was general acquiescence, or perhaps apathy, concerning the project when a very small group of militant women from a local community-action program protested the plan. The retailer did not care to get into what threatened to be a long, ugly hassle and canceled the project. Other segments of the community, who had looked forward to rising property values and a more cheery shopping area, voiced their outrage at the cancellation.

In the foregoing dispute, an established, impartial mediating body, composed of a cross-section of responsible local leaders, could probably have worked out a solution agreeable to all.

Advocacy planning. Similar in style to collective bargaining are the movements for advocacy planning.[6] They grew out of the programs of the Office of Economic Opportunity, Model Cities, and private efforts such as the Saul Alinsky-founded Industrial Areas Foundation. These movements are developing rapidly as more social and health workers, young people, public-interest lawyers, and concerned citizens become sensitized to, and involved in, the problems of the poor and powerless. Advocacy planning provides professional and technical help, whether from lawyers, architects, planners, doctors, or social scientists, so that community groups can articulate their aspirations and develop future plans.

These activities, however, have initially proven disruptive because they bring submerged issues and conflicts to the surface where inevitably they clash with established institutions and contrary public opinion. For this reason, they have received criticism and political reprisals from uncomfortable politicians at all levels.[7] But as they gain legitimacy and develop the necessary support from established institutions, they will represent another means for resolving conflicts between minority interests and the dominant culture before these conflicts ignite into actual violence.

By forming new social constituencies and particularizing issues, advocacy planning expands the views both of the involved parties and of the community at large and serves precisely the same legitimizing function as do labor unions. As the new constituency is rationalized, duly mandated leaders can emerge; then the larger society can negotiate the group's newly articulated claims with these leaders in an orderly fashion.

For instance, by organizing black groups in Chicago in the sixties, Operation Breadbasket voiced grievances through boycotting retail corporations, which, however disruptive, created a dialogue with management that prevented more dangerous, violent forms of protest and led to the legitimation of the Rev. Jesse Jackson's methods and to the more broadly concerned organization, PUSH.

Response of Government. Other software mechanisms are evolving as governmental agencies form or regroup in response to growing public pressure. One example is the Environmental Protection Agency, which absorbed dozens of conflicting smaller agencies, integrating their goals and programs.[8]

Similarly, many single-purpose agencies whose concerns are excessively "need oriented," such as the Soil Conservation Service and the Army Corps of Engineers, are now coming under increasing fire for their natural bureaucratic tendency to empire-build without sufficient regard for objective needs or realities. And still another governmental response to the call for more citizen participation has been the growth of neighborhood block associations and "little city halls," which began as complaint and service centers, and are now developing opinion feedback capabilities.

Agents of Social Change. Well-motivated citizens, whose volunteer activities are often identical to those of labor or community organizers, have tremendous potential for articulating social concerns. Through their committees and coalitions, they function as "societal change agents," analagous to the change agents used by management consultants in helping corporations adapt to new environmental conditions.

These societal change agents, however, are often negatively conditioned because they serve as shock troops in assaulting and democratizing rigid, outmoded institutions, and, consequently, they become embittered, exhausted and sometimes alienated.

We need to capitalize on our full-time citizens not only to protect them from embitterment, but to better train and direct their integrative skills as interpreters, mediators, communicators, and ombudsmen. Whether we approve or not, they will continue these activities in accordance with their moral precepts and expanding social awareness. They can organize the coalitions that will be needed to financially underpin various segments of our emerging public-sector markets, in the same way that they have already organized and articulated the "group consumer demand" that now underpins the pollution control market.

One such coalition, forged by volunteer community organizers to support the development of mass transit, will result in a political movement of sufficient power to provide huge increases in federal funds to build and restore these facilities. The industries waiting in the wings to serve this incipient mass-transit market cannot do the job by lobbying efforts alone; they will need to back the efforts of the organizers behind the citizen-action groups. This backing will provide the solid political muscle of diverse new grass-roots constituencies that include conservationists, antipollution groups, historical societies, garden clubs, women's groups, voluntary health agencies, inner-city groups most threatened by freeways, and increasing numbers of infuriated suburban commuters.

Similar coalitions of equally unlikely groups are nascent and will eventually grow strong enough to underpin the emerging regenerative economy. Helping such coalitions rationalize their goals and focus their political power is another essentially conservative activity whereby a nation can redeploy its resources peacefully as its social needs and technology change.

Essential Hardware. All the new "software," if it is to be fully effective, will require much greater access to existing "hardware," which represents the technology of social choice. Electronic communication, with its enormous potential for expanding the awareness of millions of citizens simultaneously, is already functioning as a vehicle for many of the social mechanisms previously discussed.

The new audience builder in broadcasting, for example, will not be merely entertainment, but self-enrichment and the fascination of two-way participation. What is coming to be known as "citizen access" is based on the understanding that broadcasting franchises are tools of political power and can be used to recreate electronically the town meetings of yore. Similarly, radio serves as an efficient ombudsman, in the form of increasingly popular "Call-for-Action" shows that champion individuals and groups seeking redress of grievances against unresponsive organizations.

Radio is also useful as a medium for mass psychotherapy, where people may anonymously call talk shows and voice their problems, opinions, prejudices, guilts, and fears, and obtain catharsis in an electronic, nonthreatening exchange of human experience. For the listeners, these talk shows provide new appreciation for the complexity of social problems and the need for tolerance and cooperation.

A Broadcasting Alternative. The prototypical structure in broadcasting is, of course, the public broadcasting network that grew from the need for an alternative public-sector network to augment the private commercial networks. Its phenomenal audience growth in the highest socioeconomic groups (in 1971 TV programs reached 11 million households per week, as well as millions of school children, and radio programs reached 120 million) attests to the needs and tastes left unserviced by the various market failures of commercial broadcasting.[9]

An interesting new outgrowth of public broadcasting is the proliferation of local station programing involving community leaders, geared to gathering feedback on the nature and extent of their area's social and environmental problems. These programs cover everything from drug abuse to garbage-collection problems and meet the need for mass environmental education for all ages at dramatic savings over any other method. Other new program concepts are being tested, such as "Environmental Games," that pose all the options open in an ecology problem, then simulate the effects of each alternative based on audience participation and feedback. Viewers can vote their preferences and see how many others agree with them.

An example of the possibilities involved the city of Philadelphia, which witnessed increasingly acrimonious conflicts concerning its role in the national bicentennial celebration in 1976. With little consultation as to site chosen or allocation of the money for permanent improvements, such as parks or mass transit, the Pennsylvania State Bicentennial Commission, composed largely of "chamber of commerce" interests, had managed to alienate large segments of the city by presenting a full-blown plan for a commercially oriented exposition. Maneuvering, demonstrations and threats from all sides threatened to scuttle the entire project.

At this point, Robert Lewis Shayon, TV critic and professor at the University of Pennsylvania's Annenberg School of Communications, stepped in. With Franklin Roberts, an advertising executive and Howard Anderson of the Bell Telephone Company, Shayon offered to help the local station KYW-TV in a last-ditch attempt to ventilate the situation. The station promised five-and-a-half hours of prime time if the team could produce an "electronic town meeting" in three weeks.

Twelve citizens, representing every shade of opinion, were selected, and each received a crash course on all aspects of the bicentennial plans. They debated the pros and cons of all the alternatives, while thousands of callers registered their views in heated exchanges over the air. In addition, printed ballots were circulated at banks and supermarkets and in local newspapers. As is so often the case when new communications media threaten to impinge on the usual behind-the-scenes horsetrading, Prof. Shayon encountered considerable opposition from some city fathers. Failing to squelch the show, some attempted to rig it, and others were even honest enough to admit they didn't believe in democracy anyway! But most interesting was the fallout from the program. Soon afterwards, Shayon reported, the Pennsylvania State Bicentennial Commission elected a new chairman, and the hotly contested site was dropped.

Even though the program and the associated balloting had no force of law, they defined the attitudes of the community, located a new consensus and secured the adoption of a more acceptable government and business project.

In Boston, another project with similar objectives was tested.

Conducted by MIT's Operations Research Center under the leadership of Professors Chandler Stevens, Thomas Sheridan and John Little, one of its goals was to analyze the flow of citizens' complaints, suggestions and other material received by Gov. Francis Sargent. These complaints, when profiled as to content, geographical location and time span for remedial action, can flag those areas where government agencies are malfunctioning or unresponsive, as well as identify issues and problems which cause the greatest gap between expectation and satisfaction and may even erupt into civil disturbances. The team designed a fully interactive, computer time-shared program to signal problem areas continually, expedite the complaint-handling process and route the data to the proper administrative area and was installed in the governor's Office of Public Service and the state attorney general's Citizens Aid Bureau.

Issue Ballots and Listening Posts. Another phase of the Massachusetts project has concerned eliciting citizens' opinions. Using "issue balloting" and "listening posts," the method has been tested in cooperation with the State Department of Education. The issue ballots—small booklets supplementing local newspapers—briefly summarize the education issues under consideration and formulate different sets of goals and options for future policy. A tear-off, computer-readable card is then marked for tallying and, if the citizen requests, forwarding to local legislators. The listening posts are local meetings in school auditoriums where each chair is wired to a portable system of voting button-boxes, which, in turn, are all linked to a closed-circuit TV screen in the front of the room. Instead of the usual monopolization of such meetings by persistent grandstanders, the voting boxes permit everyone's opinions to be counted. As the audience "votes" on the alternatives, its views are simultaneously tabulated and displayed on the screen.

In addition to relieving tensions, the secrecy of the procedure elicits "honest" responses. For example, in a test discussion at MIT of the subject of municipal corruption, the group was asked whether they might be tempted to practice nepotism if they were municipal office holders with discretion over tax assessments or zoning. An overwhelming percentage admitted anonymously to the temptation to "cheat a little bit." This revelation on the display screen sent a friendly chuckle of understanding rippling through the group. Television can also complement the process by wiring it into a local cable-system or a conventional public or commercial broadcasting station. Although some participants have found the issue ballots unfamiliar or too complicated, many others display sophisticated reactions, such as questioning the formulation of the issues and alternatives. Both the issue ballots and the voting boxes can handle this type of feedback, which, in turn, serves to refine the choice of questions and options.

The Hardware of Democracy. These two demonstrations of the potentials of "counter-technology" are far from isolated examples. They are the tip of an iceberg signaling a quiet revolution in public realization that computers and mass-communications technology need not be a monolithic and centralizing force. People are discovering that these machines can just as easily be adapted to more humanized, organic and decentralizing applications—wiring individual "cells" in our body-politic into the central nervous system.

Another experiment: Open Channel, a New York-based, nonprofit corporation was run by Thea Sklover, a leader in New York City's fight to require commercial cable TV franchises to make at least two of their channels available for public use. This pioneering legislation set a precedent for similar requirements across the country to open public channels for a rich, new network of interactive, community television resources.

Prototypes and Pioneers. One of the earliest projects on record was conducted in 1955 in Ames, Iowa, by the same Robert Shayon who staged Philadelphia's TV "referendum." In a series of 26 programs on school-district consolidation entitled "The Whole Town's Talking," educational station WOI-TV so successfully stimulated voter response that a special session of the state legislature was called to act on the issue. Even more amazing, by popular demand, its deliberations were also televised, and the consensus on implementing a consolidation plan was achieved.

Almost a decade later, some of the wider social potential of computers was tested in California, and, even then, the centralized, managerial bias was much in evidence. In 1965 Gov. Edmund Brown commissioned a survey by the Space General Corporation which used a series of overlays of data and five social indicators: low incomes, heavy concentration of segregated minority population, high arrest and drop-out rates, and high general population density. The now famous results zeroed in on a place called Watts. But in 1966, Ezra Krendel, professor of Statistics and Operations Research at Wharton, took the whole concept a step further. He found that by using unsolicited complaints to formulate social indicators the inadvertent impact of the analysts' own values—which may have weighted the choice of indicators in California—could be avoided. He classified over 300 complaints received by the mayor's office in Philadelphia as to content, time of response and geographical area to obtain a "snapshot" of the citizens' expectations and the actual quality of urban life and services available. Similarly, New York City has operated such a service monitoring system which evaluates the performance of city agencies.

Another approach was the brain-child of Richard A. Givens, former New York Tri-State Director of the Federal Trade Commission, who instigated a computerized study of all the judgments awarded in New York's Civil Courts, indexed by companies suing for debt collection and all collection lawyers, process servers and notaries public, as well as other data on location of suits and defendants. Under a small grant from the Fund for the City of New York, the FTC helped establish a Consumer Protection Coordinating Committee and provided computer facilities to investigate the many abuses of consumer rights in debt collection. The results showed that a small group of companies, particularly in retail furniture and credit jewelry, were responsible for most of the collections, and that a similarly small group of lawyers and process servers handled most of the claims. Using this data, the agency unearthed other abuses, such as obscure, unreadable credit contracts, defendants not properly notified of the suit or called before distant courts to increase the likelihood that they would fail to appear, thus allowing the company to be awarded on-the-spot judgments, illegal attachments of salaries and intimidation of defaulters. Many of the company officers, lawyers and process servers have since been subpoenaed by the FTC.

The conceptual development of this FTC system is toward even more sensitivity, because it does not rely on activation by an overt complaint from the articulate, energetic citizen who still monopolizes the feedback from most of the systems mentioned. Instead, it uses the computer as a tool of "advocacy planning" to expose and give voice to the grievances of the inarticulate, the poor, or the weak and apathetic—those who have nothing to lose and are exploited and often prone to violence or self-destruction. Polling, issue balloting and voting button-boxes will need further refinement to reduce the effect of the indicators' assumptions in designing the questions and formulating the issues. One safeguard most of these systems employ is to include, in addition to the "don't know" response, another possibility indicating "objection to question" or even the equivalent of "boo"!

To see how far such refinements have improved commercial polling and market research, compare them to a conventional poll conducted by the Roper Organization for the Television Information Office. One question was posed as follows: "If eliminating all commercials on children's TV programs meant considerably reducing the number of children's programs, which would you favor: (1) eliminating the commercials and considerably reducing the number of programs, or (2) keeping the commercials to keep the children's programs, or (3) don't know or no answer. Such an artificially narrow range of options naturally produces artificially simplistic results. The range of options might have included some other realistic alternatives, such as "Would you favor broadcast licensees being required to make a certain number of hours of children's programing a part of their program responsibilities under the 'public interest' section of Federal Communications Commission rulings?" Or, "Should sponsors of children's shows be limited to one-minute institutional mentions at the beginning and end of each half-hour program?" Omitting such ideas from many commercially oriented market-research polls not only reinforces the status quo but hides some aspects of the real situation which may require closer attention; i.e. if you ask a silly question—you get a silly answer!

Commercial radio and TV stations have also discovered the value of championing the "little guy" on the enormously popular "action line" shows, even though failures of their advertisers are sometimes exposed to public scrutiny. The pioneer "ombudsman" show, "Call For Action" on station WMCA in New York, has now been replicated in dozens of cities. As this sort of activity increases, innovative corporations may start listening better in the realization that even unsolicited brickbats are priceless market research.

Another experiment was designed by Dr. Amitai Etzioni of New York's Center for Policy Research. In a 14-room laboratory wired with conference-phone equipment and using light buttons and versions of Robert's Rules of Order, his team investigated the role of opinion leadership. After determining what technologies best fit various group-meeting processes and what signals best replace human kinesthetics (such as frowning or raising a hand), the hope is to see communities with cable TV attempt to set up long-distance dialogues between, say, Watts and Harlem, or Harlem and Scarsdale, N.Y.

Raison d'Être. What is the social purpose of all this feedback technology? Those within or with access to current decision systems maintain the "elitist" view that growing demands for participation are at best troublesome and unscientific and, at worst, a threat to overall stability and order. On the other hand, the new breed of "populists" seems to empathize with the powerless and those lacking expertise and information or economic leverage, and they share a faith that, with sufficient undistorted information, the common sense of the layman can equal that of the specialist.

Obviously, these systems presently aim toward reducing alienation and tapping the backlog of uncommunicated grievances causing social pressure. If this is done, the data produced can provide negative feedback loops to the trouble-generating spots in society. But counter-technology can be a potent new force for motivating interest and learning, and here most would agree with the elitists that, if our citizens are to have more power in running this complex society, they must be incited to master more information to make wiser use of their votes. Further, most of the experiments have shown that if people receive new opportunities for participation many of them mature from an initial fixation in the "complaint mode" and begin advancing constructive suggestions, as well as volunteering their spare time to work on problems eliciting their interest.

At the same time, we have also discovered that, in our information-rich society, decisions imposed by the few on the many rarely remain in force for very long. Even though discussing and revising proposals to achieve a more democratic consensus is a long, involved process, the reward is likely to be a sounder and more lasting decision.

The longer range goals, therefore, must not only be better social feedback but also more "feedforward" of relevant information for better public decision making and formulation of goals. Because the pace of social change has accelerated, we must be able to look further down the road and have more lead time in which to shift direction. To do this, we must emulate the kind of intensive discussion of alternative futures conducted in 1969-70 in the Puget Sound area of Seattle by a consortium of local colleges, civic and religious groups backed by the State Office of Planning and Community Development and matched by federal funds. Trained group discussion leaders and coordinators held hundreds of small meetings to discuss what concepts such as "quality of life" and "environment" meant for the future of the region. A local TV station produced eight shows to provide interest and the necessary information feedforward on the problems and possibilities, and polls measured awareness and attitudinal changes. Nielsen ratings showed that 10 to 11 percent of the available audience in the area—some 75,000 homes—watched the shows on such heavy subject matter as economics, social welfare, institutions and values, and land-use planning. People who became aroused by the information and the study group experiences were shown how to plug their concern into existing social infrastructure in the form of volunteer and political activity. We will review the development of these "alternative-futures" groups in Chapter 22.

A Voice for All the People. Just as Prof. John Little prepared a model of the dominant variables affecting the financing of higher education as part of the Massachusetts project described, so we can model, for example, all the alternatives for a town's transportation mix and assess outcomes for such options as: (1) do nothing and permit continued *ad hoc* growth of automobile and highway use, or (2) make more provision for safe paths for short trips on foot or bicycle, or (3) build mass-transit line for high-density area with feedforward from potential riders to determine routes, or (4) start a "dial-a-bus-on-demand" system with computerized dispatching, or (5) designate open lanes for buses on major freeways entering the town so as to lure people out of their cars with the faster trip. These simulations can be adapted as "games" for presentation on public television, and audience feedback can be profiled to change the diagrams on the screen as "votes" are recorded.

Another exciting area where the computer is being harnessed to the democratic decision-making process is in the planning of more rational and suitable use of America's remaining open land. Dr. Ian McHarg, an ecologist and land-use planner, is using computers to determine the optimum use of various kinds of land. For instance, some land is unsuitable for building; it may be too steep or the subsurface rock may be unstable so that buildings may slide, subside or wash out in rainstorms; some land is best suited to farming and grazing; some is covered with timber and must be conserved to protect a watershed—and so can be used only for recreation; and, finally, some land is best suited for residential construction or for industrial development. Dr. McHarg surveys a whole region, and—after taking into account factors such as geology, surface terrain, meteorology, population densities, transportation, water and other natural resources—superimposes on his maps different code colors representing areas best suited to building, farming, wilderness recreation, and industrial development.

Eventually, the map reveals for the entire region the best location for new towns and the best possible areas for open space, parks, farms and industry. In this way, we might plan America's shrinking land and resources to best accommodate the tens of millions of new citizens expected in the next 30 years—without risking cancerous growth of existing megalopolises, or putting up on treacherous mountainsides and floodplains ill-conceived buildings, which often cost the taxpayers dearly when their collapse is followed by demands for federal "disaster" funds. Plans like these could help indicate how we should space out our population by the year 2000, and may help to prevent future ecological debacles. Systems based on Dr. McHarg's pioneer work would help us balance the use of our resources effectively for the greatest benefit of the country as a whole, rather than for the benefit of such parochial interests as speculators.

All of these uses of computers will depend for their optimum effectiveness on how carefully and democratically the analysis is performed and the program written. If narrow, technologically oriented specialists continue to design data processing systems only with data that appear relevant to *them*, computers will continue to reflect narrow values, and produce "garbage-in/garbage-out" results. For instance, a badly-designed cost-benefit analysis of whether a city can "afford" to install air-pollution-control equipment might produce data which would "prove" that the city couldn't afford to clean up its air.

However, if data had been included on emphysema disability claims, for example, or on the correlation between absences from work due to respiratory diseases and sulphur dioxide levels in the air, the answer might have been that the city couldn't afford *not* to clean up. The best way to ensure that all these elusive variables get included in the model is to use extremely diverse, interdisciplinary teams of advisers: including, perhaps, a poet, a house manager and a child, to develop random ideas in structuring the model. Jay Forrester points out that social/physical systems, like cities, are far too complex for one human brain. The only answer we have at hand is to develop more and more comprehensive access to all of society's brainpower.

Computers can also provide a rich new store of epidemiological research data for scientists and public-health officials. At the same time, this data can provide environmental scientists with needed evidence of health effects of various polluting substances, and even pinpoint sources of toxic emissions by geographical plotting of disease incidence. Similarly, such data can provide insights into how environmental irritants interact with viruses, bacteria, nutritional states, socioeconomic status and all the elusive variables suspected in the etiology of diseases such as emphysema, cancer and heart attacks.

Publicly owned broadcasting, for instance, if adequately funded, could provide continuous public affairs programing as well as free, equitably apportioned time for political candidates and for both public officials and citizens. Higher education could be available to all via the airwaves, were we to emulate the British Broadcasting Corporation's Open University, which commenced operations in 1971. For education no longer needs buildings, but only the voluntary communion of the minds of our greatest teachers and of all who thirst for knowledge and understanding. If democracy cannot work in America, surely it cannot work anywhere!

The basic hardware of our emerging information-based democracy is already in place. We must now learn to make it serve the individual. Collecting and analyzing individual viewpoints is already common practice in the commercial world; it's done by market sampling consumer preferences, and by an increasing use of personal data banks for credit information or medical histories. We see it too in statistical studies so prevalent in behavioral sciences, and of course in the increasing use that politicians make of opinion polls to sample the views of the voters. But all these uses are potentially a threat to individuals, because they are the involuntary subjects—and sometimes the victims—of organized information gathering. These activities are commissioned by private groups to obtain only such information from individuals, or *about* them, that serves only the institution's purposes. Often the individual only vaguely understands the purpose to which the information will be put; and any study commissioned for one particular purpose, or for use of one particular group, will inevitably suffer various distortions based on the particular need and bias of its instigator. The resulting study design, questionnaire and sampling techniques are similarly threatened by distortion.

An example during the sixties was the attempt to "sell" the anti-ballistic missile system to the American people by a group that supported the President's position. They placed a full-page advertisement in the New York Times, claiming a recent poll had shown "84% of the American people were in favor of the ABM." This extreme example of a self-serving use of polling was luckily obvious in its distortion; after all, it is almost impossible to imagine that 84 percent of Americans would agree even on the merits of motherhood or apple pie! But other, more subtle and therefore more potentially harmful, uses of polling by politicians are becoming popular: for example, to "market research" voters and manipulate emotional issues to more easily win an election; to buttress the political position of a legislator or a President; to start bandwagons rolling; or to stifle dissent.

In fact, the private manipulation of information-gathering techniques is becoming as disastrous for public decision-making as Nielsen ratings have been for quality television—and such methods tend to screen out of consideration new or random ideas, which are a vital component of an innovative society. We see, too, how bewildering the political scene can become when a President invokes an unmeasurable "silent majority" as his mandate.

The battle of wills then centers on who can capture the most television coverage to display his pile of supporting telegrams, or on the number of marchers he can muster in the streets. Government by nose counting, or by polling of random samples, can lead to dangerous distortions in the democratic process.

However, from studying the various uses made of information and communication technology, we can begin to see its potential value if put to the service of individual citizens, allowing them to decide what information to transmit to their elected representatives in deciding national issues. Dr. Donald Michael, author of *Cybernation: the Silent Conquest*, has gone further in pondering how information technology might be harnessed to serve the voter and help perfect the democratic process itself. In the magazine *Daedalus* he described how a "citizen feedback" computer-planning system might work. Suppose a city's official planning body proposes to redevelop a large area of the city. At a public hearing today, only about one percent of people affected by the proposal are able to testify, and only after waiting hours for the privilege. And, in too many cases, all manner of behind-the-scenes horse trading has already taken place among various vested interests, whose votes are levered by the power of money or political alliance.

Instead, Michael proposed a network of computerized "citizen terminals." These terminals would be located in every voting precinct, and would all be hooked into the proposed plan at City Hall. All data in any way relevant to the new plan would be stored in the computer; any interested citizen would be able to visit the "citizen terminal," where there would be a technically trained operator available, as well as maps and plans. Dr. Michael thinks that laws should be passed forbidding anyone to deny the citizen access to the computer terminal, because, as he says, "Citizens with different perspectives and interests than the planners and politicians almost certainly will ask questions the professionals forgot, thereby discovering significant implications that the professionals overlooked." Eventually, one would imagine, it would also be possible for the voter to punch into the computer his preference of a range of options on the plan; the computer would flag those features of the plan most objectionable as well as most satisfactory to most people.

Is all this science fiction? If it is, it will not be because the technology is lacking, but because the decision to use it in this way will be shelved. Let us suppose, for a moment, that we did decide that every issue of local or national importance was to be decided by electronic referendum, at the appropriate local, state or national level.

The scenario might be something like this: It is an early February evening in the year 2023, and John and Jane Doe are relaxing before the television in their home communications center. The newly elected President of the United States is having her first "fireside chat" with her fellow citizens. She maps out the main issues the voters have presented to her administration, together with the widest range of options suggested by citizens from all walks of life. These options have been winnowed and tabulated by computers as to priorities. Priority number six has been flagged for resolution now to meet long-range planning goals. Priorities one through five, while of global importance, need further information input and analysis. "Priority number six," the President continues, "concerns future development plans for U.S. Region Three, which was formerly known as Appalachia; and five major options have been developed from both random voter feedback and scientific and specialist feedback, with votes from the affected region having additional weighting over the rest. The options will now be summarized and simulated on your home screen."

The first option is displayed in a series of colorful simulated maps and diagrams. It would designate the whole region as a national park, and the chief recreational playground for the two great adjacent megalopolitan regions: to the east, BOSWASH (formerly known as the northeastern seaboard from Boston to Washington), and to the west, CHIPITTS (formerly the great industrial region of the Ohio River between Chicago and Pittsburgh). The plan entails six new towns of 250,000 people each, to serve as spas and cultural meccas. Their chief industries would be leisure and tourism, health and beauty maintenance, and the performing arts. Now charts appear showing that the economy of the region would grow at 10 percent per year for the first five years, and would require capital expenditures of half of one percent of current gross national product. Then, expected influxes of construction engineering and planning personnel are shown for the first five years of building; and, thereafter, the needs for increasing numbers of recreational managers and workers, doctors, health therapists, beauticians, physical education personnel, and, of course, performing artists of all kinds.

"And now to Option Two," the President says. The second option would designate the area primarily as a natural resource bank, with a secondary use as wilderness recreation. The plan calls for filling the old mines with plastics, iron, copper, rubber and other materials salvaged from the nation's waste-disposal plants; these items would be stored until needed for recycling into production. A network of small towns would be necessary; their economies would be based largely on caretaker and inventory-control functions, while also providing for campers and hikers using wilderness areas. As each of the additional combinations of alternatives was presented, a new computer simulation would appear on the Does' screen. The President reappears and makes her formal declaration that the referendum on these development plans for U.S. Region Three would be made at 7 P.M. one week hence. She adds, "Each voter can, of course, receive their own detailed printout of the plans from the U.S. Government Printout Office by dialing 235-4707 on their computer phone terminal."

At 7 P.M. one week later, John and Jane Doe—having discussed the plans with neighbors, and at their community town-hall meeting—have made up their minds. The telecast begins and the President says, "Good evening, my fellow citizens. I hope you have all done your homework, and that those of you who are registered voters will now give America the benefit of your informed, collective wisdom in tonight's very important national referendum on the long-range development of U.S. Region Three. To refresh your memories, we will again simulate on your home screens the five alternative plans prepared with guidelines from your previous feedback. Please have your voting cards ready for the optical scanner to verify. At the end of the review of the five plans, please place your voting cards in the scanner and then punch in your choice of options, one through five, on your computer phone digit buttons."

After the voting John and Jane relax while the returns are being tabulated. It has been a grueling week of study for both of them, even though the standard work week has been reduced to the flexible time equivalent of three days—the rest devoted to household and community-based shared productive activities. Apart from the U.S. Region Three plan, they have had to study an important local education proposition involving three options on the "mix" of educational services their changing town will need in the next decade; they also have had to fulfill their voluntary community commitments. The red indicator lights; and the Does return to their home communications center. They learn that Option One for U.S. Region Three has passed.

Next month, their tasks will include determining a 10-year transportation-design mix for their own U.S. Region One, monitoring a new study course given by the University of the Air, and beginning work to establish priorities on national resource allocation for the second phase of the 25-year plan—for the years 2025 through 2050.

Utopia, or nightmarish Dystopia?

Would citizenship itself become more demanding, so that for many it will be a full-time job, and for others an overwhelming burden?

Finally, what are the political merits of all this enhanced information flow and participation? Are voters competent to handle such increased participation or would it enhance the power of the experts? Would such increased use of electronic information systems create one vast, conformist planet, with few options left for the individual or, on the other hand, might it lead to social anarchy or a tyranny of the majority, as we took the pulse of the body politic daily and reflected too accurately its passions and prejudices? No one knows whether we will prove rational enough to handle our explosive affairs. Exposing voters to powerful new methods of social decision making might teach them to better cope with complexity and perhaps even raise their awareness of the fact that there are no "solutions" to life's problems, only processes for dealing with them, some better and some worse.

We cannot stifle demands for participation: we can only make better provision that it be informed and orderly. We may be forced to increase our use of information technology to manage the complexity we have created, while devising safeguards for individual freedom and privacy—for example, the new law giving right of access to any personal dossier so that an individual may challenge malicious information or add favorable new data, such as a degree or a patent. Finally, we must also enhance outreach to all the new social feedback to provide a more sensitive signaling mechanism to business and government decision making, while assuring that the periodicity of the feedback is not so frequent as to preempt opinion leadership functions and create those wild oscillations that can lead to a "tyranny of the majority" which our forefathers so wisely avoided.

Why must we allocate such high priorities to building a vast new infrastructure of social-conflict management? Simply because in an instant-communications society it is no longer possible to solve one group's problem at the expense of another, or to allow one company to profit at the expense of social dis-economies affecting other companies or groups. The goal is nothing less than the deliberate institutionalizing of social change, so that violent revolution becomes superfluous.

We are discovering that there are just as many jobs in building mass transit, maintenance and renovation, energy conservation and the developing of the regenerative-resource economy as there are in building military hardware and SSTs.

Unfortunately, one of the many distortions in the current method of computing the GNP is that it does not include the significant economic value of a great proportion of integrative work, such as conflict resolution and that provided by parents in child raising and homemaking, or the volunteer efforts of millions in providing scores of indispensable social services, as well as cleaning up after profit-making activities that pollute the environment. The problem is to adapt our value system so as to incorporate needs, which, when fully perceived, will help raise the market value of integrative activities. We may then come to realize that competition itself can only function within a context of equivalent social cooperation and the true efficiency and value of trust, honesty and goodwill in reducing the need for costly policing of overly competitive, aggressive behavior.

Through today's advanced communication techniques citizens are learning of the need for worldwide control of technology. They see that the Volga is as polluted as the Potomac, and that Tokyo, London, and Sydney all gasp in air as noxious as that in New York. These global ecological crises are learning experiences which dramatize the fact that all of today's important problems are global problems: pollution, population, poverty, ignorance, racism, and militarism. Our children, exposed by the mass media to other cultures and space travel, are now absorbing this newly integrated ecological viewpoint just as easily as we once learned geography.

By airing these global problems, we can help others to achieve broader perspectives and even more expanded levels of awareness, so that, finally, they can perceive the true reality of our common situation—a human family stranded on a small planet, orbiting a medium-sized star in a forlorn, undistinguished galaxy.

To grasp this mind-bending reality is to experience expanded awareness that no drug can provide. Only when we share this common perception of reality can we also find a common point of view, a ground for consensus on the big problems. Whoever or wherever we are, when our individual or group self-interests are seen in a large enough perspective, we discover that they are *identical.*

[1]Speech at a meeting of Sigma Delta Chi Journalism Society, November 15, 1970, Chicago; reported in The Wall Street Journal, November 16, 1970

[2]See, for example, Lewis Coser, The Functions of Social Conflict (New York, Free Press, 1956)

[3]Laurence Lessing, "New Ways to More Power with Less Pollution," Fortune, November, 1970, p. 78

[4]Environmental Action, November 14, 1970, p. 11

[5]The New York Times, November 11, 1970, and statement by Hon. Kenneth Patton, "Regarding Regional Economic Growth Patterns," Bedminster, N.J., Town Board Meeting

[6]See, for example, Social Policy, a new journal of the social sciences, published bimonthly by the International Arts and Sciences Press, White Plains, New York 10603

[7]See, for example, Daniel Patrick Moynihan, Maximum Feasible Misunderstanding (New York, Free Press, 1970)

[8]See, for example, Elizabeth H. Haskell, "Quality of the Urban Environment: The Federal Role," a working paper of the Urban Institute, Washington, May 1970.

[9]The Corporation for Public Broadcasting, Washington, D.C.

Excerpted chiefly from "Toward Managing Social Conflict," *Harvard Business Review,* May-June, 1971, (copyright, 1971 by the President and fellows of Harvard College: all rights reserved) and "Computers in Social Planning," *MBA,* Dec. 1971. Since these articles were published much further experimenting with the social feedback mechanisms has occurred, from the Environmental Mediation Center at the University of Washington in Seattle and the Rocky Mountain Center on Environment, to those described in Chapter 22. However, our inadequate market system has precluded much democratic innovation of these systems and commercial considerations have subverted the content and foreclosed on the promise of community cable TV systems. The computer is still more of a threat to human liberty than a means of increased participation and democratic information sharing. Meanwhile, at least one systems-oriented economist, Yoneji Masuda of the Japan Computer Usage Institute, which I visited in 1973, has developed an entire theory of information economics, based on vastly increased, democratic information sharing, citizen participation and worker self-management, expounded in his paper, "The Conceptual Framework of Information Economics," *IEEE Transactions on Communications,* Vol. COM-23, #10, Oct. 1975.

"Then it's agreed—one million for research in recycling and two million to publicize it."

CHAPTER FIFTEEN

Democratizing Media

The current public interest in all forms of mass communication reflects a growing understanding of its central role in our national life. Not only do we have schools of communications at many of our colleges, but we have an increasing body of scholarly analysis of the mass media's effects on our culture and our individual psyches. More people are at last realizing too, the awesome political power that comes with ownership or control over any medium of communication, whether television, radio, newspapers, magazines, wire services, computer networks, or any other system for moving information and ideas to significant numbers of people.

Communication between all citizens and all their institutions is indeed the primary integrative force needed to turn our fragmented, uncoordinated body politic into a healthily functioning whole. The sum of all channels of communication in a society makes up its vital nervous system. The great challenge is to ensure that all the components of this nervous system are free and open conduits for the maximum possible interchange of information between the maximum number of citizens.

The channels of communication in America today are technologically advanced beyond those available to any other body politic. In fact, mass media are almost beginning to replace political parties in our system of government. They have informed and misinformed our citizens on national issues on an unprecedented scale, but in a largely unplanned manner. The mass media have shown the poor how the rich live, and have shown the rich what it is like to live in a rat-infested city slum. They have given us insight into pressing problems like "perspiration wetness," "tired blood," "bad breath," and the "blahs." They have made Americans interested in each other and whetted their appetite to communicate with each other. But the only way to do this efficiently is by using the mass media, especially the air waves—air waves that always seem to have an editor, a licensee, or a sponsor between ordinary citizens and that precious microphone, not to mention the "static" of endless commercials and entertainment programing.

Nonetheless, radio and television sets are the most efficient tools that Americans have at hand to help them understand our race relations, why our cities are decaying, what our politicians are saying, and what America's role in the world should be. For the underskilled, broadcasting could offer nationwide job training and basic education. For children, the air waves could provide more Sesame St.-type and "Headstart" programs, without the costs of special transportation or facilities. Our mass media could become a national feedback mechanism by providing a random-access conduit for all the wisdom, creativity, and diversity of our citizens.

Our mass media are only a poor shadow of what they could be—not for lack of technology, but because of our imperfect understanding of their potential power. The mass media in America are still operated on the notion that they are purely businesses whose primary concern is to make profits for their stockholders, and to provide a medium for merchandising goods. In the last decade, we have begun to learn the considerable hidden cost to society in making advertising the chief source of revenue to sustain the operations of its mass media. Since the original decision to cede the use of the air waves to private broadcasters, we have learned that if advertisers pay the cost of putting on programs, the public must pay the price of seeing only programs advertisers feel will sell their products. Instead of the justly dreaded government censorship, we ended up with censorship by sponsors and private owners.

The advertiser's desire for the largest possible audience naturally conflicts with the needs and interests of minority audiences. It also hampers the germination of new and controversial ideas, which must break into the mass marketplace if they are to gain consideration. In a sprawling country like America, coverage in the mass media is the only means of gaining a day in the court of public opinion. If minority groups cannot get coverage, their only nonviolent recourse is to beg or buy advertising. But here they must compete with giant corporate-product advertisers who can afford to pay $125,000 a minute for prime television network time. Competition for free "public service" advertising is heating up; but here again, it has been until recently the safe causes, like "Smokey Bear" or "Give to the College of Your Choice" that are permitted to get their message through.

When civic groups "sell" their ideas and programs in competition with products and politicians, who should decide how much time and space ought to be allotted to these different purposes? Just those who own or control the media? For broadcasters and regulators, this problem is already serious. Which groups deserve free "public service" time and which must pay? If a civic group, a politician, and a product advertiser all want to buy the same limited advertising time, how will broadcasters decide whose message gets on the air and whose is blacked out? For budding civic groups, the need for publicity is a matter of life and death, and a negative decision could condemn an organization to oblivion.

Similar problems have arisen in political primaries. Politicians send advance men into an area and buy up all the available time. Other candidates arrive and find themselves blacked out. And what if a local civic group had wanted air time to raise an issue that was being inadequately covered by the candidates? Some of these matters are subject to a loose set of rules (the "fairness doctrine") promulgated by the Federal Communications Commission, and now continually being challenged in court, but more often these decisions are left in the lap of business.

When a society is in ferment, as ours is today, pressure for equal access to public opinion through mass media increases as the old consensus splinters. New ideas and new minority opinion groups spring up everywhere. These new ideas are vital for the continual process of renewal and adaptation that prevents cultures from decaying. At the same time, such new ideas are necessarily disruptive and controversial, and therefore underfinanced and without institutional vehicles to promote them. The realization is now dawning on groups espousing these new ideas that in a mass, technologically complex society, freedom of speech is only a technicality if it cannot be hooked up to the amplification system that only the mass media can provide. When our founding fathers talked of freedom of speech, they did not mean freedom to talk to oneself. They meant freedom to talk to the whole community. A mimeograph machine can't get the message across anymore.

It is entirely possible that much of the recent radicalization of American politics may be due to this media bottleneck. Minority opinion groups have discovered that whereas media ignore a traditional press release on their activities, they send reporters rushing to cover a picket line or any attention-getting "happening." Once other groups caught on to this game, the media became desensitized to mere picketing, and escalation became necessary. Now to get the media's and, therefore, the public's attention, one must hold a college dean hostage, dance naked through the streets, throw a rock, or start a riot. In psychological terms, the news media have been "rewarding" and therefore reinforcing destructive behavior, by drawing attention to it and making national figures out of those who have learned what kind of behavior keeps them in the camera's eye.

At the same time, quiet, constructive behavior on the part of all those thousands who continually work to build and heal society, is punished by the negative sanction of being ignored by the media, and never reaching society's attention. Of course, there are exceptions to this generalization, and there are many responsible publications, as well as some unusual radio and television stations, that do not make a practice of exploiting sensational news. But until we recognize the dangerous tendencies of the prevalent, oversimplified journalism based on the time-honored editorial use of "rape, riot, and ruin," the radicalization of politics will continue.

Until minority opinion groups are provided with significant rights of access to mass media, and thereby, society's group consciousness, they will continue to behave in any aberrant way necessary to get attention. Just as the labor movement had to stay in the streets until it had won the right to an orderly channel of communication (in this case, a bargaining table) for negotiation and redress of grievances, so will the new political movements disrupt until the system can provide them open and orderly channels of communication.

The battle over the public's right of access to the mass media may well be the most important constitutional issue of this decade. The issue affects every segment of society from blacks who wish to be portrayed adequately in the media to antimilitary groups vainly trying to counteract the promotional budgets of military contractors; antipollution groups wishing to counteract the millions spent on defensive advertising, public relations, and lobbying by polluting corporations; or anticigarette groups trying to neutralize the millions spent by tobacco companies to promote the smoking habit. Until very recently, there have been only sporadic skirmishes fought for this right of access by a few embattled crusaders and citizens' groups. The first real change came in 1953, with the birth of educational television. But even today, public television is still underfunded compared with commercial television, and our public television stations must still largely rely on local charity to mount their programs.

Pressures to democratize media have mounted and, as always, some critics are responsible and justified, and others demagogic. Many civic groups have learned that they can challenge broadcasters at license-renewal hearings, held every three years by the Federal Communications Commission. Another response has been the explosive growth of "underground" media. Protest magazines and newspapers are proliferating and "underground radio" is beginning to flourish on FM bands held by churches and universities.

The American Civil Liberties Union worked to broaden the interpretation of the First Amendment to include the concept of the public's "right of access" to the media. Professor Jerome A. Barron of George Washington Law School advanced this concept in an article entitled "Access to the Press—A New First Amendment Right," in the *Harvard Law Review* of June, 1967. He called for "an interpretation of the first amendment which focused on the idea that restraining the hand of government is quite useless in assuring free speech, if a restraint on access is effectively secured by *private* groups." Professor Barron thinks that the cure for suppression is government regulation through court rulings and laws to force the media to give time and space to unpopular ideas.

What can be done to democratize media and permit more citizen participation? Some broadcasters have been reexamining their policies. There have been more feedback and discussion programs on local stations, including several "ombudsman" programs to help citizens get action from unresponsive government or businesses.

But efforts simply to bypass the mass media via alternative communications continue.

We must remind ourselves, meanwhile, that the present structure of our mass media was not ordained by the Almighty, but merely grew. The First Amendment should not be a cloak for our current media operators to hide behind, or to wave in our faces if we suggest anything new. We must ask *whose* freedom of the press? Just the freedom of the present owners? And if so, what about the citizens' freedom of the press, and our freedom to hear the maximum diversity of opinion on all issues?

If we succeed in freeing our mass media from some of their past patterns of operation, then we can decide what needs to be communicated and how to use communications to build our future. First, we must have faith that new information, properly communicated, can change human perception of reality and therefore our attitudes and behavior. There must be a new, mature ethic of journalism, for both electronic and print media. Current mass journalism is still largely based on the old, fragmented Newtonian vision—where humans were the dispassionate, objective observer of their world. Even though few people still believe that humans can ever observe the world objectively because they are an interacting part of it, there is still a widespread lag on the part of our mass media in perception of this integral nature of reality.

The new, post-Newtonian journalism will be less concerned with aberrant, violent happenings and manifestations. Rather, intelligent, creative reporters and editors will face up to the knowledge that true objectivity is impossible, and therefore shoulder and acknowledge the heavy burden of responsibility thus placed upon them. They will analyze the complex structures and interrelationships which lie beneath the surface events in the same way that only a handful of "little" magazines do today, and present this material simply for mass audiences. In a democracy as complex as ours, only if voters can obtain such simplified coverage of the parameters of major issues, can they hope to use their votes wisely. Mass media reporters will seek out injustices and pressures in society before they need erupt in violence or find expression in the "underground media." Just as the sensory system of primitive creatures can only signal danger or dysfunction, so our primitive mass journalism has concentrated on signaling only these inputs to our body politic. Editors will seek news of the integrative activities of people, as well as their destructive acts. Like individuals, a society needs confidence in itself, and its ability to cope with its problems. We must know of human love and courage, as well as our hates and fears.

To address adequately the need for more democratic access to public opinion, as well as to meet its huge responsibilities as our most powerful educational system, mass journalism, both electronic and print, must face up to a greatly enlarged function in a complex, mass society. If it fails, the consequences may be disastrous.

Reprinted from *Columbia Journalism Review*, Spring, 1969. Since then, alternative media activism has grown and spawned many journals, such as *Radical Software* and *TeleVisions,* as well as citizen-access books, including *Guerrilla Television,* by Michael Shamberg, and Nicholas Johnson's *How To Talk Back To Your TV Set*. Citizen groups demanding access rights have blossomed, notably Nicholas Johnson's National Citizens Committee for Broadcasting, Action for Children's Television, the Citizens Communications Center, the Media Access Project and Public Media Center.

"It's my observation that more and more consumers are looking after their own interests these days."

CHAPTER SIXTEEN

Information and the New Movements for Citizen Participation

America's emerging social values might well be termed "post-industrial" and are espoused by a growing number of more educated, politically influential citizens who are disenchanted with many existing institutions and priorities but, for the most part, still believe that their objectives can be reached by restructuring business and government machinery through constitutional means. They include environmentalists, militant consumers, students and young people, middle- and upper-income housewives newly activated by the consciousness of the women's rights movement or the boredom of suburban life, the public-interest lawyers, scientists, engineers, doctors, social workers and other politicized professionals, the joiners of political organizations such as Common Cause, the activist stockholders and the various crusaders for "corporate responsibility."

The new "post-industrial" values of such groups are to a great extent needs described by the humanistic psychologists Abraham Maslow, Erich Fromm and others as transcending the goals of security and survival and are therefore less materialistic, often untranslatable into economic terms, and thus beyond the scope of the market economy and its concept of "homo economicus." As I tried to alert business leaders at the White House Conference on the Industrial World Ahead in 1972, they constitute a new type of "consumer demand," not for products as much as for life-styles, and include yearnings which Maslow referred to as "meta-needs": for meaning and purpose in life, a closer sense of community and

"One of these could educate every kid in Cincinnati."

"One brand-new B-1 bomber costs $87 million.

Enough to wipe out the cost of public education in Cincinnati. With enough left over to fund the libraries in tne District of Columbia.

A single B-1 could pay for fire protection in Los Angeles for one year. Or finance the entire budget for the city of Atlanta.

Or pay all yearly expenses for streets, parks, and sanitation for Indianapolis, St. Louis, Pittsburgh, Hartford, and Milwaukee. *Combined.*

But what about the military benefits of the B-1?

According to a host of experts, there aren't any.

A Brookings Institution study found: 'No significant military advantages [are] to be gained by deploying a new penetrating bomber such as the B-1.'

Yet, Congress seems determined to fund the most expensive weapon in U.S. history – a 244-plane system that could cost $100 billion.

Our union wants to stop the B-1 funding.

We support a military strong enough to deter any aggressor foolish or venal enough to attack us.

But what good is it to be able to destroy Moscow ten times over if our own cities die in the meantime?"

American Federation of State, County, and Municipal Employees, 1625 L Street, N.W., Washington, D.C. 20036

cooperation, greater participation in social decision making, a general desire for social justice, more individual opportunities for self-development and more options for defining social roles within a more esthetic and healthful environment.

Ironically, these new values attest to the material successes of our current business system and represent a validation of a prosaic theory of traditional economics which holds that the more plentiful goods become, the less they are valued. For example, to the new "post-industrial" consumers the automobile is no longer prized as enhancing social status, sexual prowess or even individual mobility, which has been eroded by increasing traffic congestion. Rather it is seen as one component of a mode of transportation forced upon them by the particular set of social and spatial arrangements dictated by an interlocking group of powerful economic forces embodied by the auto, oil, highway and rubber industries. Such consumers have begun to view the automobile as the instrument of this monolithic system of vested interests and client group dependencies, which has produced an enormous array of social problems and costs: decaying, abandoned inner cities, an overburdened law-enforcement system, an appalling toll of deaths and injuries, some 60% of all our air pollution and the sacrifice of millions of acres of arable land to a highway system that is the most costly public-works project undertaken by any culture since the building of the pyramids and the Great Wall of China!

It has become expedient of late for business spokespeople to excoriate the views held by these new consumers. At best, they are seen as esoteric, at worst, un-American: but certainly a luxury not affordable by the average American family, let alone those living in poverty. And yet it must be acknowledged that these views are increasingly validated by the realities of environmental degradation, unemployment, continued poverty in spite of a climbing Gross National Product, and other visible evidence of the shortcomings of current social and economic arrangements. At the same time, some of these "post-industrial" values are surprisingly congruent with values being expressed by the poor and less privileged. Such groups—whether welfare recipients or public employees, less powerful labor unions or modest homeowners and taxpayers—seem to share the same demand for greater participation in the decisions affecting their lives and disaffection with large bureaucracies of both business and government. Environmentalists found themselves agreeing with labor and minorities that human service programs, which also tend to be environmentally benign, should be expanded rather than cut. There was also agreement that a federal minimum income program is more needed than ever, because it would create purchasing power for instant spending on unmet basic needs, such as food and clothing and, by affording the poor greater mobility to seek opportunities in uncrowded areas, would help relieve the overburdened biosystems of the cities. As another example, to recast the environmentalists' disenchantment with our automobile-dominated transportation system, we may note the very different but equally vocal objections of the poor. Over 20% of all American families do not own an automobile. For the poor, many of them inner-city residents, the cost of even an old model is prohibitive. This decreases their mobility and narrows their job opportunities, while the decline in mass transit and increased spatial sprawl permitted by wide automobile use worsen the situation. The highway building spree is too often experienced by poor and minority groups as the callous cutting of roads for white suburbanites through black and poor neighborhoods, permitting even more of the cities' remaining middle- and upper-income taxpayers to flee urban problems for greener pastures.

Therefore, although it is possible to dismiss these "post-industrial" consumers as irrelevant—and indeed they may well be less of a market for consumer goods—they nevertheless represent a new and different challenge of vital concern to business and politics as usual. Even though they are no longer willing to perform the heroic feats of consumption which have heretofore been successfully urged upon them by massive marketing barrages, their opinion-leadership roles and trend-setting life-styles will continue to influence traditional consumer tastes. This influence has been felt in the new anarchism and casualness in clothing fashions, the popularity of bicycling, the trends away from ostentatious overconsumption toward more psychologically rewarding leisure and life-styles, and reflecting the astounding growth of encounter groups and other activities associated with the human potential movement. In addition, the "meta-needs" of the "post-industrial" consumers will express themselves in increasingly skillful political activism and advocacy as they continue to find in their more holistic concepts greater congruity between their own goals and the aspirations of the less privileged. Furthermore, their growing confrontations with corporations over their "middle-class" issues, such as the environment and peace, have led them to discover the role of profit-maximizing theories in environmental pollution and of the military-industrial complex in defense expenditures and war. These insights, together with their awareness of their own privilege and their acceptance of guilt and concern for social injustice, are leading to the kind of convergence with other socioeconomic group interests so much in evidence in the movement for corporate responsibility.

Many of the corporate campaigns have been equally concerned with peace, equal opportunity in employment, pollution, the effects of foreign operations, safety and the broadest spectrum of social effects of corporate activities. Typical was Campaign GM, which simultaneously sought representation on General Motors' board of directors for minorities, women, consumers and environmental concerns. The same convergence is evident in the Washington-based Urban Environment Conference, composed of labor unions and a cross section of environmental groups pledged to stand united in the face of a growing number of corporations that attempt to prevent implementation of pollution-control laws by raising fears of unemployment, plant shutdowns or even relocating in more "favorable" states or in other countries. Both labor unions and environmentalists view such tactics as more often power plays and bluffing or the result of poor management than bona fide cases of corporate hardship.

Similarly, environmentalists and unions have worked together to reduce in-plant pollution and the ravages of such occupational diseases as black lung, or in fighting the wholesale destruction of small farms, open lands and streams through the excesses of strip-mining. This convergence is also visible in the comprehensive manifestos and social critiques of theoreticians, whether in the movements for civil rights, peace, women's liberation, or environmental and consumer protection. Most of these critiques tend to

Tomorrow morning when you get up, take a nice deep breath. It'll make you feel rotten.

It is said that taking a deep breath of fresh air is one of life's most satisfying experiences.

It can also be said that taking a deep breath of New York air is one of life's most revolting, if not absolutely sickening, experiences.

Because the air around New York is the foulest of any American city.

Even on a clear day, a condition which is fast becoming extinct in our "fair city," the air is still contaminated with poisons.

On an average day, you breathe in carbon monoxide, which as you know is quite lethal; sulfur dioxide which is capable of eroding stone; acrolein, a chemical that was used in tear gas in World War I; benzopyrene, which has produced cancer on the skin of mice; and outrageous quantities of just plain soot and dirt, which make your lungs black, instead of the healthy pink they're supposed to be.

At the very least, the unsavory elements in New York air can make you feel downright lousy. Polluted air makes your eyes smart, your chest hurt, your nose run, your head ache and your throat sore. It can make you wheeze, sneeze, cough and gasp. And because air pollution is responsible for many of those depressing "gray days," it may affect your mental well being. If you're a person who is easily depressed, prolonged exposure to polluted air certainly isn't doing you any good.

Of course, at its worst, air pollution can kill you. So far, the diseases believed to be caused, or worsened by polluted air are lung cancer, pulmonary emphysema, acute bronchitis, asthma and heart disease.

600 people are known to have died in New York during two intense periods of air pollution in 1953 and 1963. How many others have died as a result of air pollution over the years is anybody's guess.

Who is responsible for New York's air pollution problem? Practically everybody. It belches from apartment buildings, industrial plants, cars, busses, garbage dumps, anywhere things are burned.

But the purpose of this advertisement is not to put the finger on who's causing the problem. It's to get you outraged enough to help put a stop to it.

What can you, yourself, do about air pollution? Not much. But a million people up in arms can create quite a stink.

We want the names and addresses of a million New Yorkers who have had their fill of polluted air.

The names will be used as ammunition against those people who claim New Yorkers aren't concerned about air pollution.

If we can get a million names, no one can say New Yorkers won't pay the price for cleaner burning fuels, better enforcement of air pollution laws, and more efficient methods of waste disposal.

The cost of these things is low. A few dollars a year.

The cost of dirty air is higher. It can make you pay the ultimate price.

Box One Million

Citizens for Clean Air, Inc.
Grand Central Station, N.Y. 10017

IF YOU'D LIKE TO GET IN THE THICK OF THE FIGHT AGAINST AIR POLLUTION, SEND US A LETTER ALONG WITH YOUR CHECK FOR $2.00 (OR MORE) AND BECOME A MEMBER OF CITIZENS FOR CLEAN AIR, INC.

Carl Ally, Inc., New York, N.Y.

explain war, racism, sexism and all forms of social and environmental exploitation as being interrelated and stemming from current patterns of power and distribution and their roots in prevailing economic and cultural assumptions. This growing understanding of the political nature of economic distribution has naturally focused on the dominant economic institution of our time: the corporation and its political as well as economic role.

A consistent theme underlying the activities of United States citizen movements has been one of alienation from prevailing perceptions of reality. While the commercial mass media have projected subtle, but compelling, images of the kind of split-level suburban lifestyles conducive to the needs of a mass-consumption economy, the citizen movements, whether for peace, consumer and environmental protection or social equality, have marched to a different drummer. They have focused on the unresearched, often suppressed information concerning the dis-economies, dis-services and dis-amenities which we are now learning constitute the other side of the coin of industrial and technological development. They have risen naturally as social feedback mechanisms in response to the increasingly visible second-order consequences of our uncoordinated economic and technological activities and fragmented, tunnel-vision social decision making.

Citizen Movements as Social Feedback Mechanisms. It is not surprising that citizen organizations focus their efforts on modifying the goals and structures of those institutions, such as large corporations, which they believe have grown powerful enough to avoid normal social and political constraints. Not only corporations, but also government agencies, legislative bodies, labor unions and religious organizations tend to institutionalize past needs and perceptions and are ill-designed to perceive new needs and to respond to new conditions. Some measure of institutional lag is inevitable and necessary to avoid disruptive and rapid oscillations in societies. However, in conditions of rapid technological and social change, new complexities and interdependencies which are characteristic of the United States today, this institutional lag often becomes clearly counterproductive—for example, the current imbalance in the

ELECTRICITY. THE BARGAIN YOU CAN'T AFFORD.
The figures in the graph are approximate figures based on utility rates in California.
Each state has its own rate structure so figures vary, but invariably the scale remains the same.
The more electricity you use, the less you pay for it.
4.0¢ Kwhr
3.0¢ Kwhr
2.0¢ Kwhr
If you use this many Kilowatt hours per month
you pay this much for each Kilowatt hour.
100 Kwhr
500 Kwhr
2000 Kwhr

With utility bills getting higher all the time, you're probably looking for a way to save money on electricity. You thought conserving energy would help, but that only seemed to make your bills higher. Fortunately, utility companies have a bargain rate that allows customers to save as much as 70 percent on the cost of electricity. Unfortunately, you can't take advantage of that bargain.

Here's how it works: Utility rates are slanted in favor of the big users. The more kilowatt hours a customer uses, the less it costs for each kilowatt hour. Rich people living in large, gadget-filled mansions pay the lowest rate. Middle-income homeowners pay almost twice as much per kilowatt hour as the big users. And the smallest users—senior citizens living on fixed incomes, young people just starting out on their own, poor people struggling to make ends meet, working people fighting to keep up with inflation—pay the highest rate of all.

Utility companies make it cheaper for rich people to heat their swimming pools than for you to heat your home. They penalize those who conserve, and reward those who use more. They make it so cheap for big users that it's easy to be extravagantly wasteful, so expensive for you that it's hard to get by.

We're not suggesting that you use more electricity than you need to. We *are* suggesting that utility companies turn their rates around, so that those who can't afford ever-increasing costs get a break, and those who can afford it pay their fair share. That will help everyone save energy. And it'll help you save money.

Remember—lower costs for the big users mean higher costs for you. And that's a bargain you can't afford.

ENVIRONMENTAL ACTION
Room 724
Dupont Circle Building
Washington, D.C. 20036

☐ *I want to: Do something about unfair rate structures and rate increases. Stop construction of unneeded power plants. And prevent unnecessary utility company advertising at the public's expense. Please send me ___ copies of your handbook, "How To Challenge Your Electric Utility." I enclose $1.50 check or money order for each copy ordered.*

Name
Address
City
State
Zip

Conserve energy.
Turn the rates around.

Prepared by Public Media Center, San Francisco.

Public Media Center, San Francisco, Ca.

United States transportation system caused by the entrenchment of a powerful array of public and private institutions with vested interests in promoting automobile use.

Many citizen leaders realize that all institutional structures are, by definition, designed to screen out any information they perceive as unwanted or irrelevant so as to better concentrate on the purposes for which they were organized—hence, their capacity for selecting, concealing, distorting and impounding information and the resulting shortcomings of their planning and goal-setting processes. Indeed, organizational theorist Bertram Gross claims that it is impossible to measure the performance of any system independent of its structure.[1] For example, the United States Congress and its committee structure impedes information flow by slicing reality into fragments which fit its somewhat arbitrary scheme of organization. A case in point was the clash with the executive branch of government over the Congress's inability to reintegrate the welter of fragmented information needed to ascertain the total size and shape of the federal budget. Academia exhibits the same paralyzing lack of communication among disciplines and fields of research. The Executive Branch, itself, suffers the same myopia; scores of single-purpose agencies pursue their narrow goals, often addressing long forgotten problems.

A clue to the shortcomings of humanly designed structures may be found in nature. As ecologist Gregory Bateson has noted, it is rare to find ecological or biological systems which are activated by a specific need or which seek to maximize single variables. Meanwhile, information at the interfaces between many of our social problems is sparse because our society is ill-equipped to perceive, let alone research, these overlooked areas of interplay.

We have seen that all of this uncoordinated institutional activity in the United States today is based on the Cartesian view of the world which has held sway in our minds for three centuries. This has led to the growth of narrow-purpose structures and reductionism in our academic disciplines, and has in turn over-rewarded analysis, while discouraging synthesis; sustained property rights, while ignoring amenity rights; fostered unrealistic mental dichotomies, such as those between public and private goods and services; and over-rewarded competitive activities, while ignoring the equally vital role of cooperation in maintaining the cohesion and viability of the society as a whole. This tunnel-vision has clouded all our perceptions and caused us to focus on objects and entities, while ignoring their fields of interplay. It is precisely in these fields of interplay that public interest citizen movements have naturally sprung up and taken root. They have begun this vital task of filling in the information gaps on the effects of all this uncoordinated activity and painstakingly documenting our growing social costs—what British political economist E. J. Mishan calls the "bads" that inevitably come with all the goods.

We are now witnessing the collapse of policies based on this Cartesian world view. The new view now being developed by futurists is holistic, modeled on the concepts of general systems theory and analogies drawn from biological and ecological disciplines. It assumes that parts of systems can be understood and analyzed only within the contexts of ever larger macrosystems. Our current problems—whether we designate them as social, economic or environmental crises—are all part of the larger crisis of our myopic perceptions. These myopic perceptions, which were adequate for a quieter, slower moving age, no longer provide us with enough lead time to correct our course. Not only are our fragmented social structures inappropriate for describing, or dealing with, macroproblems, but the very goals of these subsystems are antithetical.

This must lead us to question a very basic assumption underlying both our economic and political systems: first, the assumption in our economic system that the aggregated goals and activities of microeconomic units will somehow add up to the public welfare. There is increasing evidence in our mounting social and environmental costs that the very opposite may be true. For example, the basic thesis of economist K. W. Kapp, expressed in his book *The Social Costs of Private Enterprise,* first published in 1950, holds that maximization of net income by microeconomic units—individuals and firms—is likely to reduce the income or utility of other economic units and of society at large. Public interest groups intuitively have understood this proposition; for example, corporations and special-interest groups can be subjectively successful or profitable by concealing or ignoring information on the external costs to society imposed by their activities.

Secondly, the same questionable assumption underlies our political system; we should have new doubts whether the competing special interests and their jostling can ever add up to even an approximation of the public interest. As discussed earlier, in many other disciplinary contexts, such as general systems theory, biology and ecology, it is considered almost axiomatic that optimizing subsystem goals is antithetical to optimizing the macrosystem of which they are a part.

The rise of the public-interest citizen movements is an expression of such new awareness—nowhere more pronounced, perhaps, than in the environmental movement in which such general systems views have become the dominant organizing principle. Similarly, as their new research enters our economic and social decision making, it diversifies the range of options under consideration, redressing the instabilities and errors caused by single-purpose-dominated policy making.

The Role of Information in Modifying Institutions and Values. Vastly increased information flows may prove to be our best hope for irrigating our impacted social system and modifying its structures, easing some of them into oblivion, while deflecting the course and redefining the goals of others. Information is, of course, the basic currency of all economic and political decision making. The quality and quantity of information and the way it is structured, presented and amplified controls all of our resource allocations. In fact, information programs and directs energy—although, as yet, we do not know the equations representing this process.

Citizen leaders know that political and social conflicts are fought with information; this understanding is clearly illustrated in the diverse information-gathering operations of Ralph Nader's many public-interest research groups. Other organizations share the same appreciation for the power of information: the Council on Economic Priorities and the National Council of Churches' Corporate Information Center research the social, rather than the economic, performance of corporations and disseminate their findings to the press, money managers and stockholders. Information is also the weapon of the new advocacy professional groups, such as the activist doctors of Health-Political Action Committee, the Scientists Institute for Public Information, the Union of Concerned Scientists, the Center for Science in the Public Interest and the Union of Radical Political Economists, as well as the legions of radicalized lawyers, sociologists, librarians, psychologists, architects, planners and even management consultants.

The strategies and tactics of these groups are based on the following, shared assumptions that new or restructured information, when deployed and amplified, can:

☐ Alter human perceptions of reality.

☐ Create changes in personal values, preferences and goals, which are later reflected in new collective and institutional goals.

☐ Explode the boundaries of academic disciplines by creating cognitive dissonances and conflicts, often leading to gradual paradigm shifts.

☐ Successfully challenge the rationality and legitimacy of resource allocations and decisions of governmental and private institutions.

☐ Strengthen the power of consumers and citizens to perceive and protect their own interests and to understand how individual interests coincide more frequently when viewed within ever larger system contexts until, when finally viewed in planetary and ecological contexts, they literally become identical.

☐ Short-circuit hierarchical, pyramidal and bureaucratic control.

☐ Illuminate the intricate chains of causality and interdependence in complex societies and their reciprocal exchanges with equally complex host ecosystems.

For example, urban dwellers must now understand the interdependencies involved in provisioning a modern city and managing its wastes if they are to be capable of making rational political judgments: suburbanites must learn how dependent their way of life is on such factors as the political mood of Arab nations, the oil needs of Japan, the viability of central cities or even the continued willingness of women to spend large amounts of time as unpaid chauffeurs. Finally, by spreading cognitive dissonance among citizens, information can release individuals from their subservience to prevailing cultural norms and create new opportunities for insight and learning.

The Alternate Media Movement. Having developed remarkable skills in gathering and restructuring information, public-interest groups also have mastered, to a significant degree, the modulation and amplification systems available in a technologically advanced society: from mass media to the informal mail and telephone networks of trust through which they channel their new information so that it acquires political and economic potency. For example, in the late 1960s activists and demonstrators began to master the staging of media events and guerrilla theater to manipulate drama-hungry news editors and reporters. They learned the commanding of media time and space through means as varied as challenging broadcasters' licenses, picketing outside newspapers and demanding hot-line, talk-in shows for the discussion of local issues and the redress of grievances; they could end-run politicians, administrators and corporations. These activities are based on the insight that in a complex, technologically advanced society free speech is an empty platitude if one is denied access to the mass-media amplification apparatus, which represents the vital nervous system of such a body politic.

In addition to opening up access to existing commercial mass-media channels, citizen groups developed their own alternate or underground media catering to the new consciousness of minority groups, women, environmentalists and communards. In the mid-1960s cost and lack of sophistication still limited underground media to the mimeograph machine and the printing press. Thereafter, underground radio began to develop as students learned how to move in on university-operated stations, and the transition to television became a gleam in every media activist's eye.

One of the most important influences during these developments was that of a New York-based group, Radical Software, which—through its magazine of the same name—linked alternate media people working all over the country and in Canada and helped them share new skills, such as videotaping with portable video cameras, splicing, editing and producing TV programs which reflected their own radical views and lifestyles.

At the same time alienated minorities of all kinds began to appreciate the potential for political power in the control of media. The battles to obtain conventional broadcasting licenses, as well as the new cable TV franchises, began. Media activists saw in cable television the possibilities for linking up communities in two-way exchanges and dialogues, covering school board and planning meetings, articulating hitherto unnoticed grievances and resolving conflicts—not to mention the longer range potential of such systems for generating revenues by providing library and data services, education, medical diagnosis and even shopping assistance. Local battles over the regulation of cable television franchises resulted in conditions being imposed upon would-be cable operators to provide channels set aside for public access.

Open Channel, the New York group founded in 1970, set out to show how the public could organize to use their newly won access channels, teaching local community groups, PTAs and voluntary service organizations how to make professional quality videotape programs interpreting their goals and activities for showing to New York's cable television audience. Other models have emerged for the use of television and print media in resolving conflicts, profiling value changes and locating possible new consensus areas, such as Choices '76, a series of media town meetings developed by another media innovator, Michael McManus, in 1973 for New York's Regional Plan Association.

Through the efforts of such innovators, media theorists—such as Marshall McLuhan, Amitai Etzioni, Chandler H. Stevens, Jerome Barron, John Culkin, Everett Parker and Donald Dunn—and activists—such as Nicholas Johnson, Peggy Charron, Albert Kramer and others; counter-advertisers Glenn Pearcy and Roger Hickey of the Public Media Center and public interest opinion researchers, citizens are gaining insight into the decision role of mass media in shaping our images and culture. They are learning also that television and computers need not necessarily be monolithic, centralizing forces, but can be used in decentralizing and coordinating modes. Instead of central data banks designed to find out about people, they can also function as random-access systems for gathering data on what people think and for funneling these opinions into political and economic decision centers.

Resource One, a radical computer group in California, developed a random-access computer network to link citizen-action groups which share its data base on resources available for fighting consumer, environmental or social equity battles. Such citizen information networks can draw on many other key data, such as the computerized files maintained by the nonprofit Citizens Research Foundation of Princeton, which cross-references all political campaign contributions of over $500, or files maintained by the League of Conservation Voters of voting profiles of every United States congressman. Similar is the New York Regional Federal Trade Commission program, mentioned earlier, which does not even require activation by a complaining citizen, on the theory that many of the most victimized people in society are too unaware or apathetic to fight for their own rights.

The falling of the last bastion against full citizen participation is, as discussed, embodied in the concept of the electronic referendum which, of course, is technically possible today. The family television set could provide the citizen with information inputs on policy options and choices, with the telephone serving as the output device whereby the votes on issues could be instantly recorded at the appropriate legislative matrix. This scenario causes nightmares for the political scientist who is cognizant of both the role of vote horse-trading between politicians as an indispensable device for conflict resolution and the role of the representative form of government in damping the daily passions of the electorate.

However, one must note that a process similar to vote trading also takes place at the grass-roots level among citizen groups and opinion leaders—in garnering support for their issues and in the forming of ad hoc coalitions—which serves an equally important conflict-resolving purpose and which would still function in referenda situations. Such electronic referenda proposals only highlight more sharply the ever increasing need for information on the part of citizens in a modern democracy. Only if citizenship became a full time job could the citizen in a complex industrial society possibly master the mountains of information necessary to make wise choices on such a daily plethora of issues.

Indeed, Edwin Parker and Donald Dunn call for the setting up of a computer-cable television information utility to be available in most United States homes by 1985.[2] Another proposal, put forward by Governor Ella T. Grasso of Connecticut when she was a Congresswoman, would make all communications by letter or phone from constituents to their federal legislators free of charge. Meanwhile, the best hope for broader access to continuing education would seem to be the model of Britain's Open University, which is conducted, with minimum barriers to enrollment, over nationwide television. For, to leave educational innovation to our schools and universities is probably as ill-advised as it might have been to put buggy-whip manufacturers in charge of development of the automobile.

Are Citizens Movements an Index of Societal Adaptability? The new urges on the part of citizens to develop their own research capabilities and to communicate more frequently with each other and the public at large also may be seen as the validation of the thesis presented by Wilbert E. Moore and Melvin Tumin in their illuminating paper "Some Social Functions of Ignorance."[3] They point out that ignorance serves the social purposes of, among other things: preserving privileged position, reinforcing traditional values and preserving social stereotypes. Such social purposes are seen today as having little redeeming worth in a society racked by the adaptation requirements of technological change. Indeed, as Gregory Bateson has noted, adaptability is a resource as surely as is coal or oil, and we need to develop a new economics of adaptability in order to understand how humans use, conserve or waste their precious stock of this commodity.

In fact, do the new citizen cadres in our midst represent a vital part of our society's stock of adaptability; if so, how can we conserve and employ their energies most productively? Gunnar Myrdal has pointed out that citizens organized for their own purposes can often serve the same functions as costly, regulatory bureaucracies.[4]

However, these citizen organizations and their research operations are often subject to severe resource limitations, even though they have proved their competence and validated their research by a much more rigorous process than that by which academic research is validated. As Ralph Nader has pointed out, any one of these independently produced studies is immediately pounced on by hundreds of paid experts in the employ of institutions which may feel themselves threatened by such information. No expense or effort is spared to fault or discredit such studies, as the Council on Economic Priorities discovered when it published its studies in 1971 and 1972 of the United States pulp and paper industry and the investor-owned electric utility industry.

Nevertheless, these studies were found worthy of publication by the Massachusetts Institute of Technology Press. The special values of these citizen-research groups is that their innovative and unorthodox approach provides priceless, if often painful, critiques of conventional wisdom and infuses the body politic with a rich new yeast of social, cultural and technological alternatives. For example, it was the determination of environmentalists which forced the consideration of research and development alternatives, such as solar energy, geothermal steam, magnetohydrodynamics and fuel cells, into the current debate on energy policies. It was the Scientists Institute for Public Information which challenged the Atomic Energy Commission's emphasis on the breeder-reactor program, and it fell to the Project on Corporate Responsibility and the National Affiliation of Concerned Business Students to develop new curricula on social performance criteria for corporate management for our university's schools of business administration. Many groups, such as the Council on Municipal Performance, already use sophisticated computer systems for their research which compares the efficiency of municipal governments in such areas as housing, crime control and waste management.

As Mancur Olsen has pointed out in *The Logic of Collective Action,* the classic economic problems of free goods and free riders enter into the prognosis for sustaining the voluntary, nonprofit activities of citizens' groups: the work and effort the citizen participant invests in protecting common property or amenity rights automatically accrues to all, regardless of whether they shared the opportunity costs incurred by their more public-spirited neighbors.[5]

Often, the result bears out the old proverb: "everybody's business is nobody's business." Therefore, society provides little motivation for the disinterested champion of the public interest and, in fact, often imposes severe financial and other penalties on do-gooders. Luckily, there are less obvious motivations at work, including those described by humanistic psychologist Abraham Maslow as ascending hierarchical needs beyond mere survival and security: knowledge, an esthetic environment, social justice, leisure and greater participation in decisions affecting one's life. These transcendent needs seem to characterize many citizen activists in this and other advanced economies in which survival is no longer a widespread preoccupation.

Value changes—based on perceptual changes, in turn based on new information—are, then, the mainspring of the new citizen movements.

Public-interest groups are now also aware of how heavily economics relies on the false assumption of adequate information availability, both in the marketplace and in illuminating such social choices. In fact, they are now beginning to end-run the economists by forcing their information into such decisions and not only successfully affecting the outcomes, but actually altering the value preferences which economists accept as immutable, given data.

An equally disturbing phenomenon for economists used to the idea that increasing information will decrease uncertainty is that the increased flow of formerly uncollected or suppressed information provided by citizen groups actually increases uncertainty, because it tends to constitute information on what is not known and what must still be researched. For example, the environmental-impact statements required of federal projects under Section 102 of the National Environmental Policy Act of 1969, more often than not contain this kind of information, which provides useful questions rather than answers.

A case in point which demonstrates the citizen groups' interest in raising the right questions involved the Congressional Office of Technology Assessment. A coalition of citizens organizations drew up criteria for opening the technology assessment process to wide public participation and scrutiny, recommending the concepts of adversary science and dialectical cost-benefit analysis; for, they understood that technology assessment is essentially a normative process involving value conflicts and, therefore, cannot be left to the technologists.

Suffice it to say that the information struggle with all the communication and structural roadblocks our Cartesian perceptions have created, will continue. Public-interest groups will continue their efforts to manipulate these fragmented concepts of rationality, because their second-order consequences will continue to activate their concern. The war of symbols between the new and the old consciousness will also continue as advertisers destroy word meaning and debase language currencies. Love is now a soft drink and a cosmetic; ecology has similarly been bankrupted of its content. Bureaucrats and politicians add to the Orwellian confusion: war is peace, bombing is protective reaction and full-employment targets shift when we do not meet them.

Public-interest groups retaliate by seeking to destroy the narrow meanings of words, such as profit, efficiency, utility and progress. When the vast value and paradigmatic shifts we are now experiencing reach some new equilibrium, we may also have discovered some higher-system-level discourse which may permit us to discuss our macroproblems in a common language. Until then, we can only hope that, in all the static and confusion of our free marketplace of ideas, the best information will win and will be allowed to be the basis for action.

---

[1]Bertram Gross. *State of the Nation: Social Systems Accounting* Ed. Raymond A. Bauer, M.I.T. Press, 1966 pp. 36-48

[2]*Science,* June 30, 1972 pp. 1392-9.

[3]*American Sociological Review,* Feb. 14, 1949 pp. 787-795

[4]Gunnar Myrdal, *Beyond the Welfare State,* Yale Univ. Press, 1960

[5]Mancur Olsen, *The Logic of Collective Action,* Harvard Univ. Press, 1965

Reprinted, with additional material, from *The Annals of the American Academy of Political and Social Science,* Philadelphia, Vol. 412 (March 1974), pp. 34–43

Life Magazine, 1902

CHAPTER SEVENTEEN

Forcing the Hand of the De-Regulators

De-regulation is a hot button in Washington these days but it is a pseudo-issue. In this growing debate, those in Congress aligned with traditional free-market enthusiasts find themselves in the strange position of potential alliance with assorted decentralists, libertarians, intermediate technology buffs and disillusioned public interest and citizen activists, over the perversion of the government regulatory process. Consumer advocates and free-market economists join in denouncing the inefficiency and plain idiocy of the tangle of conflicting regulations presided over by the alphabet agencies created in the Roosevelt Era to protect Americans from the excesses of the profit motive and untrammelled free enterprise. These regulatory bodies, now the targets of this antibureaucratic sentiment, range from the Civil Aeronautics Board (CAB), the Interstate Commerce Commission (ICC), the Food and Drug Administration (FDA) to the Federal Communications Commission (FCC), the Federal Maritime Commission (FMC), the Federal Reserve Board (FRB) and the Securities and Exchange Commission (SEC).

Indeed, the opposition seems well-nigh unanimous. In fact, the corporate lobbyists hope to reverse the 90-year-old trend toward more inclusive regulation, and thus restore freer competition to the marketplace. Whether such a turning back of the clock would reinvigorate our flabby economy, as industry spokespeople assert, is uncertain. A recent study of these supposedly offending regulations by a group at the Massachusetts Institute of Technology concluded that in the five different countries they examined, these regulations had stimulated, rather than stifled, economic growth.

Consumer vs. Producer. However, consumers, labor, social-welfare advocates and environmental activists need not be reminded that their gripes with regulatory agencies are *not* the same as those voiced by organized producers and large corporations. In fact, the opposition to the regulatory bureaucracies is deeply divided, with the chief lines of conflict lying between producers' aims and the more nearly overlapping concerns of consumers, workers, minorities, advocates of economic justice and those concerned with the safety of working and living environments. Indeed, conservative columnist William Rusher of the *National Review* sees the division in starker terms of a new polarity between producers and consumers. Rusher's rhetoric neatly distorts the picture, conjuring up visions, beloved of the right wing, of hard-working, self-righteous legions of sweating individual "producers" and entrepreneurs pitted against all those layabout, passive "consumers" of society's hard-won fruits of production—from coddled, rambunctious workers to over-fed, finicky consumers, effete environmentalists and no-good welfare chiselers.

Public interest, social welfare and environmental activists, on the other hand, are more likely to see Rusher's producer-consumer conflict in terms of giant corporations allied with the regulatory bureaucracies which have become their tools, in further oppression and disenfranchisement of the consumer. So it behooves all these groups to approach the growing de-regulation bandwagon with extreme caution. Many will be persuaded by well-meaning welfare economists that they should give the move to de-regulation their endorsement on the ground that it will increase overall "efficiency" and "productivity," and therefore redound to everyone's benefit by making the whole economy healthier. This argument is an over-simplistic trap.

Outdated Economic Theories. The catch arises, of course, because many American economists are still subconsciously hypnotized by the free-market, equilibrium models of supply and demand that they were taught at school. Thus many of them cannot see that we now live in such a highly structured society, composed of big producers and big government, that the free play of supply and demand forces now exists only in a few residual areas of our economy and in the imagination of economists. Therefore the consumer is no longer king and cannot activate the whole productive and technological system, as is presumed in the free-market model. Many of the economists who support the move to de-regulation believe that we must turn back the clock and restore this idealized marketplace. As we have learned, this can not be done, since the development of an economy and its technology is an evolutionary, not a reversible, process as economist Nicholas Georgescu-Roegen points out in *The Entropy Law and the Economic Process.*

The de-regulation move is predicated on inadequate and outdated economic theories, as well as the failure to subtract the social and environmental costs of all our current forms of production and consumption—all those dis-economies, dis-services and dis-amenities that we now know economists still tend to overlook. The de-regulation debate is therefore a red herring, to avoid looking at a much deeper problem that neither producers, industrialists or government agencies want to face. We have seen that each order of magnitude of technological mastery and managerial size *inevitably* calls forth an equal order of magnitude of government coordination and control. For example, if the nation's chemical industries create thousands of new, potentially harmful chemical compounds each year, the FTC and the FDA will be forced (with the taxpayers' money) to hire a sufficient number of chemists to analyze them and trace their effects on society.

The Future of Industrial Society. This inevitable process eventually leads industrial societies into the blind alley which I have called "the Entropy State," where their technological complexity and interdependence became impossible to model, manage or coordinate. Such a society then becomes racked with a mounting backlog of unanticipated social and environmental costs. Today, as I have tried to show, it is all of these "transaction costs" that are growing exponentially, rather than the real Gross National Product. Legislators, government departments and regulatory agencies are forced to proliferate in trying to keep up and coordinate. For example, the FCC is now being pressured by angry citizens to regulate our newest technological obscenity: automatically-dialed, recorded "junk phone calls."

More clearly than ever, the de-regulation debate is a debate about the direction of industrial society. On the one hand, will we continue the current drift toward producer-oriented, capital-intensive, centralizing technologies in the pursuit of the ill-defined goals of "efficiency" and "productivity"? And will we bear the increasing social and environmental costs, only partly measured by the ominously growing tax bill, for increasingly necessary regulation? Or, on the other hand, are we as a society and our industrial producers willing to accept the *price* of de-regulation, which inevitably must be the shift to more benign, appropriate technologies, which are more de-centralized, simpler, cheaper, which conserve capital and energy by employing more people, and which create fewer social impacts and unanticipated side effects and therefore need less regulation?

Citizen as Regulator. Such a decentralized, energy-conserving, labor-intensive production system would increase the effectiveness of citizen and consumer groups in being watchdogs of their own interests, rather than having to pay bureaucrats and regulators to do it for them. In any case, the strategy for consumer, public interest, social welfare and environmental forces to employ in the de-regulation debate is two-fold: (1) Force a discussion of these deeper issues of producer-oriented, capital-intensive technologies versus people-oriented de-centralizing, intermediate technologies; (2) Demand, as the price of their support for *selective* de-regulation that bona-fide citizens organizations, public interest research, consumer, social welfare and environmental groups be given block grants by government to protect their own interests, as is common practice in other industrial societies. Gunnar Myrdal reminds us in *Beyond the Welfare State* that this model of regulation is much more *efficient* even in narrow money terms, because citizens and consumers, if adequately funded, look after their own interests much more diligently than bureaucrats paid to do it for them. One has only to compare the cost/benefit performance in pressing the claims of American Indians of the American Indian Movement and Americans for Indian Opportunity (both citizens' organizations) with that of the Bureau of Indian Affairs, to see the validity of this statement.

In Sweden, the practice of making block grants to bona-fide citizens' organizations is well established. While in Stockholm some years ago, I asked leaders of such activist groups whether they felt inhibited by accepting such grants. Could they picket government or corporate offices without interference or reprisal? "Of course," I was told, "we would alert the press if there were such attempts at intimidation." Another model is that used in Canada to provide citizen groups with intervenor funding. In the Mackenzie Valley Pipeline Inquiry, such citizen intervenors received a total of $600,000. Native and Indian groups received $400,000 and environmental groups $200,000 in this study of the social and environmental effects of the Arctic Gas Company's plan to construct a pipeline through the Mackenzie Valley in Northwest Canada. Intervenor funds tied to projects would prevent concentration of funding on some groups and assure that it was spread equitably among whatever groups were relevant to each situation. In my article, *Politics by Other Means* (*The Nation,* December 1970), I proposed that all federal and state projects over a certain size should mandate independent funds for their assessment by citizen intervenors.

This type of broad funding mix for citizens groups might well be superior to the appointment of public advocates, whose offices might well turn into another layer of bureaucracy between citizens and the handles of power, or a stepping stone to conventional political power. Senator Edward Kennedy has introduced legislation for citizen intervenor funding in regulatory proceedings. Others propose issuing "Citizen Assessment Bonds," backed by government guarantee to provide citizen groups the funds to hire their own experts to assess public and private projects and the effects of various technologies. Some U.S. regulatory bodies, such as the Environmental Protection Administration, have made a small trickle of funds available to citizen groups for studies or public education projects. But in no sense have these small grants embodied the concept of *watchdog* activities to protect their interests, let alone as a means to reduce the cost and size of the regulatory bureaucracy itself. More often, yet another group of bureaucrats is added, to whom citizens must then be supplicant for the meager amounts squeezed from the agencies' budgets. In reality, such funds are proper compensation for iconoclastic and innovative research.

A coalition of consumer, environmental and social-welfare groups, together with decentralists, co-op and community-technology forces and media-alternative activists, with the backing of public interest professionals in law, economics, accounting and the sciences, might have the power to change the current David and Goliath battle scenes between producers and the rest of us. The de-regulation debate is now providing the opportunity to force the hand of the de-regulation enthusiasts, by demanding that the price for the support of citizens and public interest people is a set of companion measures to assure that in all de-regulated areas, adequate block grants or intervenor funds are mandated so that we can watch out for ourselves.

Reprinted from the crusading small journal, *Just Economics,* Oct., 1975. Since then, Governor Jerry Brown of California has taken a big step in the right direction. In February, 1977, in a group ceremony, he swore in 60 new members of various state regulatory boards and commissions—all citizen and consumer advocates with a wide range of expertise and including many representing minority groups. Many had been active in public-interest research and Governor Brown charged them specifically with the mandate of being "lobbyists for the people, rather than special interests." The *New York Times* called the action an unprecedented step to remove control of these boards from the special interests they were created to regulate, and a significant new turn in changing the way government works. This, as well as some of the proposals and new methods I have mentioned, may help in the continuing effort to open up such regulatory processes and to rethink their functions.

CHAPTER EIGHTEEN

Awakening from the Technological Trance

For several years public skepticism has been growing concerning the role of technology in our society. The days are now past when citizens automatically equated technology with progress. We are now all too aware of its unanticipated consequences in daily life. They range from the increasingly formidable destructive power it places at our disposal, the ecosystem disruptions it creates and the ever-larger scale of industrial operations it produces, to the individual alienation and sense of diminished power and control that so many of us now experience.

In the past, the scale of technology was smaller and its effects were still relatively localized. Many uncoordinated technologies could still coexist without impinging on each other or affecting large regions or populations. As technological mastery increased, the increasing scale of innovations, together with the cumulative effects of many small applications, began making even more pervasive impacts on populations, social structures and ecosystems. Today the unanticipated effects of our growing knowledge and the technologies it creates have seemingly outrun adaptive capabilities, whether psychological, social, organizational or political. The result is the current series of unruly crises. Whether we designate them as "energy crises," "environmental crises," "urban crises" or "population crises," we should recognize the extent to which they all are rooted in the larger crisis of our inadequate, narrow perceptions of reality. When our perception is too narrowly focused, for example, on the city or town in which we live, we tend to lose sight of all the external factors

"They have the know how, but do they have the know why?"

which affect it, such as national policies, transportation, agriculture and commerce. When one of these external factors changes, we are likely to perceive the resulting change in our city as a "crisis." For example, what was first designated as the "urban crisis" is now understood more in terms of the national policies that affect our often helpless cities: e.g., federal housing policies that, along with the Highway Trust Fund, underwrote the flight of the cities' taxpaying middle class to the suburbs; the mechanization of agriculture that drove hundreds of thousands of farm workers to the cities; etc.

Most humans are motivated by perceptions of reality that are quite limited in space and time. A majority are too concerned with tomorrow's food supply for their immediate family to worry about next week or their neighbors. Some, affluent enough not to worry about immediate survival, can enlarge their sights to concern themselves with their community and occasional longer-range problems. A lucky few with opportunities for travel can extend their concern to embrace national and even international affairs and perhaps worry about the kind of world their children and grandchildren will inherit. And some extraordinary individuals, in spite of poverty and adversity, through the power of imagination can relate to the great concerns of their time and contribute their vision to enrich our range of alternative futures.

But it is natural that most people perceive reality as their immediate environment and their personal and family concerns. And for the kind of rural, low-technology world we knew in earlier times and which is still a reality for most of the world's people, such perceptions were adequate for dealing with the problems and decisions of daily life. But now that many of the world's nations are technologically advanced, global impacts and interdependencies are the rule rather than the exception; therefore, narrow perceptions become increasingly dangerous and lead to decisions based on inadequate information concerning larger and longer-range patterns of causality. Not only do we find that "crises" sneak up on us because we were not paying attention to the significant variables, but, conversely, we do not adequately appreciate how individually rational microdecisions and actions can add up, by default, to dangerous, irrational macrodecisions. This cumulative effect of the tyranny of small decisions is seen when, on a hot summer day millions of decisions to drive to beaches result in traffic jams.

How do such human perceptions become further entrapped by our very technologies? Sometimes it is by compounding our lack of perception, as when a technology produces an unanticipated effect and we try yet another technological fix to ameliorate it, thus adding more unknown variables and increasing the impetus of social and ecosystem changes. In fact, technology has now surrounded us with a human-made environment capable of insulating perceptions from direct experiences of our vulnerability and our dependence on the primary natural ecosystems. An enormous number of arguments and misunderstandings between people of goodwill seem to occur because they did not exchange some basic information about their perceptions at the outset. The dialogue can be made much easier if each of us clarifies first where we see ourselves in time and space. Where are you in the total system? Do you most often experience yourself on a planet? As a citizen of the United States? As a member of a local community? As a family member or as an individual? Similarly, what time frames do you commonly employ? In your view, is a "long time" a millennium; a century; five years; three weeks?

Learning to examine our own space-time frameworks and the mental models they generate of the system we inhabit is now of the utmost importance for our survival. "Reality" is that selective image of the external world which, as Kenneth Boulding points out, we pull in on our own personal, perceptual "TV screens."[1] Therefore, in order to make myself more explicit, I will tell you where I think I am in space-time, and what mental models my perception has generated and by which I experience "reality." I believe that I inhabit a rather undistinguished solar system in an arm of an equally unremarkable spiral galaxy, and I also inhabit a body evolved from the elements of one of the prettiest planets in this solar system. The mental models this experience of perception generates are systemic and interactive, multidimensional, balancing both equilibrium and dynamic behavior modes, and evolving over time in some purposive manner which I believe that I shall never understand. There seem to be elements of subsystem expansion and contraction; ordering and disordering; entropy and syntropy, continuously occurring, along with energy-

/matter/information transformation. The most puzzling aspect for me is how I am a part of this process and yet seem also to experience myself as an observer of it. But let us, for the moment, leave such mind-body, subjective-objective, observer-observed paradoxes to the adventurous young physicists who are trying to deal with them as they were posed by Werner Heisenberg in his famous Uncertainty Principle.[2] Science has become a religion for all too many, while human values and ethical concerns are driven into hiding because they are embarrassingly unquantifiable and "nonrigorous." Most of the incentives in the academic world reward rather narrow, reductionist study and pseudo-rigorous examination of less and less significant phenomena.

Many distinguished scholars have called attention to these "fallacies of misplaced concreteness," as Alfred North Whitehead called such efforts of microrigor. They include, of course, Heisenberg himself in physics; Kurt Godel in mathematics; Oskar Morgenstern, Georgescu-Roegen, Kenneth Boulding, and E. F. Schumacher in economics. The torch is still being upheld in the science-policy arena by Lewis Mumford, Gerald Holton, Margaret Mead, Gregory Bateson and many others, and there are the vigorous new critiques of reductionist science by Theodore Roszak, R. D. Laing, and William Irwin Thompson.

All these humanists force us to remember that the normative nature of science is revealed in the first decision of any scientist: what phenomena to study. This choice then influences the view of reality: where we see ourselves in space-time—perhaps it's a sort of Heisenberg Uncertainty Principle at the macro, rather than the quantum, level.

I believe that human survival now requires an awareness that transcends our very natural anthropocentrism. Each great knowledge explosion has been based on a new level of expanded awareness; from Ptolemy's geocentric view of the sun and stars revolving around Earth, to the Copernican revolution, which reduced us to a subordinate position in the universe. Darwin further undermined our proud image with his theory of evolution. Much of today's new knowledge is increasingly shattering our sense of self-importance: whether studying ourselves as components of living ecosystems or as the infinitely malleable creatures of behaviorist B. F. Skinner's *Beyond Freedom and Dignity,*[3] whose profoundest emotions are nothing but electrical stimulation reproducible by brain-probing instruments. Now we learn that two more of our claims of uniqueness are being debunked: dolphins and other mammals have well-developed languages, and many other species use tools, including even the lowly ant, which loads food supplies on leaf fragments and thus multiplies its transport capabilities ten-fold.[4] We are just becoming aware of ecosystems as immanent intelligence; for example, it has been shown that grasses in typical grazing pastures are capable of growing themselves tougher and more unpalatable by increasing the cellulose content of their leaves in order to drive off excessive numbers of grazing animals.[5]

But let us not be dismayed by this new evidence of a need for greater humility. Let us instead relax, enjoy our natural curiosity and indulge the new burst of imagination and speculation it creates. Imagination, indeed, has always been one of our most important survival tools. We must now employ imagination to help us deal with the perceptual crisis that is upon us, as our species has now multiplied almost to the limits of its ecological niche on this planet. This perceptual crisis has two aspects: (1) We are experiencing an implosion, as space and resources diminish relative to our growing population. We feel the loss of frontiers, the slowing of economic expansion, urban crowding, and the evaporation of many historically defined freedoms. (2) At the same time, we experience ourselves getting smaller and less significant as old perceptual boundaries fall away. Paradoxically, as we feel *physically* confined and frustrated, we are also confronted with an expanded *mental* model of the universe. We are again facing the oldest human dilemma: a consciousness that can wander among planets, stars and millennia, but trapped in a few dollars' worth of chemicals that will degrade in a short span of years. In brief, we must again face the fact of our own death and finiteness; the old games our cultures have provided to shield us from this reality break down and become destructive and inappropriate for the new conditions, leaving us shorn of the psychological clothes to protect ourselves.

Imagination is already coming to our aid again. As physical

forms of growth are foreclosed, we are learning to make some new psychological "elbow room" in diversifying life-styles and in fashioning new images to help us expand consciousnesses for the evolutionary leap we must now make. We might imagine ourselves as a termite colony, living happily up to now for all its generations in a beam in the basement of a house. We have developed elaborate social structures and academic disciplines: termite geography, termite mathematics, physics, engineering and economics. Suddenly, our current generation has used up and transformed the interior of the beam and emerged at its surfaces. Not only does this change all the conditions within the colony and its beam, but the roof of the house seems to have blown off and the wall collapsed! Survival requires construction of a more appropriate geography, physics, math, etc., so as to incorporate the new variables and expanded boundaries and contexts.

Now let us imagine together that we are extraterrestrial visitors from one of the millions of planets in our galaxy which may have conditions hospitable to life. We are further evolved than the life forms on Earth. We zero in on our spacecraft and approach this planet. It is not important where we land; it all looks the same, a sphere of blue and white with brown patches visible below. The spot where we land is apparently called by its humanoid inhabitants "Washington, D.C." It doesn't matter, it is as good as anywhere to begin our exploration. We wander around (first taking the precaution of dematerializing ourselves so as not to scare the humanoids) and peek in a large building at a gathering of them. These humanoids are discussing their future on this planet. It's all very confusing: they still seem to be debating whether or not their planet is a spherical, finite system. We extraterrestrials know that from the vantage point of our own highly developed technology, it is not really a closed system; but from the humanoids' current levels of technology, it still is. The trouble seems to be that the humanoids have not yet internalized the learning experience that their first costly venture into space provided. They created all that hardware, flashed back from their moon TV pictures of their actual situation, but have done little, it seems, to overhaul and reprogram their educational, political and economic systems to conform with what they have learned. Still, we are encouraged to find these humanoids at least arguing and debating how to do this.

To resume our human perspective, it is indeed encouraging to know that this great debate is on the agenda. At least we are addressing at many levels our propensity to create hardware without writing the necessary software to program its orderly functioning. For example, with our technology we have created an interdependent, global economy, and now we are desperately trying to write the program of "software": monetary agreements and international rules to operate it without catastrophic breakdowns.

I have often pondered why we are so much better at creating "hardware" than "software." At one level, it is rooted in our fear of death and nonexistence. When we build cities, dams and factories, we provide for our material requirements, but we also affirm our existence and importance. These physical artifacts are so tangible that they reassure us of our own reality. Another root of our interest in hardware is that humans love to manipulate their surroundings and enjoy the sense of mastery and control they derive and the expression of self found in such creation and play. Yet another explanation may be that we would rather project our inner tensions and conflicts onto the objective world than resolve them by examining our own psyches and trying to retool ourselves. Lastly, I hope you will forgive me for wondering whether this passion for hardware is not a result of a cultural overdose of the masculine consciousness. The masculine psyche does seem more attuned (either biologically or by cultural conditioning) to manipulating external things and objects, while the female psyche by contrast seems more attuned to "software," i.e., interpersonal and social relationships and arrangements.

Technology, defined as knowledge systematically applied to human problem solving, means *software* as well as hardware. For example, the Social Security system and income-tax withholding are as much technologies as any hardware system. Lewis Mumford in the *Myth of the Machine* drew attention to our bias toward hardware in anthropology and archeology. He pointed out that when we dig for evidence of earlier cultures, by definition such remains are tangible: their hardware, whether arrowheads, axes, pots or other artifacts.

From the extent and elaboration of these artifacts, we infer their level of "civilization."[6] We often forget that many cultures may have existed without leaving a trace. They could have developed highly refined technologies, but of the software variety: techniques of conflict resolution, supportive interpersonal relationships, production systems based on elaborate barter, reciprocity and redistributive schemes, myths and taboos to regulate antisocial behavior without use of jails, clubs or physical restraints. A culture which elaborated such software techniques would have had little need for spears and arrowheads and might have had scant energies left over to elaborate its tools, and so we might assume too casually that because there were few tangible remains it was less "civilized."

In the same vein, during my visit to Japan I talked with a project director at the Japan Techno-Economics Society who was directing an effort to computer-model the value system of the Japanese people. He pointed out that from the quantities and configurations of material artifacts and technologies created by various cultures it was possible to infer a great deal about their value systems. He noted, for example, on one end of the scale the Balinese, who create exquisite music, dances, rituals, stories and clothes, but are rather uninterested in hardware. On the other end of the scale are the Americans, who are fascinated with, and produce more, hardware than any culture the world has ever known. We are even unable to enjoy leisure activities such as hiking without an incredible quantity of gear—let alone our uniquely energy- and materials-intensive hobbies, such as those involving snowmobiles, beach buggies and camping vehicles.

We know that values are the dominant variables driving not only technological but economic systems. Relationships have been established between Judeo-Christian religious beliefs and the rise of capitalism and the industrial revolution.[7] E. F. Schumacher in *Small Is Beautiful* describes the value-system that drives Buddhist economics, where labor is an *output* of production, rather than an input, embodied in the idea of "right livelihood," where work is a valuable mode of self-actualization while the product is of secondary importance.[8]

We in this culture may at last be awakening from that altered state of consciousness which Thomas Berry calls "the technological trance" and all the unthinking assumptions underlying it.[9] The most destructive of these beliefs is that we see innovation and technological progress mostly in terms of hardware and as continuous, rarely recognizing limits or the concepts of balance and paradox. This technological trance has led us on with a mirage of "efficiency" as its will-o'-the-wisp. Our technological consciousness has permitted us to conquer nature (temporarily, at least), expand our ecological niche and manage more of the variables affecting our existence. But the trade-off is that, as we proceed with this process, the task of managing the proliferating variables becomes ever more complex and onerous, until we find, as Schumacher put it, that we need "a breakthrough a day to keep the crisis at bay." We lose sight of the fact that some human and natural processes are not susceptible to increased "efficiency." Women still understand this better than men; they know that it still takes nine months to make a baby and one hundred years to grow a mature hardwood tree. While human interactions can be increased and made faster with technology, they are rarely bettered and sometimes worsened. A companion myth is that new technologies can always be "debugged" if only we wait long enough.

Let us look at some contemporary examples of this mirage of efficiency. One is the effort of officials in the U.S. Postal Service to reduce "inefficient" human labor and replace it with elaborate, automated machines for sorting mail. After reducing the human work force and adding to the ranks of the unemployed, and investing millions in capital, they find that the machines are ripping, crushing or destroying an alarming number of parcels.[10] Perhaps it might have been more socially efficient to add a million unemployed workers to the Postal Service, increasing the care in handling while reinstating the twice-a-day mail service our forebears took for granted! Another more somber example is the efforts of electric utilities to seek "efficiency" in larger and larger generating plants and in substituting nuclear power for less costly and violent technologies. For this increasingly suspect and evanescent "efficiency," they are willing to assume risks on our behalf and trade off social efficiency, since costly and elaborate police and security systems will have to be invented to contain and manage the deadly plutonium now and for thousands of years to come. This is not to mention the additional societal costs which must be paid in the loss of many cherished civil liberties. Already, consumers and citizens are in full-scale revolt against these social inefficiencies.[11]

As we now know, the word "efficiency" is fast becoming meaningless. "Efficient for whom?" is the question in the nuclear issue. In an economy with between 7% and 8% of the work force unemployed, clean, safe solar heating could provide the equivalent energy supply, while creating several *times* as many jobs per dollar invested as nuclear energy; while conservation could itself become our major new energy "source."

As discussed, our economy has overshot the mark in its substitution of capital for labor. In fact, I contend that in hundreds of production and service processes, labor has now become the more efficient factor of production and as natural resources become increasingly scarce we must employ our human resources more fully.

Former president of the British Operations Research Society, Stafford Beer, points out something obvious but crucial to our understanding of humanly designed social systems: "Institutions are systems for being what they *are* and doing what they *do*. No one believes this, which is incredible—yet true. People think that institutions are systems for being what they were *set up* to be and do, or what they *say* they are and do, or what they *wish* they were and did. The first task of the systems scientists is to look at the *facts:* what is the system? What does it *do?* If the answer turns out to be something no one wants, do not go around repeating the popular but fictitious belief in a very loud voice. Do not hire a public relations campaign to project the required image. CHANGE THE SYSTEM!"[12]

Dr. Beer went on to point out that when people become disenchanted with their institutions, whether governments, corporations, unions, churches or whatever, they express this as disenchantment with their *leaders*. "Why doesn't the leader *do* something?" etc. This, of course, is the wrong question, because the leader is an *output* of the system. The former president, J. K. Jamieson of Exxon Corp., talked of the near impossibility of shifting the course of that mammoth company, and any would-be leader who has tried to ride the tiger of such other massive systems as the Department of Defense understands the truth of this fact of institutional resistance to change.

Therefore, if the leaders are the output of the system, the appropriate question is, "Why do our institutional systems throw up leaders who seem helpless, incompetent, or corrupt?"

One of the chief reasons for our confusion over the failure of our systems to do what we designed them for is, of course, that we do not understand them and therefore cannot design or model them accurately. Any system that cannot be modeled cannot be managed. And in today's complex industrial societies most of our large institutions fall into this category.

The result is that through the filters of our myopic perception, our puny efforts at managing these systems leads to proliferating exercises in suboptimization. If you can't model the larger systems, you can take the path of least resistance and try to model the smaller, easier systems, where the numbers of variables are more manageable. Most of our societal crises today are the effects of our increasing proficiency at suboptimization.

Furthermore much of our scientific and technical enterprise is geared toward increasingly efficient efforts at suboptimization. Likewise, our nation's colleges and schools of business administration are institutions for teaching the methodology of suboptimization. We must now attempt to model the larger social contexts and interactions before developing ill-considered, short-range technological "fixes." For example, it is of little use merely to hire more police and buy more costly security hardware without viewing crime as part of the social cost of maldistribution of wealth and income. Similarly, drug addiction is a part of the social cost of maximizing profits in the drug industry. Emphysema and lung cancer are part of the social costs of maximizing profits in the tobacco industry. And there is increasing evidence that violent behavior and poor eating habits are part of the social costs of the commercial structure of television in this culture. Only such a contextual view allows us to better assess which of our "problems" are even susceptible to a technical "solution."

In addition, the often massive costs of corporate research and development programs are routinely passed along to consumers in higher prices. Yet consumers have no control over how these funds are deployed, what new technologies are developed, what social impacts they may cause and whether narrow criteria of profitability are used for such ubiquitous "R and D taxes" levied on consumers of thousands of products from light bulbs to antibiotics. Furthermore, the engineers who design projects and the companies that get the contracts are highly organized to petition for public funds to underwrite new technological developments, but the taxpayers and consumers who foot the bills do not even get wind of these public or private proposals for months. They have little information or incentive as individuals to bear the heavy financial and time costs of researching the issue and challenging such powerful forces. This particular syndrome, known to economists as the "free rider" problem, is discussed at length by Mancur Olsen in his book, *The Logic of Collective Action.* He illustrates the difficulty of challenging any policies or proposals promoted by such powerful interest groups, to which the potential payoffs are so large that their incentives to commit funds to win these prizes are always greater than the incentives of individuals to try to protect themselves by organizing and applying countervailing pressure, using their own funds.

It is no wonder, then, that technology promoters, developers and empire-building public agencies spare no expense to hire economists to prepare cost/benefit analyses to justify their plans and present them as significant advances in the public welfare. There are other more technical problems with cost/benefit analyses which I shall mention briefly. They assume that adequate information is available to all parties and they accept the existing distribution of income as a given, although these two factors can disenfranchise many citizens, such as those without economic or political power, or adequate information on costs, health effects or long-range risks of a particular technology or development. In addition, cost and benefit ratios can be completely different depending on what rate of discount is used, i.e., the assumption of what interest rates will prevail over the lifetime of the project. Such arbitrary assumptions can overstate the costs and understate the benefits or vice versa and are currently the subject of hot debate among economists.

We can thus appreciate the relativity of all such supposedly scientific methods of analysis. One case with which we are all familiar was that of the supersonic transport plane (SST). In addition to the cost/benefit analyses and their deficiencies, which were pointed out by several economists, there was a myopic attempt to focus on the speed of the plane from airport to airport, with very little discussion of the fact that similar overall door-to-door speeds could have been achieved for many more people at far less cost and risk by using the funds to construct rapid mass-transit lines from airports to center cities.

The "automobilization" of the U.S. is an even better example of the cumulative effects of myopic perception, narrow analyses and suboptimization. It has taken some 25 years for the social and environmental consequences of the auto to build up to a point where we finally noticed them—in fact, the point had to be hammered home by an energy "crisis" before we got the message. But in examining the full cost and benefits of the auto, we still have a long way to go. As Ivan Illich points out in *Energy and Equity,* we still measure time gained by speed in this fragmented way, as miles per hour *in the vehicle.* We forget the time spent in earning the money to pay for the vehicle, insure it and maintain it, which in an overall view of our lives is the real measure of our time and opportunity costs vis-à-vis automobile transportation. Moreover, Illich observes, the auto, by extending and setting in concrete our spatial living patterns and permitting greater distances between living, working and shopping, dictates that it will take more traveling time to perform such daily activities and decrees that cars shall be indispensable. By such total-system calculations, Illich estimates that our cars actually deliver us speeds of about 5 miles per hour, because fully one quarter of our waking lives are spent in performing the involuntary activities associated with this automobile-transportation system. By contrast, Illich shows, in countries without highways, people walk at an almost equivalent speed, but spend only 5% of their time transporting themselves around. Likewise, when we consider costs, we tend to look only at the cost of the vehicle, maintenance, insurance and gasoline. We often overlook the enormous tax costs of highways, police, the burdens on our court systems, the arable lands lost, crop damage, fatalities and injuries (all quantifiable or approximatable "externalities"), not to mention the less quantifiable factors, such as the explosion of cities into wasteful sprawling suburbs, the "disabling" of millions of citizens who cannot drive or afford cars, and the destruction of other transportation options, such as walking or bicycling. Nor do we measure the cost of this mode of travel vis-à-vis all the alternative modes. At an even broader perceptual level, we can discern that how much and what kind of transportation we have is based on how a culture values mobility and acceleration and how it assesses the trade-offs associated with these values. One sees also that with a different set of values, a society could just as easily measure transportation as an indicator of the level of *dysfunction* in its system.

To return to the examination of our troublesome institutional subsystems and how to contain their dangerous tendencies, we must recognize the extent to which their behavior is controlled by their programing assumptions and language. The programing language they use is that of economics, the discipline that monopolizes the discourse over all our national resource-allocation decisions. To understand the workings of any system one must also examine the assumptions and goals which program its activities. For example, the U.S. Constitution is the program set up by our Founding Fathers to provide norms for operating this social system. The judicial system then performs the comparator function of measuring behavior against these norms. As we have seen, the social system's growing economic and technological subsystems are now making it increasingly impossible to operate according to the original program of the Constitution. In fact, some of our large-scale technologies, such as nuclear power, will require abridgment of civil liberties and may be simply unconstitutional! As more public debates over resource-allocation decisions are forced into narrow economic metaphors, such as cost/benefit analysis, the subsystem goals of profit maximization and efficiency supplant the former goals of freedom, justice, equality and our collective judgments as to what constitutes a good society—rather than merely a rich one. Indeed Oscar Wilde's description of a cynic as one who knows the price of everything and the value of nothing would also fit proponents of a merely rich society.

Therefore, I have concluded that the discipline of economics itself is now the chief stumbling block to the rational discussion that our nation must have over what is valuable and how our resources are to be allocated. Imagine, for example, how different such public debate would be if we used the discipline of biology, or perhaps general systems theory, as the language of discourse. Economics, unless it enlarges its concepts, must, for the health of the social system, be subsumed by other broader disciplines. For example, the model of human behavior economics uses has not even incorporated the work of Sigmund Freud, let alone Abraham Maslow, David McClelland and other humanistic psychologists. Economics has permitted us to pursue dangerous, as well as ethically dubious, goals. It has enthroned some of our most unattractive predispositions: material acquisitiveness, competition, gluttony, pride, selfishness, shortsightedness and just plain greed. In a simple, uncrowded, bountiful frontier world, the effects of these behavior patterns were damped. In his day, Adam Smith was probably right that his "invisible hand" allowed microdecisions, however selfish, to add up to a fair approximation of the public welfare. In the crowded world of today, these human tendencies are beginning to destroy us, particularly when they become institutionalized and reverberate throughout the system, thus helping create the many "tragedies of the commons" we are now witnessing each day.

All this leads us to a new appreciation of the vital need to make careful choices and trade-offs in our national and individual decisions on resource allocations. Both business and government leaders have been guilty of masking the necessity of making such choices by their continual pumping up of consumer and citizens expectations with pie-in-the-sky promises. Politicians win votes by promising each group the pork-barrel legislation it desires, as well as the less obvious subsidies and tax breaks which have become a continual raid of the public treasury. Even in the case of necessary new programs to maintain potable water supplies and breathable air in the cities, the price tags are rarely measured in relation to other social goals.

Corporations also inflate our expectations in the more than $20 billion they spend each year on advertising, often pandering to our most infantile desires and fantasies. "You deserve the best," they tell us; "Come on down to Florida"; "Charge everything with your Funny Money Card." When depressed, we get the impression from millions of dollars worth of patent medicine commercials that we can solve our frustrations with the right kind of pills. In addition, many product advertisements obscure the trade-offs inherent in all of our consumption. They tell us the good news, but forget to tell us about the bad news—they tell us about the sparkling dishes and clothes, but forget to mention the loss of those sparkling rivers and lakes. Consumer and environmental activists have been bringing all the bad news about the inevitable trade-offs we must accept if we continue our current wasteful production/consumption patterns. If we want more of the same—energy, cars, appliances—we must expect more emphysema, strip mines, highways and pollution. If advertisers pointed out these trade-offs more truthfully, I suspect that American consumers would accelerate their embracing of lower consumption lifestyles and demand that production methods be changed to reflect the new scarcities.

Therefore, it is vital that citizens inform themselves and that voluntary sector organizations, environmental and consumer groups and alternative, Ralph Nader-inspired information-gathering groups become an integral part of the process of defining technological options and helping shape the agenda of our science-policy decisions. Meanwhile, there is a role for every well informed layperson, and that is forcefully to ask the right questions: "Have all possible options been adequately explored?" "How will the costs and benefits be distributed among different groups and individuals?" "What are the social and environmental impacts and the future consequences?" "Will the new technology or project create irreversible changes?" "Can the goal be reached by any other means?" and, if not, "Should the goal be further examined in light of other goals and priorities?"

As we have all discovered with environmental impact statements, it is no good for the public to assume their objectivity, especially when they are prepared by the same agency that is promoting the project. Similarly, in assessing new technologies we cannot afford to assume the objectivity of even the most prestigious scientific panel or the seemingly most unimpeachable organization preparing such technology assessments. The perceptual and organizational biases are too endemic and often below the threshold of consciousness. Therefore, public participation in every phase of such assessments is the best way of providing a rigorous watchdog function to spur such scientific assessors into broader perspectives and more thorough analyses. Notices of all technology assessments to be contracted by public agencies should be published in advance in the Federal Register and should be freely available to all interested publics. Diverse voluntary organizations, especially ones representing those particularly impacted by new technologies, such as labor and consumers, and those concerned with environmental impacts, must, with the help of their own teams of volunteer experts, critically examine every phase of such assessments. They must review the study design and the assumptions used in modeling the problem and examine the composition of the scientific case team chosen to determine whether significant perspectives and disciplinary skills have been omitted. Among such often omitted considerations are those of occupational health to ascertain effects on workers; political science to determine, for example, whether a technology will have centralizing or decentralizing effects; ecology to investigate total energy-conversion efficiencies and whether natural systems could perform the same task without resort to massive new "hardware"; and welfare economists to examine impacts on the distribution of wealth and income. But even such specialized studies will not be enough. Voluntary sector groups must follow the project closely and demand opportunities to critique the work in progress to ascertain whether new options have been discovered and adequately explored, whether major new uncertainties have emerged and are being given proper attention. For, as Joseph Coates of the U.S. Office of Technology Assessment has pointed out, if technology assessments are done well, they are bound to turn up some bad news. But only vigilant public participation can assure that such findings will not continue to be suppressed.

Systems require balance to achieve homeostasis: competition balanced with cooperation, selfishness and individualism with community and social concern, material acquisitiveness with thirst for knowledge and understanding, rights with responsibilities and the striving for love, justice and harmony. As we study more deeply the homeostatic systems in nature, we infer not the absurd caricature interpretations of social Darwinism to rationalize our greed but rather the interdependence of all living things. We learn, too, that natural systems never maximize single variables, such as "profit" or "efficiency." We can infer too, that maximizing behavior on the part of any individual or firm is shortsighted and destructive of the larger system. An extreme example is the whaling industry, which is still maximizing its catches, in spite of the knowledge that it is only a matter of a few years before most of its prey will be extinct.

We now realize that we must learn humility if we are to face these complexities we have created. We sense the truth that only the system can manage the system and see the airy arrogance in some of our concepts of "management" and "administration." We examine anew the easy assumptions that sociotechnical systems are even susceptible to manipulation by legislation, just as ancient kings learned that they could not affect the behavior of natural systems by royal decree. We marvel once more at the ingenuity of "primitive" cultures, whose most obvious characteristic is the relative absence of government, because social controls have been internalized.[13] We are, indeed, at a crossroads faced with our own sociotechnical complexity. We can take the path of stepping up the computer power to model these complexities, or can try to disentangle some of the unnecessary interlinkages and the overcoherent technologies themselves and by such decentralizing of means, reduce the numbers of interacting variables that must now be managed.

We also realize that hard choices and trade-offs must now be made. It is no longer just a matter of budget priorities between education, transportation, health or more private consumption; or between R and D priorities, public and private investments, capital or labor intensive production, energy alternatives. These new trade-offs are visible in every societal subsystem, from governmental and corporate organizations to the educational system, where maintaining capital plant and equipment must be traded off against flexibility and adaptability, while similar choices must be made in teacher training and curricula. At the personal level, educators and professionals must also deal with these new trade-offs: whether to specialize further or to expand their horizons into interdisciplinary studies, even at the expense of "rigor" as academically defined and rewarded; and whether to trade expanded consciousness for greater secular power and emoluments. We see that such goals conflict, because knowledge has become the servant of power in too many cases, and our educational enterprises have too often turned out intellectual mercenaries, whose lances are for hire to justify policies of entrenched bureaucracies and interest groups, rather than merely to search for the truth.

At last we see that science is not neutral, nor is technology, and its pretensions to "value-free" objectivity are now debasing the currency of public debate and preventing us from making adequate social choices. Technology now creates its own social configurations and we must ask to what extent the continued drive toward big-bang, capital-intensive technologies simply concentrates power, wealth and knowledge in fewer and fewer hands, while making the rest of us poorer and more powerless and actually increasing overall human ignorance. At least, it is now clear that the "free market" is not working to direct technological innovation to consumer demand, as it should. If it were, we would not now have a debate raging about the "appropriateness" of technology, which has spilled out of the market-choice arena into the realm of social and political choice. All this was predicted in 1944 by Karl Polanyi in his study of human production and exchange systems, *The Great Transformation.* Polanyi pointed out that leaving resource allocations to a free market system would merely suboptimize the social system while leading to rapid environmental depletion. He demonstrated that free markets, far from being derived from some natural order or human behavioral laws, as Adam Smith thought, were created by carefully designed human planning and software. The conditions thus created for the operation of free markets were bitterly contested and legislated over many decades. As discussed in Chapter 2, the British social legislation that effected land enclosure had the intent and consequence that both land itself and the labor of the dispossessed peasants became commodities for sale. This laid the groundwork for the industrial revolution. The increase in efficiency of production was won at a terrible price in social dislocation and inefficiency. In the larger scale of history, market systems are a mere blip associated with the rise of industrialism and actually have been a rare aberration in societies.

This brief discussion of market failure is necessary if we are to assess technology properly and try to simulate its likely second-order consequences. Each major technological innovation redistributes power, destroys some jobs and creates others, rearranges population patterns and creates new ranks of winners and losers. Technologies do not arise in a vacuum. There is always a force field of institutional vested interests whose interactions may tend to promote or suppress technologies. For example, the institutional and financial commitments to nuclear power have starved solar energy for decades, as James C. Fletcher, head of the National Aeronautics and Space Administration has pointed out.[14] Therefore, to keep technology assessments intellectually honest, their review panels must appoint representatives of potentially impacted constituencies, such as labor, consumers and environmentalists, a mode of operation I have encouraged at the U.S. Office of Technology Assessment (OTA).

The technology assessment debate can also be focused around whether technologies are producer-driven or consumer-responsive. As the shortage of capital and inflation force tougher social choices, cost-benefit and risk-benefit analyses become more difficult. A typical producer-driven technology is the computer-automated tomographer costing about $500,000 to diagnose rare brain diseases undetectable by other procedures. A large output of such devices seems unjustified, yet promoters sold great numbers to hospitals, which are recovering their huge costs quickly by adding this diagnostic procedure to their normal health check-up routines and charging extra fees. With limited resources, can a society permit capital to flow into this type of questionable technological proliferation, when, perhaps, health dollars spent on disease-preventive education, or on other more vital but less glamorous equipment, might pay off many times greater benefits?

We must perforce study the technology/society interface and the baffling problems I have just described. As our "headspace" and conceptual backdrop expands, we must never forget that most humans learn experientially. Today, in many colleges you will find that the engineering, chemistry and physics students have steered their professors out of the lofty classroom-based conceptualizing and into the open air. There they are doing hands-on experiments with small-scale, renewable-resource, appropriate technologies: whether methane gas production from bioconversion, solar-energy collectors, wind generators or hosts of other "soft" as opposed to "hard" technologies, as summarized in Figure 12, page 366.

I have seen the sheer joy in the faces of physics professors working with their hands constructing these vital, experimental, ecologically-sustainable technologies, whose great merit is their power of integration. Today, all of us must learn to study whole systems with our whole integrated selves. We see too, in the resurgence of interest and delight in the crafts, the limitations of other human satisfactions inherent in the welter of cooky-cutter, mass-produced goods. I have mentioned the hundreds of small new magazines of appropriate technology, entrepreneurship and self-sufficiency. All are highly recommended, together with the writings of a home-grown alternative lifestyle futurist, Murray Bookchin, who anticipated the alternative technology movement in *Our Synthetic Environment* published in 1962.[15] The study of alternative technological modes is also the study of how to integrate them into human cultures. A useful summary of such issues is *Technology as a Social and Political Phenomenon,* by Philip Bereano.

The dreams of technology-based hedonism, where machines would work and people would be trained for leisure, were, I believe, premature and based on inadequate ecological models. They will remain beyond our reach until we learn to control our population, reduce the impacts of our technology, and share our resources more equitably. It is time for the science and technology of industrialism to realize its conceptual limitations. The stakes have never been higher for human survival. We need a new scientific paradigm, more suitable for a science that stands on the brink of nuclear disaster and genetic manipulation, a science that has the potential of enhancing human evolutionary possibilities or turning us into a race of what John Leonard calls, "bionic junkies." This new scientific approach must be self-reflective. The old, innocent view—"scientist observing phenomenon"—will no longer suffice. Today we must pull back one "photo frame" and include a new composite view: "scientist observing *her or himself* observing phenomenon." This reflective paradigm exists in psychology, where those seeking to practice psychoanalysis on their fellow creatures must themselves submit to a prior psychoanalysis.

Now, perhaps, we should urge all scientists to undergo psychoanalysis, in order to understand their own deeper motivations, impulses and ego needs. Physicist Gerald Holton has commented on the personality types of people who choose scientific pursuits in his penetrating article, *"Scientific Optimism and Societal Concern: Notes on the Psychology of Scientists,"* in the Hastings Report of December, 1975. Or perhaps we should call a moratorium on giving Nobel Prizes in highly controversial and dangerous areas of research, such as nuclear physics, or recombinant DNA, so as to assure that incentives in this area are purely those of questing for truth, rather than personal ego gratification. Perhaps William Irwin Thompson is right when he notes that we need to rediscover the Pythagorean tradition in science rather than continue to pursue today's Archimedean science with its seeking to control nature through prediction and experimentation. Thompson notes that the Pythagorean approach embraced a mystical view of science that was integrated with art and religion. Fritjof Capra, in *The Tao of Physics,* draws the same conclusions and shows how physics has progressed in this century to an ever more metaphysical world view, more consistent with the direct, experiential insights of Eastern religious traditions than with the atomistic, mechanistic view characteristic of the now receding industrial age.[16] Dr. John Todd of the New Alchemy Institute typifies the new reverence which now must inform our scientific enterprise. He sees evolution not as pre-programed, but as a continual sacred dialogue with our planet. Radical horticulturalists Alan Chadwick and John Jeavons of California share this view and their methods produce astonishing yields. Such augmenting of natural productivity and reintegrating our activities into biospheric processes is already leading to a design revolution and a rethinking of many problems of production and of energy and materials management. Architects now design houses with "passive" heating systems, i.e., constructing and positioning the house to take advantage of natural solar and wind conditions that will reduce or eliminate the need for a heating unit. Production of fertilizers may not *require* factories, but can be approached by recycling animal and human wastes or by genetically engineering plants to augment their own nitrogen-fixing capabilities. Our own capabilities and imaginations will be stretched by our current crises. Millions of us are transcending our old fragmented viewpoints and rising to new levels of human awareness and many academic fields are undergoing creative ferment. Time is short; but we can all do no less than play our part in this human evolutionary struggle.

---

[1]Boulding, Kenneth, *The Image,* Ann Arbor Paperback, Univ. of Michigan Press, 1956

[2]Heisenberg, Werner, *Physics and Beyond,* Harper Torchbook, 1971, p. 96. Heisenberg's Uncertainty Principle states that the accuracy of measurement is limited in principle, i.e., as one tries to measure smaller and smaller particles or phenomena, the act of observation itself affects the object or process under observation. Heisenberg (1901-76) founded quantum mechanics, for which he received a Nobel Prize in 1932, and was director of the famous Max Planck Institute for Physics from 1946 until 1970.

[3]Skinner, B. F. *Beyond Freedom and Dignity,* Bantam paperback edition, 1971

[4]*New York Times,* April 4, 1976, "Two Zoologists Find Ants Using Tools"

[5]*Science,* "The Strategy of Ecosystem Development," Eugene Odum, April 18, 1969

[6]Mumford, Lewis, *The Myth of the Machine,* Part One, *Technics and Human Development,* Harcourt Brace Jovanovich, 1966, p. 23

[7]Weisskopf, Walter, *Alienation and Economics,* E. P. Dutton, 1971

[8]Schumacher, E.F. *Small Is Beautiful,* Harper & Row, 1973, p. 50-58

[9]Berry, Rev. Thomas, "Future Forms of Religious Experience," Information Paper #50, Futures Planning Council, Episcopal Diocese of California, 1976

[10]*New York Times,* March 26, 1976, "Postal Officials Say They Miscalculated on Parcel Damage"

[11]*New York Times,* Feb 3, 1976, "Three Engineers Quit G.E. Reactor Division and Volunteer in Anti-Nuclear Movement"

[12]Beer, Stafford, keynote speech before the Canadian Operations Research Society, Ottawa, Canada, 1974

[13]*The Ecologist,* "The Family Basis of Social Structure," Edward Goldsmith, Jan. 1976

[14]Fletcher, James C., Administrator of the National Aeronautics and Space Administration, speech to the National Academy of Engineering, Washington, D.C., Nov. 10, 1975

[15]Bookchin, Murray, *Our Synthetic Environment,* Knopf, 1962

[16]Capra, Fritjof, *The Tao of Physics,* Shambala Press, Berkeley, Calif., 1975

Based on the Damon Lecture given at the annual convention of the American Association for Industrial Arts, Des Moines, Iowa, April, 1976.

CHAPTER NINETEEN

Technology Assessment

In my role as a member of the Advisory Council to the U.S. Congress Office of Technology Assessment (OTA) since its establishment as an official scientific and technical research unit in 1973, I have shared in the intellectual excitement of this developing field of policy studies, its models, methods and underlying philosophies. The ideas expressed in this chapter grew from my interpretation of the many discussions and debates since the nineteen sixties in academia, in science policy circles and in Congressional hearings, as well as from my involvement in OTA.

I soon realized that it was more than a clash of interests. I was involved in the seesaw battles between the declining industrial culture and its opponents in the emerging ranks of the alternative-technology, the consumer and the environmental-protection movements, the citizen activists for social and economic justice and the counterculture; and the pitting of institutions against networks, a war of information and symbols. It involved nothing short of a major paradigm shift in the society and, as clearly as the new global resource constraints, augured the decline of industrialism itself and of its preoccupations with instrumental rationality, materialism and technological determinism. The entire 300-year period of the scientific enlightenment and the idea of "progress" that had flowed from the logic of Aristotle and René Descartes (itself a reaction to the dogmas of the Medieval church) had run its course. Thus, the discussion of technology assessment (TA) perforce became a discussion of the underlying philosophies of the "objective," "value-free" Western scientific method itself.

The evolving concepts and methodologies of technology assessment seem to be polarizing around two conflicting philosophies: (1) technology assessment should develop as an essentially "value-free" scientific discipline and (2) technology assessment is a normative process which must be recognized as rooted in and responsive to the dynamically changing values of the society on whose behalf it is conducted. Let us briefly examine the worldview and resulting paradigms associated with each approach. Firstly, the philosophy that sees technology assessment developing as a value-free scientific discipline appears to rest on Cartesian premises: that wholes can be understood by examining their parts: which in turn favors analysis over synthesis and leads to an overreliance on "objectivity," empiricism, quantification, and all the other paradigms associated with reductionism, still best summed up by Alfred North Whitehead as "the fallacy of misplaced concreteness."

The second philosophy that sees technology assessment as an essentially normative process recognizes that the notion of "value-free science" is an oversimplification that conceals the often unacknowledged value assumptions which influence the choice of phenomena to be studied and account for the idiosyncratic models and methods employed.[1] As Heisenberg's Principle demonstrates, the act of observation can change the object or phenomenon under examination. The "scientific" view of technology assessment is opposed by the "normative school" on additional grounds: its linear, extrapolative, quantitative modes of thought have led to policies which exclude the dynamic role of human values, and which often result in short-term, narrowly conceived, economic "maximization" of subsystems at the expense of larger social and ecological systems. And, as Garrett Hardin has noted, such policies also lead to technological determinism, "An implicit and almost universal assumption of discussions published in professional and semi-popular scientific journals is that the problem under discussion has a technical solution. A technical solution may be defined as one that requires a change only in the techniques of the natural sciences, demanding little or nothing in the way of change in human values or ideas of morality."[2] The fragmented policies of suboptimization and the values and perceptions that guide them can continue and be rewarded with positive feedback (i.e., profits, increasing technological mastery, and managerial control) until these subsystems begin to stress their social and ecological host systems. Then the feedback, often after dangerous lag times, turns negative. Indeed, as Alexander Christakis notes, "this often leads to a process which Gibbon described in the *Decline and Fall of the Roman Empire,* when the very forces of decay and collapse are legitimated as the proper behavior."[3]

Our particular perceptual myopia, which has focused on products not processes, on objects and entities rather than their broader fields of interplay, has led us to reward analysis and punish synthesis, to overvalue competition while ignoring the equally valuable role of cooperation, and to artificially define "problems" and devise "solutions" to fit them. As brain researcher Robert Ornstein has noted, the linear, rational modes of cognition are located in the left hemisphere of the brain, and accordingly produce the inductive, empirical, and often reductionist modes of thought and analysis that currently predominate in Western cultures. This predominance has led to an undervaluing of the cognitive functions of the right brain hemisphere, which processes data simultaneously and in spatial, rather than linear modes.[4] The right brain hemisphere controls intuitive, imaginative, artistic, and deductive thought processes, such as those insights which create the great hypotheses of science that can lead to pattern recognition of formerly inchoate data, and its eventual empirical validation or rejection.

A new synthesis of these two modes of human thought and perception is sorely needed in science policy formulation.[5] Pretensions of "value-free objectivity" in science are now threadbare, and yet science policy makers are naturally confused by the current, healthy reassertion of normative, nonquantifiable considerations as well as ecological concerns. These value shifts are now being elucidated by the many visible disruptions wreaked by technologically, economically, and bureaucratically deterministic cultures. It is precisely in the larger social and ecological host systems and in the overlooked fields of interplay between all our ill-fitting, disparate technologies and institutions, that public interest advocacy groups and social movements have sprung up. They have coalesced around the many dis-economies, dis-services, and dis-amenities we see industrialized societies are now producing. Such citizen movements for racial and social justice, environmental and consumer protection represent an inevitable and vital social feedback mechanism to correct the course of our society. These groups are reasserting values too often ignored because they cannot be rigorously stated or quantified in the prevailing empirical, reductionist terms. It is for this reason that such citizens groups realize that only political pressure can get the reductionists' attention sufficiently to reopen the debate over ends and values rather then means.

I believe that the current polarities in the philosophies of technology assessment may be seen as a part of this important and continuing debate between holism and reductionism. The scientific, technological, and professional groups would like to simply get on with the job of overhauling the methods of systems analysis, technology forecasting, cost-benefit analysis, and planning, programing, and budgeting systems and adding the requisite number of previously unnoticed variables to these analytical models, so that the composite result will emerge as technology assessment. On the other hand, the citizens groups and potentially impacted parties are determined that technology assessment shall be a holistic mode of discourse where the vigorous articulation of pluralistic values, ethical norms, and societal goals will allow all TAs to present the broadest range of technological and societal options to the electorate for democratic debate and resolution. By this definition, therefore, the question of public participation in technology assessment becomes tautological. In the view of citizen leaders, there is no such thing as technology assessment without public participation—indeed, they maintain that citizens groups *invented* technology assessment, rather than the technologists, who are now seen as opportunistically running off with it for intellectual fun and profit.

How shall we heal this polarity in philosophy when it runs throughout our still dualistic, reductionist, Cartesian culture? Technology assessment is a new opportunity to try to devise a disciplined method of incorporating both modes of thought, perception and research—just as many of the breakthroughs in science itself have successfully harmonized the brilliant, intuitive hypothetical insight with the equally necessary disciplined empirical process of its validation. If we are to achieve such demanding goals for technology assessment both the scientific and the normative schools will have to acknowledge their interdependence and approach each other and the process with mutual respect and openmindedness. The cognitive inputs into technology assessments of both scientific and value-articulating personnel should be accorded equal validity and respect, and rewarded equally in monetary terms, in spite of our existing cultural biases of academic accreditation.

All this will be difficult, not only because of traditional, Tower of Babel barriers to discourse between disciplines and between science and the humanities, but because great shifts of power hang in the balance. This recognition produces paranoia on all sides, as new priesthoods threaten old ones clinging to the thrones of the domain of science policy. Where the ends of knowledge are power, science becomes the servant of the powerful. The current debate over humanizing science and technology seeks to reexamine the goals of knowledge. Scientists, technologists, and professionals are essential to this debate, but they must be able to disengage from their operational roles sufficiently to see clearly where the limits of their technical competence end, and where their values carry no more weight than those of any other citizen in a democracy. It is for this reason that François Hetman in *Society and the Assessment of Technology* noted that all technology assessments should include at the outset an inventory of vested interests in the promotion or suppression of any technology to be assessed and an analysis prepared of the force field that such vested interests may create.[6]

Since citizens groups are still fearful that TA may simply become a new process of technology ratification or technology promotion, there will need to be: (1) a demonstrated intellectual commitment to open-mindedness and pluralism of approaches and (2) a continuous watchdog function performed by citizens groups monitoring the entire process. Methodologies for this unwieldy but necessary process are developing and will become refined with further experience. I do not underestimate the difficulties involved in developing TA as an open, participative process. But we should not shrink from the task merely because it is difficult. It is precisely because public policy has too often tended to follow similar paths of least resistance that we now face our multiple crises of suboptimization. A key concern will be that of developing political theories of equity and representation to legitimate such new forms of democratic processes. It now seems obvious that the channels for political expression and value-conflict resolution that sufficed in a simple agrarian democracy are no match for the greatly enlarged load that they must bear in a complex, mass industrial society.

We have reviewed some key political and philosophical rationales for public participation in technology assessment processes. Following from these rationales, voluntary groups of consumers, taxpayers, labor unions, minorities, environmentalists and poor people will inevitably demand to be involved in structuring the questions and formulating the policy options that constitute the process of technology assessment. (For purposes of definition and brevity, I will call "impacted parties" all those organized and unorganized citizens, taxpayers, consumers, workers, minorities and the poor who do not function primarily as instigators of technological change nor have significant financial or intellectual investments in the promotion of specific technologies. It should be noted that this definition runs counter to the prevailing production-factor analysis used in traditional market economics, as discussed in Part 1.)

Many still believe that public participation in TA is implicit in the fact that OTA serves Congress. In this view, Congress's use of and deliberations over the TA reports itself constitutes voter participation. While there is little contention on the part of impacted parties that TA is limited to structuring the questions and formulating the policy options while Congress preserves its role of deciding and legislating, there is some fear that a key issue may become submerged: that the crux of the TA process and the decision making that follows lies in the manner in which the policy questions are structured and the choices formulated. The growing awareness that information and the ways in which it is structured, analyzed, presented and amplified are key components in all political, technological and resource-allocation decisions have led to the determination of impacted parties to participate in the TA process itself. Therefore impacted parties are interested in being involved in the entire TA sequence: shaping the questions and selecting the priorities for assessment, designing the studies, examining the assumptions used in modeling the problems, making up the project teams, serving as members of advisory panels, reviewing ongoing draft reports and final results, bidding through their organizations for contracts to perform TAs.

Support for this broad interpretation of public participation in TA comes from many quarters. Senator Edward M. Kennedy, chairman of the Technology Assessment Board (TAB) has stressed a commitment to the fullest possible public involvement in TA processes through such means as full public disclosure and the mechanism of advisory panels and exploration of a variety of other avenues. The report of the National Academy of Science Panel recommended that TA processes be open to the widest possible range of responsible influence, including surrogate representatives of weak, diffused interests. And in testimony presented to TAB by Dr. Guyford Stever, the director, and Dr. H. Kenneth Gayer and Joseph F. Coates, then members of the National Science Foundation, emphasis was laid on the need for broadly based assessments so as to achieve credibility and for involvement of public interest groups in the conduct of TAs throughout the whole process of conception, organization, execution and final evaluation.

The question remains: "How is this participation to be channeled in orderly, logical and legitimate ways?" The Technology Assessment Act itself provides the basic prerequisites, particularly in providing for full disclosure of information, except if security statutes would be violated or TAB finds it advisable to withhold information in accordance with provisions in section 552(b) of Title 5 of the United States Code. In all other cases OTA conducts its activities in accordance with the rules of the House and Senate on public access to hearings and mark-up sessions as provided for in the Legislative Reorganization Acts of 1946 and 1970, the Freedom of Information Act (PL 89-487) of 1966 and the Federal Advisory Committee Act (PL 92-463) of 1972. In addition, the Technology Assessment Act provided for setting up the Technology Assessment Advisory

Council (TAAC) as a mechanism for involving members of the public in the activities of the office, to compensate for the elimination from H.R. 10243 (as reported by the House Science and Astronautics Committee) of TAB members who were not Senators or Representatives. At the first joint meeting between TAAC and TAB members on January 24, 1974, many TAB members stressed their conception of TAAC as a vigorous, independent body which could increase the diversity of views needed to develop TA as a broadly gauged policy research tool. And in the "Summary of Legislative History of OTA" in the staff study, *Technology Assessment for the Congress* (prepared for the U.S. Senate Committee on Rules and Administration), it is noted that "The continuing value of traditional adversary processes for supplying information and disclosing truth also will apply to Technology Assessment."

There are many additional methods of channeling public participation into TA. Some are easily implemented and have been initiated already, others will require experimentation and considerable trial and error. As we proceed with such experimentation, it should be remembered that as much can be learned from failure as from success. There is still some very understandable fear that either TA processes will be rendered disorderly or "unscientific" by such experiments or that the hopes of the public may become focused too heavily on TA as a new panacea for redressing past errors and humanizing technology. In addition, there are important questions how such new participation is to be legitimized and incorporated into the theories of political science, public administration and legal doctrines. In addition, while the requirements of broad information disclosure have been provided for, a remaining obstacle may be the claims of proprietary interests, patents, etc. In these probably inevitable clashes over what one might call "property rights" versus "amenity rights" in information, it seems clear that the public's right to know of potential impacts of new technologies must prevail over private rights to develop such technologies in secrecy pursuant to claims of market competition. Presumably, many of these issues will have to be decided by the courts.

Implementation of policies of full disclosure and broad dissemination of information on TAs, both before and after the fact, presents no unusual problems, and building up two-way communication with citizen/consumer, labor, public interest, professional and impacted groups is crucial to the intellectual honesty of technology assessment studies. Experiments such as the Public Interest Group Advisory Panel of citizens set up by Arthur D. Little, Inc. to review its Solar Energy TA contract for the National Science Foundation are useful, but they suggest that public participation should not be left in the hands of the contractor, but, as a general rule, should be overseen by the contracting agency or contracted for separately. In the case of OTA, this principle seems to be established, and contractors to OTA will not manage public participation in their work. This role will be assumed by OTA itself, by the advisory panels and by participation of TAAC and other oversight mechanisms.

Likewise, implementation of broad dissemination procedures for Requests for Proposals (RFPs) presents no problem. Notices of RFPs could be expanded from the current publication in the *Commerce Business Daily* to include, for example, *Consumer Reports* or a national newspaper such as the *Christian Science Monitor,* the *Wall Street Journal* or other publications. If, as Dr. Gayer suggested in his testimony before TAB, public-interest groups are to be considered as potential TA contractors or subcontractors, they must have access to RFPs simultaneously with traditional contractors. The methods employed by the Federal Environmental Protection Agency might be reviewed in this regard. EPA maintains very large mailing lists of public-interest and citizens' organizations, sends out regular *Citizen Bulletins* and frequently awards contracts to such organizations. In addition, wider dissemination of RFP notices might dispel the apparent "East Coast bias" of using Washington-based media exclusively, regional interest in TA might be stimulated and the infusion of new ideas and new bidders might enrich TA processes, which are recognized as more of a creative art than purely scientific.

OTA could assist those impacted parties that identify themselves as desirous of participating in various phases of the TA in question and begin the process of soliciting candidates and curricula vitae of scientific and non-scientific personnel proposed by such groups for appointment to advisory panels or as consultants or reviewers of the

TA as it is conceived and progresses. It will be through such means that OTA can build a large and valuable file of such "impacted party representatives" and their diverse skills and disciplines, and in the process create a network for increasing public understanding of OTA's role in illuminating science policy choices for Congress and the public.

One method of broadening public participation in TAs is the public hearing. It may be preferable on grounds of political theory to have such hearings conducted by TAB itself, or by an appropriate committee of Congress, i.e., composed of elected representatives of the voters, rather than by the office itself as an executive proceeding. Too often, agencies of the Executive Branch have been ceded the responsibility of conducting public hearings, which can become mere charades as administrators try to balance interests and resolve social conflicts more appropriate to the legislative process. Often such hearings result in concealing social conflicts on "utilitarian" or technical grounds and violate the canons of political and social-choice theory as well as the dictates of common sense. Such public hearings, if conducted by TAB or other appropriate committees of Congress could be particularly helpful in the crucial process of developing priorities in the possible candidate issues for technology assessment. Such broad public input on priorities could provide an educational experience to the electorate, and satisfies the important need for participation at the key stage of formulating the questions that the technology assessors will be asked. Such an open process of developing TA priorities is also likely to contribute more than any other single step to reducing the suspicion with which our scientific and technical establishment is now viewed by the public and to closing the distance between them. It might further reduce this sense of alienation if the public were to receive early notice of such hearings in accordance with the anticipatory thrust of TA itself. In conjunction with such hearings, experiments might be performed, such as polling public attitudes, using Delphis, computer conferencing and audience-feedback programs via public television. If the TA process is to fulfill its promise as an early warning system, it will need to experiment with eliciting "feedforward" from the public on what issues may be emerging, before they engulf the legislative process and become too "hot" for careful TA-type evaluation.

It is because normal political channels have become overloaded that demands for participation have become so urgent and that each promising forum or new arena, including the Office of Technology Assessment, is immediately politicized. Until more sophisticated concepts and methods of participation evolve, we must proceed by trial and error, employing and testing all our available hardware and software of conflict resolution, whether expanding the use of collective bargaining and mediation panels or using two-way cable TV for electronic town meetings and much more of the underutilized computer and communications systems, which could provide the basic hardware of democracy in an electronic age. Part of the rapidly developing "software of democratic participation" are the imaginative concepts of public interest law and economics and that of "advocacy science," as set forth in the statement "Criteria for Technology Assessment," prepared by a coalition of voluntary sector groups which in 1975 consolidated as the National Council for the Public Assessment of Technology in Washington, D.C. Dialectical cost benefit analyses will also be necessary to assure the comprehensiveness of technology assessment methods. In this way, costs and benefits, rather than being averaged out per capita and thus concealing serious potential conflicts, will be prepared not just for those who stand to benefit from the promotion of a technology, but also for those groups in society such as labor or consumers, that may stand to lose. It will also be necessary to place on all technology assessment advisory panels, task forces, and case teams technical and professional or nonprofessional people who reflect specific concerns, such as those of labor, consumers, the poor, or the environment. These people, while acting as stop-gap "human surrogates" for previously omitted variables, will help scientists still enmeshed in the false perceptions of "scientific objectivity" to appreciate their own often subconscious advocacy, their perceptual and cultural biases, and the unacknowledged normative nature of their disciplines and methods. As many of our most brilliant minds in every discipline from quantum mechanics to philosophy attest, reality is not fragmented into convenient segments or neat Cartesian dualities. Only an integration of subjective and objective styles, inductive and deductive reasoning, rational cognition and ethical striving can hope to develop technology assessment into a tool of policy synthesis for humanizing and redirecting the goals of knowledge.

---

[1]See for example the interesting discussion of various paradigms used by science in "The Ways Behind The Hows" by Ian I. Mitroff and Murray Turoff, *IEEE Spectrum,* Vol. 10, No. 3 (March 1973), pp. 62-71 (Institute of Electrical & Electronic Engineers, Inc.)

[2]Garrett Hardin, "The Tragedy of the Commons," *Science*, Vol. 162 (1968), p. 1243

[3]A. Christakis and P. Jessen, "Policy Planning for Humankind," *Fields Within Fields*, No. 11 (Spring 1974), p. 43

[4]Robert Ornstein, *The Psychology of Consciousness* (Viking Press, 1972)

[5]A brilliant exposition of this problem is made by Laurence H. Tribe in "Technology Assessment and the Fourth Discontinuity: The Limits of Instrumental Rationality," *Southern California Law Review*, Vol. 46 (June 1973), pp. 617-660

[6]François Hetman, *Society and the Assessment of Technology* (Paris: OECD, 1973), p. 85

Based on material prepared for the Sub-Committee on Methodology of the Technology Assessment Advisory Council of OTA and published, in part, in *Public Administration Review,* January, 1975.

CHAPTER TWENTY

The Decline of Jonesism

We have noted the preoccupation with hardware, technique and the material means of our lives—all of the "How" questions that have so concerned industrial cultures. As industrialism exhausts its logic, the pendulum swings again—to a reexamination of the "Why" questions.

The cultural assumptions and economic arrangements associated with the game of "keeping up with the Joneses" are in for a drastic reevaluation. In a world of growing scarcity and interdependence, this sport—so popular among the middle and affluent classes in rich countries—appears to have reached its limits. Jonesism is becoming a steadily less viable motivating force for economic growth, since such mass-consumption strategies in already affluent nations require an unfair portion of the world's diminishing resources and prevent needed economic growth in the less affluent countries. The result is envy and greater international tensions in our global village, because only a small percentage of the world's growing population can hope to achieve such high levels of consumption.

There are many levels of Jonesism. Governments indulge in an international version of it with their state airlines and big-bang technologies, such as rockets. There is corporate and organizational Jonesism, typified by the proliferation of glittering glass skyscrapers, thick carpets, self-conscious art collections in the managerial aeries, sleek cavalcades of limousines and corporate jets—all sensitively tuned to the needs of status, prestige, and display. Perhaps most interesting of all is consumer Jonesism, exquisitely interrelated with and reinforced by all the other varieties. It is there that Jonesism begins and there it may end, as our individual perceptions and values change.

Let us now take a closer look at Jonesism in all these forms and examine the new backlashes which are now unleashing counter forces.

A brief glance at the history of Jonesism may help set the stage for our inquiry into its present and probable future expressions. The conspicuous consumption of surplus has been a continuous theme of human behavior since the building of the pyramids. What is relatively new today is that the excesses of kings and emperors now provide a general model for emulation by large numbers of people. Before the industrial revolution, a level of mass consumption beyond that necessary for subsistence was almost unthinkable. Surpluses, when they occurred, tended to be relatively small and unpredictable, dependent on cycles of harvests and military plunder. The distribution of surpluses was, therefore, hardly a central preoccupation and was often imaginative and idiosyncratic, whether dispensed in feasting and potlatches or used for conspicuous displays of the rulers' power or the groups' prestige, or for religious and artistic expression.

It was not long after industrialism began generating larger and more predictable surpluses that the first critics of the early expressions of Jonesism began to voice doubts. Marx pointed to the social hardships that industrialism created and focused on the inequities in the distribution of the surplus that was produced. As American industrialism developed into the opulence of the Gilded Age, Henry David Thoreau, rejecting the pursuit of consumption, presented his fellow citizens with the alternative values expressed in his famous book *Walden*. Scandinavian-born economist Thorstein Veblen took up the torch and railed against the vapid excesses of the rich in his book, *The Theory of the Leisure Class*, published in 1899. Even economist John Stuart Mill and the father of modern growthmanship, John Maynard Keynes, predicted an end to the continual increase in the production of goods and the eventual satiation of human wants. Underlying all such critiques of materialism were the teachings of major religions, which had always admonished against the seeking of earthly riches and celebrated the search for spiritual values and the golden rules of cooperation, simplicity, humility and love. The futility of Jonesism was perhaps best expressed in the famous text, *Desiderata*, found in a Baltimore church, dated 1692, which stated simply, "If you compare yourself with others, you may become vain and bitter; for always there will be greater and lesser persons than yourself."

Jonesism Afflicts Poor Nations. Now let us turn to the contemporary phenomenon of Jonesism and its expression in international, national, organizational and individual behavior. International Jonesism is as old as the nation state, and its symbols, such as public monuments and buildings, new cities such as Brasilia, well-equipped military forces, high technology, and membership in the United Nations, are all familiar.

It has widely been assumed that all nations could eventually pass through the stages of economic growth, as described by economist Walt W. Rostow, and reach the mass-consumption phase, and even the heights of sociologist Daniel Bell's "post-industrial state." But this kind of international Jonesism seems less viable in a world with a population now roughly four billion (and set to reach some six billion in twenty-five years) and with increasing scarcities of resources. World-wide inflation is making military and public-works expenditures more burdensome while adding little increase in national security.

In addition, Western-style industrial development often seems to create problems more onerous than those it seeks to ameliorate. While the need for development in poorer countries is indisputable, capital-intensive growth makes these nations dependent on rich countries for both capital and technology and renders indigenous workers inadequate, without costly training, for all but the lowliest jobs.

With such losses in national self-sufficiency come other problems—such as migrations from rural areas to explosively growing cities and often, greater maldistribution of wealth and unemployment. New and expensive infrastructure must be provided: railways, roads, communications, sewage treatment and publicly subsidized housing. And as indigenous consumers awaken to advertising transplanted from the affluent world, they often are lured into spending their meager incomes on marginal goods, such as patent medicines, candy or such inappropriate substitutions as costly infant formula for mother's milk, often with dire nutritional consequences.

Intermediate, labor-intensive technology is a viable development alternative. Capital-intensive industries can never provide enough jobs, while simpler, less expensive technologies can create rural self-sufficiency, raise individual productivity, stop the cancerous growth of cities, reduce social inequities and the dependence of poor countries on rich ones. Following the concepts of Schumacher, Illich, Bookchin, Dickson, Lappé, and others arguing for development based on Mahatma Gandhi's idea of production by the masses—not mass production—many non-governmental organizations are planning to introduce them into the U.N. Conference on Science and Technology in 1979. The traditional goals of international Jonesism may have been modified by China's labor-intensive, rural model, which is now exciting great interest because it employs the resource that all less developed countries have in abundance, the labor and skills of their people.

Jonesism in Wealthy Countries. Now let us turn to Jonesism as it is practiced in the affluent nations of Western Europe, North America, and Japan, where it is encouraged as a key motivating force in domestic economic growth. A typical example of this form of Jonesism occurred in the recession of 1958-9 when President Eisenhower approved a saturation advertising campaign that used the slogan "You Auto Buy Now." The social costs of Jonesism and the continual force feeding of mass consumption through advertising and planned obsolescence were first widely proclaimed in such books as J.K. Galbraith's *The Affluent Society*, Vance Packard's *The Waste Makers, The Hidden Persuaders* and *The Status Seekers*, and Marshall McLuhan's *The Mechanical Bride*. These critics all noticed that our society, while well supplied with cosmetics, plastic novelties and tailfinned automobiles, seemed to be generating social and environmental costs, such as polluted air and water, decaying cities, disrupted social and community patterns—as well as drug addiction, crime, excessive mobility, rootlessness and other effects which Alvin Toffler later summed up in *Future Shock*.

The idea that the world's resources might not always be equal to the task of fueling helter-skelter material consumption has come as a rude shock to Keynesian economists. The successful Organization of Petroleum Exporting Countries (OPEC) may be considered as writing on the wall; already other producing nations are following suit in exacting higher prices for their resources.

Even if infinite substitutions can be found—as most economists hope—prudence now dictates a less wasteful economy geared more to filling the basic needs of poorer citizens than encouraging the overconsumption of the middle and affluent classes. The continual pursuit of superaffluence in rich countries, as Robert Heilbroner speculates in his book, *An Inquiry into the Human Prospect*, may lead to increasing sabotage and international blackmail of rich countries by Third World terrorists as well as increasing the probability of wars. Barry Commoner noted in *The Closing Circle* this growing moral dilemma of rich nations, whose economies use the lion's share of the world's resources and create most of the pollution.

The coming world food shortage juxtaposed with increasingly rich meat diets of affluent consumers highlights this dilemma. Since meat production requires large quantities of feed grains, the consumers in affluent nations eat many times more grain per capita than people in the Third World. A reduction of meat consumption would release grain wastefully used to fatten livestock for direct human consumption by the world's needy. Lastly, not only must we start leveling down our own consumption of all resources, but we must also prepare ourselves for a geographical redistribution of production to other countries with indigenous resources, large, eager labor forces, and relatively unpolluted environments.

Businesses Will Tell Customers Not to Buy. How will the decline of Jonesism affect corporations, which have played a key role as employers, producers, advertisers and cultural standard setters in reinforcing this human tendency? For one thing, corporations themselves will likely have to begin helping consumers get down off the Jones trip. Having raised the Frankenstein monster of demand through the over $20 billion they spend each year on advertising, the corporations now increasingly find themselves unable to deliver the goods. The electric utilities have been the first hit because of their primary role in powering other production processes as well as consumer uses. Because their primary inputs of coal, oil and natural gas are now in short supply and are now priced more realistically (including a greater measure of their social and environmental costs) the utilities are squeezed between soaring construction and operating costs, and consumer resistance to rising rates. When it became clear to them that they could no longer increase capacity profitably, they began to reduce their advertising and promotion and started asking us not to use their product.

These "de-marketing" campaigns run by utilities were a forerunner of things to come. When the energy crisis came into focus in the winter of 1973-4, the oil industry found itself in similar straits. After years of gasoline price wars and frantic sales promotions, the American people had almost come to believe that unlimited supplies of cheap gasoline were their God-given right. Overnight the slogans changed to "Save Gasoline," "Oil is Precious," "Don't Be Fuelish." Philip Dougherty of *The New York Times* stated that no issue since the end of World War II had generated as much advertising as the energy crisis.

The result was that the public was furious. Consumer groups claimed that the companies had created the crisis in order to reap windfall profits. Environmentalists charged the companies with using the crisis to accelerate wasteful, crash programs for exploitation and control of new energy sources and to scuttle the environmental legislation of a decade. The furor led to angry hearings in the Senate and cries to break up the companies and reduce their jealously guarded depletion allowances.

Not surprisingly, the whole debacle gave a boost to the fledgling movement for counter-advertising, as civic groups tried to obtain media time under the Fairness Doctrine to air their own views. Consumption of oil, gas and all energy had become, like cigarette consumption before it, a controversial political issue rather than a simple matter of merchandising.

In the future, more and more types of consumption will become controversial and will embroil the producers in counter-advertising campaigns and political turmoils, because consumption creates as many external costs as does production. In some cases, consumption of specific items will seem irresponsible, just as the wearing of furs of endangered species is viewed today. Similarly, whole classes of products that are excessively resource-intensive will be ostracized in the same way that many people today reject large, overpowered cars and throw-away bottles. Corporate advertising playing on psychological insecurities and motivations for status and visible success symbols typical of the Jones syndrome may be called to account.

A story in *Forbes* documented the saturation of the U.S. cosmetics market and the now declining profitability of cosmetics companies. Baroque proliferation of ever more improbable beauty products has finally led to rejection of such "bondage," as one corporate survey response termed it, and the return to favor of natural appearances. Already two professors of marketing at Northwestern University, Philip Kotler and Sidney J. Levy, are trying to brace corporate executives for the unfamiliar world of de-marketing, and have published an article in *Harvard Business Review* entitled "De-Marketing, Yes, De-Marketing." Advertising executive Jerry Mander of San Francisco believes that if advertising is to survive as a major industry it had better get into the business of selling social issues and promoting public service organizations.

For some industries faced with imminent supply scarcities, such as those in petrochemicals and aluminum, the writing is on the wall: De-market, or else! For when consumers have been led to expect life with limitless quantities of cheap paper products and cooks have been hooked on aluminum foil, the reaction to the withdrawal of these items is anger. The problem for companies is that we have all become so inured to the daily advertising barrage and its thousands of petty deceptions, that it is hard to believe companies which say that their supplies have been cut off or that prices have become too high to permit profitable manufacture of some products.

The only answer for such beleaguered corporations will be to let others be the messengers who bring the bad news. This is the role that environmentalists and consumerists have been playing for years. Corporate advertising has proclaimed the good news while avoiding the bad news: it hailed the labor-saving appliances and pushed junk foods, while overlooking their effects: over-weight, unhealthy, flabby consumers. While advertisers have touted the advantages of fast cars and energy-gulping appliances, environmentalists have reminded us of the price we pay in more nuclear plants, strip mines and emphysema. Therefore, the most viable strategy for corporations as they become caught in supply binds will be to divert a portion of their advertising budgets as grants to appropriate voluntary organizations, whether consumer, environmental or other community groups, so that these organizations can mount the needed de-marketing campaigns and serve as credible messengers to bring home the bad news. Such groups would not shirk the task of explaining the new need for hard trade-offs and choices, because they are hardened to controversy and themselves advocate changes in lifestyles, values and economic arrangements.

As such lifestyle changes are mandated by declining energy and resource availability, some corporations may not survive the shifting patterns and may follow the buggy-whip makers into oblivion. Others equally heavily dependent on resource-intensive consumption may have the political power to force the taxpayers to bail them out in the manner that several aerospace, airline and utility companies are attempting today. Those that negotiate the transition to the steady-state economy will be those willing to serve public-sector needs, such as mass transit and recycling, and to minimize resource use by emphasizing durability rather than obsolescence. Corporations may also have to be content with modest profit margins because the companies will have to internalize more of the social costs of production and consumption. And as noted, energy for transportation will be more realistically valued, and overcentralized production by giant corporations will become less efficient than smaller regional and localized manufacturing serving de-centralized markets.

Now let us turn to the individual and examine the basic motivations for the Jones syndrome. Sigmund Freud identified narcissism as a crucial factor in human behavior and its many expressions in display and status or recognition seeking as a validation of the individual's significance and very existence. We all desire in our own way to write our particular version of "Kilroy Was

Here" on the course of human events and to etch some mark on the natural world around us. Norman O. Brown in *Life Against Death* and Ernest Becker in *The Denial of Death* both suggest that all of our history can be interpreted as the saga of human striving to validate the importance of our existence and to overcome our fears of death and nothingness by our frantic manipulation of each other and of nature.

But the expressions of such drives as the material acquisition and overconsumption of modern-day Jonesism is a fleeting phenomenon associated with a relatively brief 200-year span of industrialism, which may now be waning. In 1937 Karen Horney warned of the psychological toll of Jonesism on the American citizen. In her landmark study, *The Neurotic Personality of Our Time*, she noted the characteristic neurosis produced by cradle-to-grave competition to keep up with the Joneses is associated with three dilemmas: aggressiveness grown so pronounced that it cannot be reconciled with Christian brotherhood, desire for material goods so vigorously stimulated that it cannot be satisfied, and expectations of untrammeled freedom soaring so high that they cannot be squared with the multitudes of responsibilities and restrictions that confine us all.

The Revolt Against High Consumption. In 1950 David Riesman examined the personality traits and sociology of an increasingly abundant and mobile society in *The Lonely Crowd: A Study of the Changing American Character*. He mentioned that as social change accelerated, values shifted so rapidly that individuals, rather than relying on the "gyroscopes" of their own inner-directed principles, began to rely on their "radar screens" and become other-directed, constantly shifting course to conform to the tastes, opinions and values of their peers and society. As opportunities for material gain and upward mobility in consumption styles increased, other more traditional roles and niches which conferred status, dignity and respect in other ways fell before the single, dollar-based success standards of Jonesism. In 1954 David M. Potter in his book, *People of Plenty*, provided an insightful summation of the cultural, social, economic and individual drives that contributed to our distinctive American culture.

In the mid 1960s the rumblings against the tyranny of Jonesism became overt, built on these earlier insights and the emerging consciousness of the "beat generation" rebels such as Jack Kerouac, Allen Ginsberg and others, whose lifestyles were sensitively examined by sociologist Robert Jay Lifton. The student revolt against what they saw as the frivolous, meaningless acquisitiveness of their parents, which exacted an exorbitant price in corporate conformity and loss of self-fulfillment and personal growth, soon led to the development of the flourishing counter-culture of today. Overconsumption and status-seeking goals began to be viewed as obscene, and the advertisers of deodorants and grooming products, playing on deep psychological fears and insecurities, contributed to the backlash of the shaggy, gloriously unwashed "hippies." Old clothes and artifacts have become the counter-culture's symbols of status, and riches in discretionary time are prized more highly than money income. Environmentalists soon joined in the movement for simplifying lifestyles as they understood the destructive role of resource-intensive production and consumption on the environment.

Swedish economist Staffan Linder's popular treatise, *The Harried Leisure Class*, pointed out the anomalies of Jonesism, in that as affluence increased, time became more scarce. Poor people had plenty of time, but the affluent came to resemble hamsters on an exercise wheel with no time to enjoy the hard-won fruits of the rat race. Linder showed that consumption takes time in the same way that production does, a point often overlooked by economists, and that the time required to use and maintain our cars, boats, swimming pools, campers, and gear for skiing, tennis, golfing and hiking provides an upper limit to consumption. As time becomes scarcer for the harried overconsumer, such delights as the leisurely unfolding of affairs of the heart, tranquil reading, unhurried contemplation of where we are going and why, give way to revolving-door sexual encounters, singles bars, speed reading, business-related vacations and the tyranny of the clock.

In his book, *Alienation and Economics* (1971), Walter Weisskopf reached similar conclusions and pointed out that for humans, the real dimensions of scarcity are not economic but existential. Time and life are the ultimately scarce resources, because of our mortality. Such psychic needs are similar to those described by psychologist Abraham Maslow: peace of mind, love, self-actualization, community, and time for leisure and contemplation. Kenneth Boulding has highlighted our propensity to believe that the rights of private property ownership permit us to use up rather than merely to use resources, even though their utility to us is not thereby enhanced nor diminished if we reuse and recycle them. Now our concerns are also moving beyond property rights into consideration of amenity rights, which are often violated by the consumption activities of snowmobilers, beach buggy riders, hi-fi buffs, and transistor radio lovers.

There are many signs that the need for getting down off the Jones trip is now widely understood. The energy crisis spurred the trend toward small cars, bicycles and mass transit. Appliances are now rated for their energy economy and people are rediscovering the pleasures of home food growing and physical exercise. The popularity of survival manuals such as *The Whole Earth Catalog, Living Poor With Style*, and *Diet for a Small Planet* is not only due to their educational content on how to reachieve basic self-sufficiency when the crunch comes, but also because they articulate the values and pleasures—not to mention the psychological relief—in store for those who kick the Jones habit.

Can we humans mature sufficiently to transcend Jonesism's basic motivations or must we be content with redirecting these human drives toward less materialistic and self-destructive goals? Perhaps it will be sufficient to alter the goals and symbols of success so that our narcissism is expressed in self-actualization and in reintegrating our self-images, so that competition may be channeled into enhancement of physical fitness and well-being, and acquisitiveness may reemerge as striving for knowledge and higher levels of consciousness. Status and recognition can be achieved in other ways; for example, many societies confer medals and symbolic status for social achievements. The British dispense knighthoods and peerages while the Russians and Chinese offer symbolic rewards for heroic feats of production and service to the people. In satisfying needs for display and expression, we could learn from many so-called primitive cultures in their imaginative use of body adornment, color, dance ritual and festivals to objectify and celebrate our collective emotional and spiritual yearnings. If such outcomes were the result, then getting down off the Jones trip might open up new vistas for cultural exploration.

Reprinted from *The Futurist*, October, 1974.

CHAPTER TWENTY-ONE

Citizen Movements for Greater Global Equity

As the new ranks of aware citizens in overdeveloped countries learn to kick the "keeping up with the Joneses" habit, their growing planetary awareness raises their concern for greater global justice. The role of international nongovernmental organizations has long been viewed as beneficial to the goals of world peace and social development.[1] The proliferation of such private, transnational groups, whether religious, professional, scientific or devoted to cultural communication and exchange, is a familiar phenomenon, and even though most of these groups are based in industrial countries, they are also on the increase in other parts of the world. Many such groups were founded by women who, although domiciled in rich countries, shared the concerns of women everywhere for better health, living and working conditions for their families and the nurturance of their children. The United Nations International Women's Year observance in Mexico City spearheaded a new level of women's concerns. It helped raise the consciousness of economic and social development planners to inequities that, because they fall hardest on women, lead to the exacerbation of many other problems; from food production, malnutrition and population to unemployment and family breakdown.[2]

It is interesting to note that these nongovernmental organizations formed over the past fifty years by women, their proliferation in many countries and subsequent convergence on world problems and the restructuring of policies to address them, are a prototype for international action. They have served as models for many similar

THE FOURTH WORLD
Gulf
ITT
NORTHROP
UNITED BRANDS
EXXON

efforts, for the environmental movement, as well as groups concerned with broad problems such as peace, population, food, health and shelter. The role of nongovernmental organizations in ventilating official government debates over these global functional issues has proved beneficial in the United Nations-sponsored special conferences on the human environment in Stockholm (1972), population in Bucharest (1974), and food in Rome (1974), and enriched the habitat conference in Vancouver (1976) with desperately needed unorthodox, "lateral" thinking and new ideas. The concept of having a continuous "forum" or "tribune" operated by nongovernmental organizations at the United Nations itself holds similar potential, at a time when international policies on a host of issues need innovation and reconceptualization.

While the role of international nongovernmental organizations has been studied frequently, much less attention has been paid to indigenous citizen movements in the industrialized countries and their potential for creating a more equitable world order, as these countries make the inevitable transition from growth to steady-state economies. Particularly, I will focus on how the extensive research developed by these citizen movements for consumer and environmental protection, economic and social justice and corporate accountability can be made available to developing countries, so as to improve their ability to assess the wider range of actions as prospective host countries to new enterprises. As these industrial countries, the United States, Japan and those of Western Europe have reached high levels of development, they have produced not only extensive infrastructures and material goods but also broadly educated populations, assertive consumers and workers, all of whom demand not just material well-being, but better working conditions, health care and living environments and insist on greater participation in not only political but economic, decision making. At the same time, economic growth in these countries (as measured by Gross National Product) has not adequately addressed social costs and has led to increasing levels of environmental pollution and social disruption and maladjustment.

All of these factors are now inevitably slowing real economic growth in all highly industrialized economies and contributing to its redistribution to other regions of the world. In fact, it is these nonmaterial limits to growth, generated by such internal factors, as well as the external reordering of world trading relationships implied by a new economic world order, that are now slowing growth in industrialized countries, long before actual resource limits are reached. Contrary to official Western views, it is likely that the demands for a new economic world order will be as beneficial for overdeveloped countries as they will be for developing countries in the rest of the world. Many citizens are now aware that overconsumption of even important nutrients can be fatal—and overconsumption of material goods can exact a heavy price in the loss of other social and environmental amenities. In addition, such a new economic world order, where industrialized countries will be forced to pay more realistic prices for their raw materials, will help them to adjust more smoothly to the advent of actual depletion of many resources. If no such political pressures were present to reduce their supplies of cheap resources, many industrial nations might have collided headlong with such scarcities, thus causing even greater disruption, since market prices alone cannot provide sufficient lead times for readjustment. Expressing a view widely shared by environmentalists, Dr. Howard T. Odum of the University of Florida's Energy Center, has said:

> In 1972, the Organization of Petroleum Exporting Countries deserved the Nobel Prize, since they did more than all the schools, universities and leaders to educate Americans, Western Europeans and Japanese people about the need to squeeze the waste out of their economies.[3]

Yet it will take much time for the lesson to sink in, since most of these countries are still officially committed to the concept of force feeding industrial growth by deficit financing, tax cuts and all the other Keynesian remedies for stimulating aggregate demand for goods and services. It seems that the higher the echelons of national leadership at which such debates are conducted, the more devoid of new ideas they become. While leaders intone banalities, less exalted NATO country representatives have met privately as individuals and discussed "off-the-record" in highly realistic terms the bankruptcy of

Keynesianism.[4] But due to official silence on such matters, these strategies by industrial nations have become conventional wisdom and creative challenges rarely emanate from the academic community, but rather from citizen and public-interest research groups and labor unions.

And yet it now seems clear that attempting to pump up their whole economies in order to ameliorate unemployment and mask inequities in distribution is too costly in resource consumption and simply leads to supply bottlenecks and raising inflation rates. Similarly, deflating their whole economies can no longer reduce inflation which is now structural, and due to large producers' ability to increase prices and reduce production; to the rapidly mounting social and environmental costs now falling due; to many government regulatory agencies that must be supported by taxpayers, even though in the United States many of them have become little more than producer cartels. Furthermore, the sharply declining productivity of capital investments and its role in inflation is due to new forces that can only be understood from beyond the discipline of economics and are rooted in the inexorable laws of physics. This declining productivity of capital investments, visible, for example, in the energy sector, is due largely to a declining indigenous resource base. For example, a British forecast noted that in spite of the billions of pounds being invested in the North Sea, these new oilfields will maintain peak production for only a short time, and that Britain would only achieve self-sufficiency for the five year period 1980-85.[5] Thus as more and more capital must be ploughed back into obtaining resources from ever more degraded and inaccessible deposits, there is a sharp fall-off in productivity in all industrialized countries. We see this in their increasing rates of inflation and balance of payments deficits. The U.S. trade deficit is estimated by Treasury Secretary Blumenthal to reach $30 billion in 1977, four times the 1972 deficit of $6.4 billion.

Clearly, the problems of most Western industrial economies, as well as that of Japan, are now structural, with inflation as a permanent feature. Indeed, these mature industrial economies are now characterized by levels of technological complexity and social and economic interdependence that produce the typical "viscosity" and unresponsiveness described by Adolph Lowe in *On Economic Knowledge*.[6]

As I have argued, these mature industrial societies may not see the bright, clean, knowledge and service-based future predicted by Daniel Bell as "the post-industrial state," but rather, may reach a climax in their potential which I have called "the entropy state." Today in the United States, we see large segments of GNP and employment rest on wasteful, resource-intensive private goods, such as automobiles, washing machines and refrigerators, and on nonviable technologies, such as aerosol cans and products based on polychlorinated biphenyls and vinyl chloride. Some 25 per cent of United States water supplies are polluted beyond government standards[7] and incidence of cancer has risen to almost epidemic proportions,[8] as well as incidence of other diseases of industrialism, such as heart disease, hypertension, obesity, addiction and mental illness.

All of these social factors are, indeed, the real limits to growth in industrial countries, as well as our conceptual limits in managing growth. These conceptual limits go beyond our inability to model and therefore to manage these sociotechnical systems we have created.

Most of the citizen movements in North America, Europe, Japan and Australasia have organized around these mounting social costs, which are now seriously distorting these economies and rendering them almost pathological. These so-called "externalities" have now reached above the thresholds of sensory awareness of millions of ordinary citizens. Official reassurances that all is well are met with increasing disbelief as they are told that they must now surrender some of their precious civil liberties and gamble with their children's futures to be "served" with more nuclear-generated electricity. These aroused citizens are by no means all mindless young technophobes or radicals. Well-dressed, clean-shaven, middle-class businessmen and their suburban wives comprise the major forces in California fighting against nuclear power. Hundreds of thousands of middle-class mothers are bringing massive pressure to ban commercials and violent programs from children's television. Hundreds of highway projects, federally funded to stimulate the economy, have been stopped by skillful, determined citizen opposition, while similar citizen efforts have diverted such government funds to less wasteful, mass transit and the development of clean, safe, nonpolluting energy sources, such as solar, methane conversion, wind and water power. Such citizens in all industrial countries are beginning to see, in spite of the mystifications of economists, that capital-intensive technologies reduce employment at the same time that they increase structural inflation and bind workers into greater dependence on giant corporations and vast technological enterprises.

Many are unconvinced that we must consume even more of the world's resources and apply more Keynesian adrenalin in order to keep other industrial and developing countries from falling into recessions. In a resource-short world of rising population, such short-term, unsustainable strategies are becoming addictive and self-defeating. Only a redistribution of production, rather than consumption, can in the long run assist developing nations, so that they may add more value to their own resources and develop their own markets. As the gap between rich and poor countries widens many are persuaded by such educators as Ivan Illich and Frances Moore Lappé, who claim that only by renouncing industrial expansion can the industrial nations help to bring food and population back into balance in the developing countries.[9] Many citizens now see what their leaders have not yet recognized: that capital-intensive technologies tend to centralize power, wealth and knowledge in fewer and fewer hands at the expense of increasing aggregate human ignorance and social and economic inequality. They also understand that traditional market economics seriously underestimates social costs and that prices do not reflect these costs, while markets are becoming ever less useful in allocating resources, since they cannot deal with future scarcity situations and only operate on effective demand, rather than actual human needs.

Naturally, at first these citizen movements organized around simple, tangible social costs, such as air and water pollution. Others organized around issues of peace in Vietnam, social and racial justice, sexual equality and consumer concerns. However, in the past five years there has been a broad convergence which has moved beyond fragmented views of isolated problems and the belief in the efficacy of "add-on" technological fixes, to more holistic and systemic concerns.

Most of the new awareness concerns the role of large oligopolistic corporations in controlling resource allocations and their influence on political decisions, as finally revealed widely during Watergate investigations. For example, formerly conservation-oriented organizations such as the Sierra Club, the Wilderness Society and the Conservation Foundation now also concern themselves with economics; national church organizations study corporate social performance and accountability, while Ralph Nader's initial concern with automobile safety has grown into a million-dollar conglomerate enterprise funded by small, individual contributions, which covers a range of systemic concerns from the drive to control corporate behavior by federal chartering, to battling for tax and regulatory reform and funding citizen and student-activist groups across the country. Many of the newer citizen groups reflect the new broader concerns: the Council on Economic Priorities, the Investor Responsibility Research Center, INFORM and the hundreds of stockholder proxy campaigns that take place every year at corporate annual meetings, all highlight the unaccountability of corporations, while the Movement for Economic Justice and the Exploratory Project on Economic Alternatives and similar local groups focus on economic alternatives, and the Peoples' Business Commission advocates worker control of corporations. Others focus on broad functional areas, such as alternatives in transportation, energy, communications, self-sufficient technologies, rural development and community land ownership, while for the first time since 1880, the United States population is now migrating back to rural areas.[10]

Meanwhile, the phenomenon of citizen challenge and ridicule of "experts" is forcing paradigm changes in a host of academic disciplines. Theoretical crises are evident in sociology, psychology, economics and even physics. Thousands of dissident academics, scientists, engineers, lawyers, economists, accountants and management consultants have now joined the ranks of citizen movements or do voluntary "moonlighting" work for their causes. Radicalized professionals modeled after Nader's public-interest lawyers and activated by the nuclear and antiballistic-missile debates of the 1960s now include such organizations as the Scientists Institute for Public Information, the Federation of American Scientists, the Union of

Concerned Scientists, the Center for Science in the Public Interest, Accountants in the Public Interest, the Public Interest Economics Foundation, the Committee for Social Responsibility in Engineering, as well as such international professional groups as the International Society for Technology Assessment and many similar activities in Western Europe and Japan. All of this research on social costs and social and environmental impacts is a rich resource waiting to be exploited by developing countries and which may enable them to avoid some of the costly social pitfalls of economic growth.

The extent of the political strength of such movements in the United States is extensive. Even by 1973, the Harris Poll found that 87 per cent of Americans agreed that "it is good to have critics like Ralph Nader to keep industry on its toes."[11] Another study by Ronald Inglehart, *The Silent Revolution*, reveals a wholesale shift toward "post-materialist" values in all mature industrial nations, and toward greater concern for quality of life, pursued with skillful political activism.[12] Another measure of the numbers of such social change advocates is the explosive growth of the publications that cater to their concerns and the commercial success of such journals as *The Whole Earth Catalog,* and *Ms.* as well as black journals such as *Ebony, Black Enterprise* and *Essence* and a host of regional publications. Radical social critics such as Illich, Nader, Schumacher, Jacques Ellul and Lewis Mumford, have become international best-sellers along with Richard Barnet and Ronald Muller's *Global Reach* and the now famous studies of the Club of Rome. A skeptical advertising campaign launched by the Center for Growth Alternatives and the Public Media Center, under the theme "We Can't Grow On Like This," has been aired in free public service time by 800 radio and 150 television stations. Economist Robert Heilbroner in his best-seller *The Human Prospect* notes the growing skepticism, formerly the concern of élites, as to whether material acquisition leads to greater satisfaction. A new report, entitled *Another Sweden* by Gorhan Backstrand and Lars Ingelstam of the Swedish Secretariat for Future Studies, suggests new goals for their country which would contribute to a more just global distribution of resources; reduce inequalities in Sweden itself; be influenced by what visitors from the Third World consider to be the most provocative aspects of affluence

WE CAN'T
GROW ON
LIKE THIS.

We've always operated on the assumption that bigger is better. But is it?

Like the dinosaurs, societies and economies can grow too big for their own good.

America is fast approaching that point. The natural resources we need to live – clean air, water, land fuels, metals – are getting scarcer. Some are on the verge of extinction. Others are becoming prohibitively expensive.

At the same time we're wasting tremendous amounts of these precious resources. And our wastes pollute our communities, our nation, our world.

We need to learn to use our resources efficiently and economically and to share them better so that everyone gets a piece of the pie.

We need to conserve the raw materials that jobs depend on, because if we deplete our resources now, things will be that much tougher later.

We need to put people to work *doing* things instead of just making things. The things we *do* make have to save resources instead of wasting them. We can build mass transit instead of freeways, rebuild our cities instead of spawning new suburban sprawl, put people to work cleaning up our environment instead of despoiling it. Harsh prescriptions? Maybe. But ones that will assure a more prosperous future.

For a better tomorrow, let's stop using resources like there's no tomorrow.

CENTER FOR
Growth Alternatives

1785 Massachusetts Avenue NW
Washington, D.C. 20036
202/387-6700

A television/radio public service advertising campaign is available on growth and resource consumption. To order, write Public Media Center, 2751 Hyde Street, San Francisco, CA 94109, or call 415/885-0200.

Public Media Center, San Francisco, Ca.

in Swedish life and set not only a minimum income, but a maximum level of *per capita* consumption as well.[13] Traditional food aid and CARE programs have now been attended in many industrial countries by new soul searching as to their own implications in the world food crisis. Awareness has been raised as a result of this and the United Nations debates on the new economic world order, as to the maldistribution of the world's wealth and the extent to which the overproduction, overconsumption and overpollution of the over-developed countries has created or exacerbated the underproduction and underconsumption and population problems in developing countries. There has also been a growing realization that the low birth rates in industrial countries may have been achieved by exploiting resources from their former colonies. In addition, high *per capita* consumption levels in rich countries and their greater impact on carrying capacities are too often overlooked in gross population figures. Some measure of the sincerity of these concerns can be demonstrated in the fact that in spite of official advice from former United States Agriculture Secretary Earl Butz that Americans should increase, rather than curtail, their meat consumption, thousands of conferences and church groups have urged abstinence from excessive meat eating, and sales of meat substitutes such as soy beans and meatless recipe books such as *Diet for a Small Planet*[14] are soaring. Direct political pressures from such voters helped modify the free-market views of the Ford administration and restore commitment to food reserves and food aid.

What further potential have these citizen movements for spurring beneficial redistribution of the world's production and consumption patterns? First, it must be acknowledged that inflation itself is a factor in these shifting values of American and European consumers, who now are searching for rationalizations for their increasing inability to "keep up with the Joneses." Many have discovered the psychic relief of dropping out of the consumption "rat race" and instead, opt for more leisure, as I have pointed out. Developing countries need ample warning that mass markets for consumer goods in industrial countries can no longer be relied upon to assure steady or increasing markets for raw materials, and that strategies must be developed to widen and expand purchasing power in their own developing regions. An indication of the shrinking mass market in the United States is the bankruptcy of the W.T. Grant department store chain and the falling sales of another retailing giant, Sears, which has depended on three factors, population growth in the suburbs, high home-building rates and automobile sales. Now inflation is ruling out the single-family home, the population is getting older and the mass market is losing some of its old magic.

Citizen movements for corporate accountability tend, sometimes unwittingly, to accelerate the shift of production away from industrial toward developing countries, as they force corporations domiciled within their reach to internalize more social costs. In some cases this effect may be offset by corporations' desire to deal in countries with a historic respect for private capital and relative political stability. But on balance, developing countries will be asked to host more and more multinational manufacturing facilities as they search for exploitable resources, cheaper labor and lower standards of social performance. This presents both pitfalls and new opportunities to developing nations, which already have learned that they have greater leverage than heretofore, in exacting better shares of profits and control of such new facilities. Hosting multinational enterprises will only prove beneficial if such developing nations take full advantage of their new bargaining power by driving the hardest possible bargains so as to assure that their people, resources and environments are not exploited for quick profits, and that the capital-intensive technologies that these companies favor do not aggravate unemployment and urbanization problems, and provided social costs can be realistically anticipated and minimized.

In order to assure the best chance that developing countries can take advantage of their new opportunities *vis-à-vis* multinationals without compromising autonomy over their own peoples' goals, the new United Nations Center on Transnational Corporations has been set up to provide information to any developing country in negotiations with such a multinational company.[15] Funded at less than $1 million, this center is an important beginning, but what is needed is nothing less than a full-scale international databank on corporate accountability (IDCA). Such a computerized databank might need to be operated under an independent charter, with some ties to the United Nations, but with a board of governors composed of distinguished individuals from a variety of professional fields, with two-thirds being appointed from developing countries. The IDCA might be domiciled in a country with a long history of neutrality and a stable commitment to a free press. The IDCA would collect and maintain files on all aspects of the governance, operations and social performance of corporations operating as multinationals. All legal charters and regulations applying to such corporations in their home countries, as well as international legal standards promulgated by such international organizations as the European Economic Community, would be accessible, for example, to the different structures and responsibilities of boards of directors, the various composition of such boards as between inside and outside directors, as well as those directors representing major constituent groups in the society, e.g., consumers, labor or the general public, as is the case in some countries of Europe. All legal regulations would be collated from countries concerning the rights of stockholders, workers, consumers, the rights of minorities, the requirements of disclosure of information concerning not only financial affairs, but also complainant lawsuits, government agencies' rulings on, for example, antitrust and monopoly behavior, violation of consumer, worker or stockholder rights, or environmental, health or safety regulations.

Not only could such a databank be compiled from official sources in all countries chartering or domiciling such multinationals and their affiliates, divisions and subsidiaries, but it could be cross-referenced with the growing resources of high-quality, independent studies of corporate social performance being prepared by public-interest research groups in many industrial countries, such as the Council on Economic Priorities, the Investor Responsibility Research Center and the Scientists Institute for Public Information in the United States, the Jishu Koza in Japan, *Ciudadano* magazine in Spain and a host of similar organizations in the United Kingdom and other European countries. The IDCA could access consumer testing and reporting, environmental impact studies and many other private and academically based research teams now documenting social costs data, and the theoretical journals on corporate social policy such as *Business and Society Review*. Multinationals themselves could file information with the IDCA on programs and achievements in social performance, such as providing new health and recreation benefits or superior employment conditions or in rehabilitating of despoiled mining sites or achievement in controlling pollution or hazardous wastes.

A core staff would verify and update all information inputs to the IDCA and assure academic quality of the research, while keeping references on major areas of scientific disputes. Imaginative subprograms could make the IDCA valuable to a variety of users, but particularly to developing countries in their dealings with multinationals. For example, a drug company wishing to manufacture and sell infant formula or certain nonprescription drugs might be checked out concerning its performance record as experienced by other host countries, and rulings on the particular drugs could be scanned for restrictions imposed on their use, their prices and research on possible contraindications. Or a corporation wishing to set up a normally polluting manufacturing process, such as paper making or metal smelting, could be assessed before the fact, by examining its prior performance in other countries, and the IDCA could also be queried as to the state of the art in controlling harmful air or water pollution from such processes, as well as adverse health environmental effects likely to be encountered otherwise, and the full social costs, benefits and trade-offs involved. Similarly, the IDCA could index and cross-reference parent and holding companies, divisions, subsidiaries, joint ventures and partnerships; interlocking directorships, etc., so as to elucidate such often obscure financial relationships, and make this data accessible and tailor it to conform to the needs of users, whether prospective host countries, agencies of governments, United Nations agencies, the press or academic and nonprofit research organizations.

Multinational corporations are now seen as major actors on the world stage, surpassing the power of many national governments and states. The legal tools necessary to subordinate such private, profit-making enterprises have not yet been fully developed by national governments, so as to assure the preeminence of the human and social goals of their people. Since it is even more unlikely that a suitable body of coherent international law will soon emerge to regulate the activities of these corporations, an IDCA would seem like a sensible first step. Establishment of such an international information system would tend to bring some order into this uncharted area of international relationships, and provide a solid data foundation for future regulatory needs and on which public opinion could be informed on a fuller basis of the societal costs, benefits and trade-offs of corporate operations. If such an IDCA were established, the models for its research and operations are now available through the pioneering efforts of citizen movements and public-interest research organizations in industrial countries, as well as similar independent and collaborative efforts in developing countries. In the meantime, a start at collecting some of the available research on corporate social performance can be made, and a library of such material, hopefully, will be assembled by the new United Nations Center on Transnational Corporations, and made available not only to developing countries, but to concerned citizens and stockholders in industrialized nations as well.

New citizen movements in overdeveloped countries are now coalescing around ecological awareness and focus on the need for more labor-intensive, appropriate technologies, which can increase human autonomy and productivity with modest capital inputs. Such technologies geared toward full employment and decentralized production which can increase the incomes and welfare of rural and poor people are now increasingly relevant to the poorer citizens of over-developed countries. In the U.S. a National Center for Appropriate Technology with an initial $3 million appropriation is operating in Butte, Montana, advising and funding community groups. The United Nations has long recognized the usefulness of such economic development and strategies, and in 1971, the *World Plan of Action for the Application of Science and Technology to Development* noted that "developing countries with ample reserves of unskilled labor and the need to find greater use of their primary materials, require technologies which would subsitute labor and local materials for capital and foreign exchange." Through organized citizen pressure of United States-based environmental and science-

Fig. 12. Characteristics of "Hard" v. "Soft" Technologies

Hard technology society	Soft technology society
Ecologically unsound	Ecologically sound
Large energy input	Small energy input
High pollution rate	Low or no pollution rate
Nonreversible use of materials and energy sources	Reversible materials and energy sources only
Functional for limited time only	Functional for all time
Mass production	Craft industry
High specialization	Low specialization
Nuclear family	Communal units
City emphasis	Village emphasis
Alienation from nature	Integration with nature
Consensus politics	Democratic politics
Technical boundaries set by wealth	Technical boundaries set by nature
Worldwide trade	Local bartering
Destructive of local culture	Compatible with local culture
Technology liable to misuse	Safeguards against misuse
Highly destructive to other species	Depends on well-being of other species
Innovation regulated by profit and war	Innovation regulated by need
Growth-oriented economy	Steady-state economy
Capital-intensive	Labor-intensive
Alienates young and old	Integrates young and old
Centralist	Decentralist
General efficiency increases with size	General efficiency increases with smallness
Operating modes too complicated for general comprehension	Operating modes understandable by all
Technological accidents frequent and serious	Technological accidents few and unimportant
Singular solutions to technical and social problems	Diverse solutions to technical and social problems
Agricultural emphasis on monoculture	Agricultural emphasis on diversity
Quantity criteria highly valued	Quality criteria highly valued
Food production a specialized industry	Food production shared by all
Work undertaken primarily for income	Work undertaken primarily for satisfaction
Small units totally dependent on others	Small units self-sufficient
Science and technology alienated from culture	Science and technology integrated with culture
Science and technology performed by specialist elites	Science and technology performed by all
Strong work/leisure distinction	Weak or nonexistent work/leisure distinction
High unemployment	(Concept not valid)
Technical goals valid for only a proportion of the globe for a finite time	Technical goals valid for all people for all time

Source: Robin Clarke, "Biotechnical Research and Development (Britain)," reprinted in David Dickson, *Alternative Technology,* Glasgow, Scotland, Fontana Press, 1974.

policy groups and the alternative technology networks, United States foreign-aid and development assistance strategies are being challenged and reexamined. Former promoters of "big-bang," capital-intensive technology projects such as the World Bank have now come around to the new thinking and target more of their assistance directly to rural aid and appropriate technology,[16] while the United States Agency for International Development has set up a new $20 million fund for similar projects. Similar citizen efforts to reexamine capital-intensive technology and shift emphasis to technologies that are capital, energy and materials conserving and geared to full employment are growing in most industrial countries. Their conceptual basis is illustrated in Figure 12, describing the characteristics of capital-intensive versus labor-intensive technologies and the different social and political configurations they generate.

It is now becoming apparent that an ecologically sound society is necessarily a full-employment society and that it is only inadequate, partial economic analysis that appears to set the goals of economic and ecological efficiency in a false conflict. The need for a shift to labor substitution for capital in many production processes, even in the United States, is now clear in the declining marginal returns we are experiencing from our capital investments. We have noted how prevailing tax policies skewed the capital/labor ratio toward inefficient capital intensity and have favored large corporations and massive, centralized technologies that are now becoming vulnerable to breakdown and sabotage. Such tax policies have also created and intensified structural unemployment and structural inflation, while increasing the concentration of power and wealth. Many, including the author, now favor repealing further tax subsidies to capital and instead shifting to a system of tax credits for employment, rather than investment, as proposed by economists Berndt, Kesselman and Williamson of the University of Wisconsin[17] and promoted by many environmental organizations.

Citizens groups are now challenging economists who justify economic growth and technological complexity and scale on narrow notions of "profit," "efficiency" and "productivity." Capital-intensive innovation is usually justified on the basis of the need to stay competitive in foreign markets. However, investment subsidies are still masking the true scarcity and rising costs of capital and resources, while projections of historical increases in labor costs *vis-à-vis* capital and materials are still masking the new efficiency of labor.

Such new citizen challenges to traditional economic theories in industrialized countries and how they result in suboptimal social policies, overgrowth of corporations and lead to dis-economies of scale, may help prevent similar waste and inefficiencies from occurring in developing countries. Hopefully they may benefit from our mistakes and avoid some of the more destructive stages of industrialism and its grosser social costs. Over-developed economies may be transformed by internal political action. This politics of reconceptualization is perhaps most advanced and articulated in New Zealand, where the new Values Party and its intelligent, globally-aware manifesto, *Beyond Tomorrow* captured 5% of the vote at their last election.

Such new issues are now being forced into policy debates in all industrial countries by citizen movements and their sympathetic professional advisers. Their challenges to corporations as rational resource allocators, to the former hegemony of elitist science policy making, the irrationality of narrow, suboptimizing economic analyses and the myth of "objective, value-free science," are now forcing reconceptualization of knowledge and more holistic redefinition of national goals. Thomas Kuhn has studied these reconceptualization phenomena in the scientific world and how they lead to changes in paradigms. Such paradigm changes are often great unifications of formerly disjointed fields of knowledge; they do not necessarily add new knowledge, so much as increase the efficiency of existing knowledge. Such a restructuring of knowledge is not only occurring in our academic disciplines, but among the general population in the thousands of mass, spontaneous, adult-education efforts that citizen movements represent. The ecology movement, the women's movement and those for social and economic justice and corporate accountability are in a very real sense efforts to expand awareness of new interdependencies and represent societal learning; while the spread of the human potential movement is based on the experiential understanding that we humans are social creatures and need each other and our web of familial and community relationships. Many groups promote such awareness in the United States, from the older World Federalists to newer efforts, including the Institute for World Order, Planetary Citizens, New Directions, Global Education Associates and their fine, new book, *Toward A Human World Order,* by G. and P. Mische,[18] as well as the Worldwatch Institute, headed by Lester Brown, whose studies have received worldwide attention. The spread of such understanding through proliferation and linkage of all these citizen, nongovernmental organizations may promote planetary awareness and the understanding that individual self-interest and narrow group self-interest, when seen in a large enough context, is identical with the self-interest of the human species. For the first time in human history we are aware of the indivisibility of human destiny on this planet.

[1]See, for example, Johan Galtung, "Non-territorial Actors and the Problem of Peace," in Mendlovitz and Baldwin (eds.), *On the Creation of a Just World Order*, New York, N.Y., The Free Press, 1975.

[2]"International Women's Year Studies on Women," Elise Boulding Series of six papers available from the University of Colorado, Boulder, Colorado, 1975

[3]Lecture by Dr. Odum at the University of Florida Energy Seminar, 14-17 October 1974

[4]Bilderberg Meetings, Cesme Conference 25-26 April 1975, Cesme, Turkey

[5]*New York Times,* 1 December 1975

[6]Adolph Lowe, *On Economic Knowledge*, New York, N.Y., Harper Torchbook, 1970

[7]*New York Times*, 13 May 1973

[8]*New York Times*, 28 November 1975

[9]Frances Moore Lappé, *Food First;* and Joseph Collins, *Beyond the Myth of Scarcity,* Houghton-Mifflin, 1977.

[10]William N. Ellis, "The New Ruralism," *The Futurist*, August 1975, p. 202

[11]"The Deteriorating Image of Business," *Bell Magazine* (American Telephone and Telegraph Co.), January-February 1975, p. 10

[12]Ronald Inglehart, *The Silent Revolution,* Changing Values and Political Styles Among Western Publics, Princeton University Press, 1977.

[13]*The Internationalist* (London), September 1975, p. 29

[14]Frances Moore Lappé, *Diet for a Small Planet*. This unusual recipe book contains a detailed, well-documented analysis of the destructive effects of colonialism on world food production and trading patterns. Published by Friends of the Earth, Ballantine Books, 1971

[15]The United Nations Centre on Transnational Corporations, directed by Finland's Klaus Sahlgren, operates as a unit of the Secretariat reporting to the Secretary General. It has begun by collating a manual-filed bibliography of information on transnationals and its second stage will consist of a similar bibliography of regulations, national policies and codes which will be published in March 1976. Later it hopes to computerize these files and move toward researching the characteristics of some 15,000 companies operating transnationally, with emphasis on the 1,000 giant firms. The centre will execute research at the request of the General Assembly, for example, in preparing reports pursuant to the General Assembly's recent resolution on bribery and corruption. (Information provided by N. Wang, Director of the Centre's Research, 4 December 1975)

[16]Robert McNamara, "Greening the Landscape," *Saturday Review*, 24 December 1974, p. 25

[17]Berndt, Kesselman and Williamson, "Tax Credits for Employment Rather than Investment," 1975. Institute for Research on Poverty, University of Wisconsin, Madison, Wisconsin (United States).

[18]*Toward a Human World Order,* Mische, Gerald and Patricia, Paulist Press, New York, 1977

CHAPTER TWENTY-TWO

Pluralistic Futurism

The growing interest in futurism in this and many other countries is welcome evidence that the human species is not inevitably programed on self-destruct. We have always been better at creating hardware and building large sociotechnological systems than we have been at writing programs to manage the new interlinkages they create. The lag time has varied, but many of us share the sense that it now must be shortened. The best current example of this lag is that we have now created an interdependent world economy and are now desperately trying to update the world's trading rules and rewrite the "program" to manage this larger system.

Thus, futurism is an evolutionary response to the feedback we humans generate by our own technological and social activities. Interest in futurism is exploding because we are encountering an explosion of unanticipated feedback (much of it negative and alarming) from the second-order consequences of our actions.

Naturally, this interest is growing most rapidly in industrialized societies most perturbed by these effects: the U.S., Japan and the countries of Europe. Some futurist efforts are organized internationally, within groups such as the Organization for Economic Cooperation and Development, headquartered in Paris; the United Nations special conferences on environment, population, food, employment, habitats, Law of the Sea and desertification. In fact such U.N. efforts can force such futurist concerns onto the political agendas of even reluctant member nations. Newer cooperative efforts, such as the International Institute of Applied Systems

Analysis are sponsored by the U.S., the Soviet Union and other countries and headquartered in Austria.

Many of these same industrialized countries have their own research and planning activities, some sponsored by their governments, such as France's Commissariat General du Plan, Canada's Institute for Research on Public Policy, the Royal Norwegian Council for Scientific and Industrial Research and the U.S. Environmental Protection Agency. Some futures research is academically based, such as the Netherlands Sociological Association's group on Prospectivism, the Swedish Society of Futures Studies, Yugoslavia's Science and Society and the Japan Techno-Economics Society, which also actively concerns itself with technology assessment. Other futurist activities have sprung from voluntary initiative, such as France's Association Internationale Futuribles, Political and Economic Planning in the United Kingdom, the World Future Society, Resources for the Future and Worldwatch Inst. in the U.S., as well as the London-based International Planned Parenthood and the Club of Rome. In addition, all multinational corporations must plan and, therefore, have widely varying futures research efforts as well as employing futurist consulting organizations, such as Ted Gordon's Future Group, the Hudson Institute, Forecasting International and the Stanford Research Institute.

Specialties in Futurist Research. Even though the hallmark of futurism is holistic, rather than a narrow approach, inevitably all futurist activities focus their interest in specific areas, either to remain consistent with their own world views, such as the worsening population/resource ratio that occupies the Club of Rome, or to advance ideas and cultural communication, such as the ecumenically oriented associations, including Futuribles, the Rome-based IRADES, the International Creative Center of Geneva, Belgium's Mankind 2000 and the U.S.-based Earthrise and the Committee for the Future. Some serve government need for assessing technology, such as Congress's Office of Technology Assessment, the MITRE Corporation and the National Science Foundation. Others concentrate on economic and social planning, such as the Council for National Living in Japan, which has developed social indicators to update their Gross National Product indicators and the Nomura Institute, which concentrates on economic forecasting. Other organizations focus on attitudinal and market research, such as the Russell Sage Foundation, Yankelovitch, Skelly and White and the Institute for Life Insurance Trend Analysis Program. Lastly, there are increasingly sophisticated efforts to research viable political futures, such as the World Order Models Project of the Institute for World Order and the Goals Research Project of the Club of Rome, and, most importantly, the planning efforts of the countries of the Third World which are developing their own models of a new economic world order and advancing them at the United Nations and other international forums.

There is no need to further augment this brief sampling of futurist programs and groups, since there are existing directories such as that published by the U.S.-based World Future Society. The Library of Congress now has its own Futures Group and can assist in locating these resources. The Center for Integrative Studies at the University of Houston, Texas, has completed its third Survey of Futures Studies covering over 1,000 institutions and individuals around the world, and the American Management Association has issued a new futures directory, Exploratory Planning Briefs, 1974-75.

But these rich and accessible resources on the futures field by no means exhaust the possibilities. Naturally, their attention is focused on the larger, more visible operations and those individuals who publish widely on the subject of the future. But futurism is much broader and more diffuse, since like other infant research fields, such as technology assessment, environmental impact analysis and energy modeling, it is still in its formative stages. Anyone can still be a futurist. There are, as yet, no degrees or licenses needed; the only requirement is that of self-proclamation.

This is good in that it encourages people to transcend older, narrower fields, to experiment and communicate across traditional disciplines. It also provides new opportunities for the native intelligence of the less formally trained and the self-educated. In fact, prior academic training can often block the creativity and imagination so necessary to conjecture about the future and its discontinuities. But, unfortunately, such an open field invites opportunism and commercial exploitation. As more funding becomes available for futures research, we must all guard against an all-too-familiar breed of "consultants" who proclaimed themselves urbanologists in the sixties, environmental experts after Earth Day and who can pour their old wine into any kind of new bottles, whether labeled "technology assessment," "futures research" or whatever the next hot ticket turns out to be.

Formalism Kills Creativity. And yet, paradoxically, attempts to professionalize and license futures research and dispense degrees in the subject are also premature. Formalism will destroy precisely the creative imagination that makes the field more of an art than a science. Too often taxonomy itself can become the enemy of thought. Technology assessment is another case in point, being an experimental field, with most practitioners still rather low on the learning curve. These new, more inclusive research methods, together with environmental-impact analyses, must embrace much greater uncertainties. In fact, the poor decision makers in government who commission such research in the hope of reducing these gray areas often discover that these studies actually increase uncertainty—by defining more systematically what is still unknown!

However, well-performed technology assessments and futures studies can extend the frontiers of many quantitative methodologies. But in the last analysis, rigorous quantification becomes impossible, since unconscious biases in scientific methods and in the very framing of research questions produce a distortion; greater certainty about sub-systems but loss of larger perspectives. The conflicting values around such issues as technology and alternative futures involve weightings and trade-offs between our often intransitive goals. Therefore, these forms of futures research cannot exclude social, political and even moral questions.

In fact, we must now acknowledge the difficult truth: that values, far from being peripheral, are the dominant, driving variables in all technological and economic systems. It is for this reason that the Committee on Anticipatory Democracy expressly endorses the concept of open-ended futurism and the widest possible participation of all segments of our society both in formal futures research and in formulating alternative futures. Citizen feedback and "feedforward" are essential in helping our big, unwieldy institutions adapt to our future needs. Or, as noted earlier, Warren Bennis states that, "Democracy becomes a functional necessity whenever a social systems is competing for survival under conditions of chronic change."[1]

This may account for the spectacular rise of nongovernment citizen feedback/futures research and planning efforts, such as those now being conducted in many states, for example, California Tomorrow, Massachusetts Tomorrow, New Hampshire Tomorrow, the Northwest Environmental Communication Network, and Rhode Island 2000 (proposed by Earthrise), to name a few. All of these broad-based efforts to envision and discuss alternative futures for states and regions often utilize mass media and rely heavily on existing infrastructures of voluntary groups of all kinds. Many state-sponsored programs are run along similar lines, including the Utah Process, Hawaii 2000, Ohio 2000, and the Commission on Minnesota's Future. They range from wide-open, participative models, such as Alternatives for Washington, to the much more elitist model typified by Goals for Dallas, which has done little more than amplify the voices of the already well-represented and consolidate the power of the politically adept. Not surprisingly, it produced the biggest municipal-planning disaster of the past decade, the Dallas-Fort Worth Airport.

As we continue to monitor and utilize these types of futurism, we will become more aware of the extent to which these efforts range from narrow elitism to sincere attempts at broad citizen participation. These regional futures efforts all indicate the malfunctioning of existing political channels, devised for an earlier, less hurried age, which are now severely overloaded. Gleaming in many a futurist's eye are the ultimate processes of political participation: the instant electronic referendum, which generally strikes panic in the hearts of political scientists and legislators alike. However, as I have pointed out in Chapter 14, this computerized "hardware of social choice" is available and we had better begin considering both its potential and its problems more seriously.

Futurist Defined. If futurism is so widespread, who then, is a futurist?

Many work in programs not specifically designed as futures studies. A good example is the new field of energy modeling which has developed in about 30 countries and is already more accurate and predictive in resource-allocation decisions than economics, whose preeminence it may replace.[2] And yet energy modelers such as Howard Odum of Florida's Energy Center, Bruce Hannon of the University of Illinois' Center for Advanced Computation and Stephen Berry of the University of Chicago do not call themselves futurists or their programs futures research. Meanwhile, many in the "futures" field of planning, technical and economic forecasting have turned out to be better historians than futurists. In fact, the recent survey of the Center for Integrative Studies shows that most of the key centers working on futures studies are relatively small and underfinanced. The survey confirms that much of the pioneering work is done by individuals and groups who are not in programs labeled as futures studies and that many such innovative individuals tend actually to be squeezed out as soon as their efforts gain validity.

Some of the greatest futurists have been artists, such as the Dadaists, who ridiculed formalism, and the Cubists, who taught us a new way of seeing. One of the greatest works of futurism in this century, and still ominously forewarning us, is George Orwell's novel, *1984,* written in 1948. My candidate for the best U.S.-published futurism of the past five years is Californian Stewart Brand's *Whole Earth Catalog*. Other futurists don't write about the future so much as they imagine it and create it. Activists in consumer and environmental protection have surely been as much futurists as those who monitored the trends they were creating. Was not Rachel Carson a futurist? Public-interest law and public-interest research, pioneered by Ralph Nader, has spread to the sciences, economics, accounting and even operations research. These public-interest research organizations, often run by the brightest, more creative and socially concerned scientists, can provide society with a rich new resource for futures studies only now being recognized by research managers in the academies, the National Science Foundation and other executive branch agencies.

Those who work for peace, social and economic justice have created new value orientations in our society which those researchers in the social indicators field have been obliged to incorporate into their models. Was Martin Luther King not a futurist? Activists for corporate accountability and ethical investing have helped create new courses at hundreds of business schools on the social responsibility of corporations, as well as many new careers in the companies themselves.

Environmentalists created the concepts and the social impetus for technology assessment and environmental-impact analysis. Health activists began the movement of professionals toward preventive, rather than merely curative medicine, important as that is. Appropriate technology activists and alternative lifestyle experimenters provided the political push that is finally leading to the development of energy technologies, including solar, wind, wave, thermal gradient, biomass and methane conversion systems.

And yet some of the key innovators in all these momentous social changes of the past decade are not listed as "futurists," largely because they were too busy doing the innovating to write about it. For example, the Council on Economic Priorities (on whose board I am proud to serve) is rarely listed as a futurist organization, or its founder, Alice Tepper Marlin, listed as a futurist. And yet the Council pioneered methods of comparative analysis of corporate social, rather than economic, performance, which are now used by futurist consulting firms as well as Wall Street security analysts.[3] Similarly, Dr. John Todd of the New Alchemy Institute is rarely listed in futurist directories, and yet he is performing vital, underfunded biotechnical research which may soon help us maintain our agricultural yields with vastly reduced fossil-fuel inputs.[4]

And the media-access activists, Nicholas Johnson, Nick de Martino, Peggy Charron, William Wright, Albert Kramer and Michael Shamberg pioneered interactive media concepts, public-cable channels and citizens' rights in broadcasting, but are not numbered among the academic futurists who study the effects they set in motion. And lastly, the millions of American women, who pioneered perhaps the farthest-reaching concepts of feminism, have created an enormous demand for consulting futurists to study the ramifications through virtually every institution in our culture, from the nuclear family to the multinational corporation.

Futurism an Open Field. This is why futurism, just like every other field, will always entail politics, rather than congeal into a "value-free" scientific discipline. Even the other scientific "objective" disciplines are finding this stance more threadbare as the public becomes more knowledgeable. Economics is now clearly revealed as a normative rather than a scientific discipline, and each day physics becomes more metaphysical. Therefore futurism must always remain an open field, since in so many cases, social innovations, and those discontinuities that futurists must try to forecast, emanate from those not-too-tightly institutionalized. Often the clearest vision comes from those who are the most alienated from the norms of their culture and not suffering from its blind spots. For we must remember that institutions are always screening out reality and most of our citizens are now enmeshed in them and find their imagination and self-development constrained. In fact, the larger and richer the institution, the less likely it is to be adaptable to the future.

So the rise in anticipatory democracy is welcome and necessary, however untidy it may appear to the traditional academic researchers. And when futures research projects are commissioned, I hope that government will scan the entire field and choose from the full array of potential contractors and resources, including public-research organizations and individuals. The synthesizing potential of such groups is illustrated by the efforts of the Cousteau Society to re-draw the U.N. Declaration of Human Rights to move beyond its myopic concern with our own generation and to include protection of the rights of future generations.

Lastly, what of the future of futurism? Much future study began simply, with linear extrapolation from the past and other myopic methods dealing narrowly with either specific issues or areas, technologies or institutions; for example, simple demand projections or straight-line technological and economic forecasting. Thus they extended only time dimensions without extending space dimensions. We now have many superior methods, such as those reviewed by the National Science Foundation, which extend both time and space dimensions and are thus more inclusive and contain fewer blind spots. It is a most interesting exercise to review some of the recent predictive efforts that exhibited such blind spots, such as the 1966

President's Commission on Technology Automation and Economic Progress, which overlooked the magnitude of structural unemployment problems; Herman Kahn's study[5] of the Japanese economy which overlooked its resource dependency on other countries, as well as linear projections of demand for electric power which vastly overestimated growth.

New Paradigms. The future of futurism is in creating new paradigms, those different pairs of disciplinary spectacles we put on which enable us to see some patterns while distorting or obliterating others. Old paradigms and models of reality are now breaking down in many disciplines, notably in economics. We always know when such a paradigm break is occurring, such as the market-equilibrium theory in economics, because the list of exceptions to the approved theory grows longer and longer until its adherents grow dogmatic and embarrassed. The future direction of futurism can be detected in new organizational forms, such as networks which are metaphysical, rather than hierarchical structures; in new modes of perception, such as Kirlian photography; in expanding awareness of human potential now evidenced in biofeedback and research on the last great frontier, the human mind. As we humans seem to be heading for an evolutionary cul-de-sac, the most creative and intuitive will still be trying to tell us that the way out of the box is to redefine the problem: to look up from our preoccupation with retooling the world and try changing our goals and focusing on our evolutionary task of retooling ourselves. Futures research is societal learning. We are all in it together.

---

[1]Warren Bennis, *Changing Organizations*, McGraw-Hill, New York, 1966

[2]Further information can be obtained from the International Federation of Institutes of Advanced Study, Sturegaten 11, Stockholm, Sweden

[3]The Council on Economic Priorities, 84 Fifth Avenue, New York, N.Y. 10011

[4]The New Alchemy Institute, P.O. Box 432, Woods Hole, MA 02543

[5]Herman Kahn, *The Emerging Japanese Superstate,* Prentice Hall, 1970.

Diane Schatz, permission RAIN magazine.

CHAPTER TWENTY-THREE

The Emerging "Counter-Economy"

I have shared my thoughts with you concerning the many signs, both physical and metaphysical, that industrial cultures are breaking down. But I want to emphasize that the breakdown of an old culture can also signify a needed breakthrough. Times of crisis, as the Chinese say, are both times of danger and opportunity. From ecological theory, we know that all biological systems (including human societies and those abstractions they call their "economies") involve continuous cycles of entropy and syntropy: the breaking down and building up of structure and the constant recycling of the detritus that releases the nutrients for new growth, synthesis and evolution. So let us now look at what is being born: the emerging, regenerative, "counter-economy" now beginning to grow amid the old industrial systems.

This basic model of the entropy/syntropy cycle and the irreversible evolution of all natural and biological systems is crucial to our understanding of the particular subsystem we call our "economy" and in helping us see current economic difficulties in longer time perspectives as the onset of the decline of industrialism. This decline will undoubtedly prove uncomfortable, as it already is for the millions of unemployed in mature industrial countries. As I have tried to show, the decline will only affect the unsustainable modes of production and consumption it has fostered, but with leadership and foresight, the adjustments can still be made without severe consequences, provided the discomfort is meted out to all groups equitably and the poor and powerless are protected from bearing the brunt of the dislocations. Meanwhile the declining system is already releasing "nutrients": capital, management and human energy and initiative which are spurring the development of this already visible counter-economy, now beginning to flourish in the interstices of our existing institutions.

I shall try to sketch the contours of this new counter-economy in the hope that those with computers and access to funds and academic resources will flesh out a picture of this regenerative economy's potential and relay it onto our mass-media scene for public debate. While economists struggle to recycle themselves in order to address these new conditions and unfamiliar variables, it seems to those whose vision has remained unclouded by economists' mystification that this transition is obvious, that it can be inferred from extremely simple metaphors, e.g., "There is no such thing as a free lunch." "Nothing fails like success"; "Growth can be cancer." Indeed, average citizens in these societies have learned to tune out their leaders and mass media and are well on their way to understanding the true situation, in spite of the obfuscations of legions of intellectual day laborers and the divinations of "experts."

In fact, it is fairly self-evident that these mature industrial societies could not continue expanding at past *rates*, simply because such rates are always in relation to the size of a *base*. Any citizen knows that as a base grows, the *rate* of its expansion must sooner or later decline; whether one is looking at the rate of increase of today's shares in IBM or Xerox, compared with their past spectacular performance, or the *rate* of growth in the size of oil tankers, airplanes or human settlements. In fact, the only current exception to this rule appears to be bureaucracies, but they too, may decline like the over-centralized, unsustainable technologies that gave rise to them. And yet, I still find a great deal of hand wringing and rubbish talked in Washington today, about falling *rates* of economic growth (GNP-defined) and falling *rates* of technological innovation and "productivity" (inadequately defined), where the *base* for calculating such rates—the giant U.S. socio-technical system—is the largest on the planet! Surely we know by now that human cultures have a habit of rising and then declining as they exceed some resource limit, run out of technological adaptability or simply lose creative cultural steam.

As mentioned in Chapter 12, Karl Polanyi may well have been correct in 1944 in *The Great Transformation* in characterizing as a rare aberration in human history, industrialism and the package of social legislation which installed "the free market system" in England and gave rise to the Industrial Revolution. Up to that period humans had normally employed two other major production and resource-allocation systems: reciprocity and redistribution. We may now have to countenance the proposition that industrial culture and our own petroleum age shown in Fig. 4 may be a brief episode in human history. Yet this may not be an unmitigated disaster, or even a disaster at all. So I am not impressed when U.S. rates of technological innovation and "productivity" are compared with official horror to the higher rates of Japan (with a postwar base about an order of magnitude smaller than our own). I am not upset when warned that new "science and technology gaps" are widening and Congress is urged by science and high-technology-promoting groups to appropriate ever more tax dollars to save us from this fate. Their underlying assumption in all these exhortations is that the health of the scientific and technological enterprise, as currently defined and constituted, is coterminous with the health of the country as a whole. I and many others reject this proposition.

It is therefore my contention that since such a vast sea change in the structure of mature, industrial countries is now under way, much of their intellectual paraphernalia has also been rendered obsolete. Apparent paradoxes abound, which are usually signals that paradigmatic shifts are required. The continuing crisis in economics, particularly that we have seen in macroeconomic management, is accompanied by crises in sociology, psychology, and even physics. As discussed earlier, two insistent paradoxes concerning advancing technological complexity are now almost unavoidable in any academic field of policy relevance in an industrial society: (1) Such complexity systematically destroys free market conditions, making laissez-faire policies ever less workable while at the same time we humans have not yet learned how to plan such societies; (2) It destroys the conditions necessary for democratic political governments to function, since legislators and even heads of state, let alone the average voter, cannot master sufficient information to exert popular control of technological innovation, while the hazardous nature of new technologies often requires societal regulation and policing that erodes or abrogates civil liberties, so that complex technologies such as nuclear power are *inherently* totalitarian.

Human societies are only now beginning to review systematically strategies (some long forgotten or in the cultural traditions of such subcultures as the Amish or so-called "primitive" societies) to deal with the destabilizing social effects of technological changes and innovations.

Such complexities are beyond human capabilities and even our most advanced modeling methodologies at present, for, as mentioned in Chapter 9, these choices are involved with the evolution of our species. Most species make such choices over eons, and unconsciously, through their genes and the processes of natural selection and ecological succession. These natural, evolutionary choices in the development path of species also involve these meta trade-offs and the "economics of flexibility," i.e., "spending" flexibility now versus "storing" future flexibility. The mode nature chooses involves profligate sacrifice of individual members of a species before a set of adaptations is selected for genetic "hard programing" of the genotype itself. These concepts are discussed at greater length by Gregory Bateson in *Steps to an Ecology of Mind*, as he brilliantly generalizes from such evolutionary genetic concepts to human and social systems. He notes, "In all homeostatic systems, higher systems of control must lag behind event sequences in the peripheral circuits... so as to permit stochastic processes of trial and error to experiment with adaptations." Timing is all. If genetic adaptation to change is too rapid, this may only maladapt the species for the next set of environmental perturbations. The paradox is exquisite: "Nothing fails like success." That humans are now at the point in their own evolution where these formerly genetic, evolutionary choices are emerging as conscious, cultural decisions is evident in the nature of all these meta-level policy choices which we have seen that industrial societies face today. Far from being faddish, the counter-culture and citizen-protest movements of the past decade that are forming the nucleus of the emerging counter-economies based on self-reliant, decentralized, ecologically harmonious lifestyles, are deadly serious and must be explicitly documented and reinforced, since they represent the best repositories of social and cultural flexibility during the decline now underway in many mature industrial countries and the coming contraction in its system of world trade.

The evolutionary dilemma summarized by the aphorism, "Nothing fails like success," can be restated in terms of anthropology as the Law of the Retarding Lead, which holds that the best adapted and most successful countries have the greatest difficulty in adapting and retaining their lead in world affairs, and, conversely, that the backward and less successful societies are more likely to be able to adapt and forge ahead under changing conditions. L. S. Stavrianos uses these principles to argue in his book, *The Promise of the Coming Dark Age,* that for the crowded, ecologically depleted planet of today, Western societies must flounder; on the other hand, there is the possibility of synthesizing unique development strategies of their own by societies adapting from the Chinese, communal model of self-reliant development and mass-participation, from the Yugoslavian model of worker self-management, and from selection from the counter-culture, citizen-action, community-control models now developing in the counter-economies being born in the Western industrial countries.[1] Stavrianos, Forrester, Goldsmith, Illich,[2] I and others have argued that if Western societies are to become regenerative and sustainable, they indeed may have to learn from these so-called less developed countries, as well as relearning much traditional wisdom and skill from their own pasts in order to reinforce, augment and culturally reward the emerging counter-economies in their midst.

In the face of the current transition of industrial societies and the efforts at reconceptualizing the new situation, it is not surprising that levels of cognitive dissonance are increasing. In the U.S., in spite of a decade of predictions of their transience, we see the durability of the counter-culture, the environmental movement according to Harris and Opinion Research, now stronger than ever,[3] consumer and public-interest advocacy, the women's liberation and minority-rights movement, as well as the diffusion of the new values and new lifestyle options brought about by these trends, while statistics document the reverse migration from cities back to rural areas.

Another indication of the general disaffection with the "business as usual" approach to government was the election, without massive corporate and special-interest campaign funds, of President Jimmy Carter by a coalition of those groups for whom the U.S. economy was not working: labor, the less affluent, small business, small farmers, environmentalists, blacks, women, younger voters and others.

In any period of cultural transition, the dominant organs of a society often increase their efforts to reassure the public, while their leaders privately express doubt and fear. This is not surprising, since it is precisely these institutions of government, business, academia, labor and religion, as well as their leaders, which are in decline and whose power is threatened and eroding. The information gathering and disseminating media, the statistics and the indicators are all geared to measure the society's wellbeing *in terms* of the wellbeing of these existing institutions. Therefore, the growing shoots of the society go unmeasured and are overlooked and will remain insufficiently monitored and studied as possible new social models. We cannot afford to wait until the conceptual wreckage of industrialism is sifted and composted. We need to study the counter-economy at the same time that we are examining our now inappropriate statistics. As we drown in useless data, we must remember that there is a natural hierarchy in information-systems: from the smallest bits of raw data to models which pattern it, to concepts that inform the models—to goals that give purpose to the concepts—and finally, the values from which the purposes flow and which *drive the entire information-system and the culture itself.* New perceptions and paradigms can generate more realistic models and provide more appropriate statistics, and the new efforts in this area are cause for encouragement.

It is premature to envision a planetary counter-economy, since much pragmatic experimentation will be needed before the ancient and now irrelevant dogmas and conflicts over capitalism versus communism can be transcended. Both are based on technological determinism, inadequate notions of industrial "efficiency" and "progress" and display an anthropocentric blindness to ecological dimensions and short change human potential. Economics itself has become an inoperative category, an academic relic useful for accounting in micro-transactions, whose terminology is necessary only in order to communicate with those socialized into its conceptual habits, who, unfortunately, still exert an enormously dangerous measure of influence in our resource-allocation systems. Some of the most interesting and significant manifestations of the counter-economies in the U.S., Canada, the United Kingdom, the Scandinavian countries, the Netherlands, Japan, Australia and New Zealand, include:

☐ The growth of counter-media and alternative publishing (a measuring rod of the counter-economy): for example, in the U.S.: *Prevention* (nearly 2 million circulation), *Organic Gardening* (1 million), *Rolling Stone* (approximately 1.5 million), *Mother Earth News* (300,000), and *The Whole Earth Catalog* and *Epilog*; the proliferation of regional magazines dealing with ecological lifestyles and appropriate technology; the some 80 publishing ventures operated by feminists; the rise of the black press; and the hundreds of small, often cooperatively owned, book publishers and distributors.

☐ The alternative marketing enterprises: for example, the *Alternative Christmas Catalog*, which offers, instead of materialistic goods and junk gifts, a vast selection of "psychic gifts," such as subscriptions to counter-culture magazines and newsletters and memberships in various counter-culture, public-interest and citizen organizations; the growth of organizations marketing rural crafts, such as quilts, embroidery, clothes and toys to urban department stores, often on a nonprofit basis; the alternative merchandising media now offered to small rural businesses and crafts by the burgeoning counter-culture media and their inexpensive advertising rates and well-defined audiences; highly professional alternative opinion-survey firms, such as American Research Corporation and Hart Associates; public-interest advertising agencies, the best known of which in the U.S. is the Public Media Center of San Francisco, which does not take ordinary commercial clients but sells citizen organizations and their social causes.[4]

☐ Another new mode is development of alternative marketing groups such as Oxfam's "Bridge" in Britain, which catalogues and links small, rural producers of handcrafted goods and art in the Third World with affluent, concerned consumers in Britain, thus setting up new trading links that operate in a nonprofit, people-to-people mode, avoiding profit-motivated trading channels and multinational enterprises.[5] Another growing mode is that of staging both rural and urban "fairs," where various sectors of the counter-economy can nucleate and cross-fertilize; these feature new lifestyle speakers, book stalls, booths for local citizen organizations, commune-made arts and crafts. An example is the Toward Tomorrow Fair in Amherst, Massachusetts, an annual event which features some five acres of alternative technology exhibits by small businesses in solar, wind and bioconversion, flushless toilets, do-it-yourself housing, home and garden tools; the fair, which attracts some 30,000 people, has become an institution. Similar fairs and festivals, such as the Cousteau Society's "involvement days" and the many concerned with improving nutrition and holistic health, draw similar crowds. They augur new lateral linkages and networks insulated from traditional industrial merchandising and based on emerging value systems impervious to the old, materialistic Madison Avenue "hard sell."

☐ The growing interest in household economics, i. e., the economics of *use*-value, rather than *market*-value. Professor Sol Tax has proposed in *The Center Report*[6] that we begin looking through the other end of the telescope at the possibilities of the *family* as the basic corporate unit and, where necessary, alter our tax structure and laws to favor this smallest unit of economic and social organization rather than large corporations, as at present. In 1780, over 80% of all Americans were self-employed, many in household economies.[7] Canada's Vanier Institute for the Family is dedicated to the concept of household economics and the promotion of local self-reliance; to decentralizing economic power and exposing the absurdities of national economic statistics that do not recognize the enormous productivity of households, simply because they are not, and probably never should be, a part of the cash-based economy. In *Home, Inc.,* Scott Burns estimates that if our national statistics were to include the value of households and the work performed by men and women in them, the total would equal the entire amount paid out in wages and salaries by every corporation in the U.S.[8] Burns notes that government statisticians only value the household when it breaks down, i.e., they know the cost of welfare, aid to dependent children and social services, and thus could compute, negatively, the value of viable households. In addition, he notes that while income tax laws allow corporations to deduct and depreciate items of capital equipment, householders are forced to treat their own productive assets, e.g., sewing machines, ovens, freezers, yogurt makers, home tools, as if they were *consumer* goods![9] A survey in 1969 by Ismail Sirageldin measured the total value of all goods and services produced by the household sector in 1965 as about $300 billion.[10] The increasing protest at the statistical blackout perpetrated for so long on the household economy is, of course, being spearheaded by women, who have been consistently ignored by economists' definitions of "productivity" and "value," as well as excluded from the GNP and their rightful access to retirement security. Many women fight these terrible injustices by going outside the home and competing successfully in the market economy, while others, together with concerned men, fight to restore the proper place of the family in our economic life and strengthen its role in the vital nurturing and socializing of the young and in maintaining intergeneration cohesion. Many others in the counter-culture work to enlarge the definition of the family to include communes and intentional families of all kinds, based not only on sex roles but also on work and companionship. Most of our economic statistics are devised to plot the market system, rather than to trace the full and real dimensions of our total economic system. So far, the sparse academic efforts to overhaul the GNP to include the value of the household economy and subtract the soaring social costs of market activities—such as Tobin's and Nordhaus' Measure of Economic Welfare (MEW) and, more recently, Samuelson's Net Economic Welfare (NEW)—have been minor adjustments that have never been promoted by the economics profession in spite of general acknowledgment of the glaring errors in GNP measurements of economic growth.[11]

☐ The growth of the various movements for alternative technology: These movements are flourishing in all mature industrial societies, particularly Britain, the Scandinavian countries, the U.S. and Canada. There is almost a surfeit of studies and statistics related to the characteristics of dominant modes of industrial scientific and technological development, since technological "productivity" and technology transfer are issues of keen interest in national and international trade. However, the emerging economies based on more culturally and ecologically appropriate technology have been almost totally ignored by most agencies of government in the countries where these movements exist. One can only infer the size of this sector from advertising lineage in alternative media and best-seller status of its gurus such as Schumacher, whose posthumously-published new book, *A Guide for the Perplexed* was eagerly awaited. While obituaries in the dominant press were respectful, a groundswell of memorial services and salutes occurred in many countries. Magazines also grow apace in this field, such as Canada's *Alternatives;* Britain's *The Ecologist, Undercurrents, Resurgence, Appropriate Technology;* and Australia's *Earth Garden, Grass Roots* and *The Powder Magazine.*[12] In the U.S. are *Foxfire, Journal of the New Alchemists, Co-Evolution Quarterly, Rain, Science for the People, Workforce, Shelter, Self-Reliance* and scores of others, many of whose ideas emerge after a several-year lag in the more cautious, Washington-based journal, *The Futurist.*

☐ The rebirth of populism and the cooperative movement, neighborhood and block development, "sweat equity" urban renewal, land trusts and the increased bartering of skills and home-produced goods and services: The Co-operative League of the U.S.A., in its 1975 review, stated that more than 50 million Americans belonged to cooperatives; these included co-op banks to provide credit that commercial banks deny, 22,879 credit unions, 2,034 insurance plans with over 7 million members, about 1,700 nursery schools, about 8,000 farmers' marketing coops, 999 rural electric coops serving 6.5 million customers, 258 student housing coops, 241 rural telephone coops serving 750,000 subscribers, 102 fishing coops with 7,098 members, 125 memorial societies serving about 500,000 people to provide dignified burial rites, and (the most recent type) 223 food coops serving about 577,000 consumers.[13] A bill to establish a national cooperative bank came close to passage in the 94th Congress and will, probably, pass during the Carter Administration; it will help counteract the discrimination coops have suffered at the hands of commercial, multinational banks. Many local organizations are campaigning to set up state-owned development banks modeled after the successful state Bank of North Dakota, in order to make credit available to small-scale farmers (whom the U.S. Department of Agriculture had proposed to consign to oblivion by dropping them from its statistics), and local community development corporations and co-ops. The Massachusetts Community Development Finance Corporation may be one of the first in the field; already authorized, with an initial appropriation of $10 million, it will buy stock in enterprises owned in common by the residents of any area in Massachusetts.[14] Meanwhile, grass-roots community organizations are springing up; many are based on the spectacularly successful model of ACORN in Arkansas, a coalition of poor homeowners, farmers, sharecroppers, workers and urban residents of all races, whose solidarity has shown how local satrapies controlling millions of dollars of tax money can be captured by determined voters who study their own local political systems carefully and how utility rate increases can be fought (and also reduced, as in California and other states by the Lifeline coalitions). Horrified at the success of these grass-roots efforts, utilities have launched emergency counterattacks in the form of stepped-up lobbying and advertising (charged to their customers' bills), while deposed political hacks in Arkansas and editorials in Little Rock newspapers talked of "the Lilliputians who banded together and tied down Gulliver."[15] Not only is it costly, but it demands intellectual creativity to design new statistics on this cooperative, neighborhood-based political and economic activity, since few official measures are available and status quo institutions and media often are successful in playing down such activities or blacking out reporting of them in commercially controlled media.

☐ The rising worker-participation and self-management movements: These are more active in West Europe and Canada than in the U.S. or Japan. Western Europe's political traditions make for easier dissemination and cohesion of efforts to gain more influence over the quality of working life. In the U.S. the labor movement has, up to now, seen more material rewards in traditional bargaining strategies for higher wages. This attitude is changing as under inflation real incomes have remained fixed for protracted periods and faltering growth can no longer assure workers of steady employment, let alone a larger share of a growing pie. In Japan, worker demands for self-management have been headed off so far by management's commitments to lifelong employment and all-embracing welfare benefits. But as the Japanese economy is weakened by its raw-material and energy dependence and reliance on unrealistic levels of exports, corporate executives are adopting Western-style mass layoffs; the radicalization of the Japanese labor force will result, with uncertain outcomes. Western European countries, such as Sweden, Norway, Germany, France and Britain, are making almost daily accommodations to the pressure from unions; examples include Sweden's Meidner Plan to shift substantially the ownership of Swedish industries to their workers, Germany's worker-management parity on boards of steel companies, and the milder forms of job enrichment that are becoming commonplace.[16] U.S. management is well aware of the radical implications of these experiments. In some cases, as workers have successfully organized production, achieved productivity gains and worked out salary differentials in their own committees, they gain the confidence to ask the taboo question, "What are all these managers doing, and aren't they just featherbedders?" At this point, many of the experiments are quietly discontinued, since the workers are challenging "the divine right of management." In Britain, this issue, which underlies the ancient war between labor and capital, has erupted into class warfare. As mentioned, the model for much worker self-management experimentation is Yugoslavia, where the concept of private property is similar to that of our own Founding Fathers.

☐ The resurgence of ethnic and indigenous peoples all over the planet: This phenomenon requires further study. Such populations are beginning to recognize the inability of metropolitan centers to meet their needs and to determine that the exploitative relationships from which they suffer must be resisted openly and severed. From the extraordinary secret "summit" meeting, described earlier, which was held in Trieste in 1975 by the oppressed ethnic minorities of Europe, with its vision of a "Europe of Peoples," to the American Indian Movement in the U.S. and the demands of Canada's native residents of the Yukon and Northwest Territories for the return of their lands,[17] these manifestations are widespread. Pioneer research in this field is being conducted by Elise Boulding, whose book, *The Underside of History*,[18] is must reading, while *The New Internationalist* reviewed these global struggles in its December, 1976, issue. If these ethnic peoples are able, as seems likely, to forge stronger links with the growing world feminist movement and with citizen-action movements in the industrialized countries, the prospects for their contribution to a saner, more just world system are exciting. Many of the goals of these ethnic peoples are similar to those of the counter-culture and citizen-action movements: curbing the excesses of the profit system, creating new principles of balance and accountability of both corporate and bureaucratic power and wealth, making science and technology serve democratically determined social and human goals, decentralizing decision making and public access to information and government, ending discrimination based on race and sex, and drawing on the wisdom of traditional cultures to redesign lifestyles harmonious with each other and the natural environment.

☐ The global ecology movement and the feminist movement: Both are providing unique roles in social transformation, since they operate in all industrial societies within their mainstream. The male-dominated industrial societies find it impossible not to deal with either, as the wives of corporate and government bureaucrats become their uncomfortable social "consciences," while many male executives must face their own sons and daughters to defend ecologically or socially destructive daily decisions.

☐ The building of new coalitions in industrial countries between formerly fragmented citizens' groups: I have noted in earlier articles how formerly disparate local groups operated in the 1960s in the U.S. Some fought air pollution in New York and California, some tried to clean up local rivers or stop roads or airports, but with little knowledge of each others' activities. After Earth Day in 1970, there was a convergence and quantum leap in conceptualization of their fragmented problems of "pollution" as protection of the environment and the planetary biosphere. Similar convergences occurred around the concept of corporate accountability. Consumer and environmental advocates and members of civil rights, women and student movements joined with antiwar groups and with media-access, freedom-of-information and counter-culture forces; all had found that the targets of their grievances were, among other things, the practices of large corporations.[19] Today's coalitions bring together into a new holistic conceptualization of the decline of the industrial system itself, many of these existing forces along with new and older elements of the labor movement, rural voters, small business people and farmers, grass-roots and neighborhood groups, and appropriate-technology advocates. An example of the workings of such new coalitions was a conference which I participated in organizing on Working for Environmental and Economic Justice and Jobs. It was held in May, 1976, and was attended by 300 leaders from labor, environmental and social justice organizations to push for labor-intensive public projects and private investments and to support the goals of a full-employment economy and reordered corporate and government priorities.[20] Since then, similar conferences have followed in Ohio, Colorado, California, New York and other states. Environmentalists for Full Employment documents in its report, *Jobs and Energy,*[21] the many more jobs available in the alternative, renewable-resource economy. An international coalition of environmentalists, labor, small businesses, solar technologists, architects, designers, artists, writers, church people and others espousing the emerging post-industrial values now sponsors Sun Day, a planetary celebration of the Sun observed annually on May 3rd. These coalitions are nothing more than group-consumer demand (that cannot be expressed in the marketplace) for mass transit, solar and wind energy, recycling and bioconversion industries—the growing edge of the regenerative, sustainable economy. Their political efforts will begin to shift the pattern of investment to underpin these new sectors of the economy, just as they created the pollution control industries and the $50 million existing solar industry in the U.S., as described in Chapter 14.

☐ Lastly, the most convincing statistics so far on the dimensions of the emerging counter-economy in the U.S.: These are contained in the 1976 Report of the Stanford Research Institute's Business Intelligence Program, Guidelines No. 100, which estimates that between 4 and 5 million adult Americans having reduced their incomes drastically and withdrawn from their former slots in the dominant industrial, consumer economy, have now transformed their personal lifestyles to embrace what the SRI researcher Duane Elgin terms "voluntary simplicity." SRI estimates that another 8 to 10 million American adults adhere to and act on some, but not all, of the basic tenets of the "voluntary simplicity" approach, which embraces frugal consumption, ecological awareness, and a dominant concern with personal, inner growth. SRI claims that these statistics on the adoption of material frugality and "psychic riches" are important, because they may augur a major transformation in the goals and values of Americans in the coming decade.

A Harris survey taken in May, 1977, seems to confirm the value shift seen by the Stanford group. It found that by 79% to 17%, the public would place greater emphasis on teaching people how to live more with basic essentials "than on reaching a higher standard of living." By 76% to 17%, a sizable majority opts for "learning to get our pleasures out of nonmaterial experiences" rather than on "satisfying our needs for more goods and services." By 59% to 33%, a majority would stress "putting a real effort into avoiding those things that cause pollution" over "finding ways to clean up the environment as the economy expands." A lopsided 82% to 11% would concentrate on "improving those modes of travel we already have"; only 11% would emphasize "developing ways to get more places faster." By 77% to 15%, the public comes down for "spending more time getting to know each other better as human beings on a person-to-person basis," instead of "improving and speeding up our ability to communicate with each other through better technology." By 63% to 29%, a majority feels that the country would be better served if emphasis were put on "learning to appreciate human values more than material values," rather than on "finding ways to create more jobs for producing more goods." By 66% to 22%, the public would choose "breaking up big things and getting back to more humanized living" over "developing bigger and more efficient ways of doing things." And by 64% to 26%, most Americans feel that "finding more inner and personal rewards from the work people do" is more important than in "increasing the productivity of our workforce."[22]

In other countries, such as Canada, where there is a constitutional mandate for government leadership (rather than the still prevailing U.S. view of government as "traffic cop," responding but not leading), the voluntary simplicity theme emerges officially. The Canadian government, through such agencies as the Science Council of Canada and the Advanced Concepts Group of Environment Canada, has for many years supported research on changing goals and values of

Canadians and their implications for science policy and social and economic strategies. For example, Environment Canada contracted with the New Alchemy Institute to develop an alternative strategy and working models of energy-conserving, regenerative technologies for Prince Edward Island and Nova Scotia; the Science Council of Canada has conducted studies on an alternative course for Canada under the rubric of "the Conserver Society" and released a comprehensive report, *Human Goals and Science Policy,* authored by Dr. Ray Jackson in 1976; while a broad survey of Canadians' attitudes was released in 1976 by Environment Canada, *Canadians in Conversation About the Future,* by economist Cathy Starrs.

In addition, the Canadian government commissioned a massive study, released in December, 1976, by McGill and Montreal universities, entitled *The Conserver Society: A Blueprint for the Future?*[23] It specifically disputes the current assumption that Canada can continue as a wasteful mass-consumption society and examines five different scenarios for the future:

1. *The Status Quo Scenario* (Doing More with More)
2. *The Growth with Conservation Scenario* (Doing More with Less)
3. *The High Level Stable State Scenario* (Doing the Same with Less)
4. *The "Buddhist" Scenario* (Doing Less with Less—rejecting mass consumption)
5. *The Squander Society Scenario* (Doing Less with More—waste and extravagant consumption)

The last of these scenarios is a caricature of the worst features of the U.S. society: waste, manipulative advertising, planned obsolescence and self-destructing goods. Scenarios Two, Three and Four require mandating specific government policies: conservation, durability of products, recycling, use of renewable resources, phasing out of all but informational advertising (i.e., banning all persuasive, exaggerated and manipulative commercials encouraging consumption), as well as full-cost pricing (i.e., "internalizing" in production and including in prices all the social costs that we have discussed, at least all those that are quantifiable). Other specific changes the report discusses include more part-time and flexible-time jobs and work schedules to relieve personal pressures and public congestion; and many other ways of more intensively using our existing capital, productive facilities and other assets, such as vacation homes, which might be shared along with cars, laundry equipment, garden and household tools, etc., under new arrangements.

One of the most interesting aspects of the counter-economy—and one of the greatest services it is currently performing—is that of creating new images of the future, new alternatives, in technology, work, lifestyles, family arrangements and societal roles. As Elise Boulding has noted, a nation that loses its power to image its future will become directionless and falter. The misty outlines of possible, achievable and viable alternative futures can now be recognized in many industrial countries. Fiction can help again; for example, Ernest Callenbach's underground best-seller, *Ecotopia,* portrays such a viable, coherent alternative future based on some of these new values and has actually spawned a political movement in California and Oregon.[24]

Those for whom the status quo is still providing comfortable livings and meaningful lives are naturally impatient at the citizen movements, the public-interest advocates and the counter-economy's fumbling efforts to articulate its visions with cohesion, to innovate technology, to gain access to media, markets and credit sources and government contracts. Futurists' scenario building and hypothesizing and explorations of values are also scoffed at for their lack of "rigor" by comfortable academicians well rewarded for their orthodoxies. Yet even learning in the future will be less institutionalized and more entrepreneurial, whether in the renewed interest in apprenticeship and interning, or in the learning that will more often occur in the workplace as a result of developing more communal and enriching production modes.

The process of reconceptualization is a vital, if unremunerative, activity, and since it requires that existing models be critiqued and conventional wisdom be demythologized, it is more often punished. My hypothesizing in this book concerning the emerging counter-economy, which may be overlooked due to our conceptual blinders and statistical conventions, can be similarly critiqued. I have gleaned few facts to support my hypothesis—indeed, if I had found much hard data, I would be a historian rather than a futurist! I confess that I am a member of the counter-economy myself. The past decade of my life has been spent in citizen-action movements, public-interest groups and in social change. I am self-employed and already operate a family corporation. We have reduced our material expectations and our corporate income in order not to serve those corporations whose values we reject. We have no income security or group pension plan; but on the plus side, we do not need to commute to work but live by the natural cycles of sun and seasons, and best of all we have the freedom to think and write what we please, without institutional constraints. The decline of industrialism, while upsetting to many power wielders, is a heady, if insecure, time for many of us. To share our sense of awakening and rebirth, old intellectual, emotional and financial investments will have to be written off, while personal risks will be required. Yet even those who choose to remain industrialism's managers or intellectual servants to the old system will no doubt have to lower their own goals or material acquisition and accept less job security, tenure and pension benefits. There is, after all, a very appropriate personal trade-off between striving for greater secular power and wealth and ego gratification, and taking the path toward expanded consciousness. Those of us in the counter-economy have opted for the latter—and consider it a bargain.

[1]L.S. Stavrianos, *The Promise of the Coming Dark Age*, W.H. Freeman, San Francisco, 1976

[2]See, for example, Jay Forrester, *World Dynamics,* Edward Goldsmith, "The Blueprint for Survival," *The Ecologist,* Jan. 1972

[3]Harris Survey, Dec. 1975, *Chicago Tribune,* and Opinion Research Corp., Princeton, N.J., Survey, Dec. 1976

[4]Public Media Center, because of its low overhead and salaries, offers campaigns at approximately one-tenth the cost of conventional advertising agencies. Some of its clients include Ralph Nader's Critical Mass, Americans for a Working Economy, and the Center for Science in the Public Interest, all of Washington, D.C.

[5]Bridge Alternative Marketing, Oxfam, 274 Banbury Rd., Oxford, England

[6]*The Center Report,* Feb. 1976, Center for the Study of Democratic Institutions, Santa Barbara, Calif.

[7]*Commonsense II,* Bantam Books, 1975, p. 20

[8]Burns, Scott. *Home, Inc.,* Doubleday, 1975, p. 6

[9]Ibid., p. 35

[10]Ibid., p. 11

[11]*Financial Analysts Journal*, "Limits of Traditional Economics," by Hazel Henderson, May-June, 1973

[12]Rivers, Patrick, *The Survivalists,* Methuen, London, 1975

[13]*Co-op Facts and Figures,* Co-operative League of the U.S.A., Washington, D.C., 1975

[14]*Self-Reliance,* June 1976, p. 8, Washington, D.C.

[15]*Acorn News,* Nov. 1976, Vol. 6, #8 Little Rock, Ark.

[16]*New York Times,* July 8, 1975

[17]*The New Internationalist,* Dec. 1976, p. 22-23

[18]Boulding, Elise, *The Underside of History,* Westview Press, San Francisco, 1977

[19]*Business in 1990,* Proceedings of the White House Conference on the Industrial World Ahead, Nov. 1972, "Redeploying Corporate Resources Toward New Priorities," Hazel Henderson, U.S. Govt. Printing Office, Washington, D.C.

[20]*Environmentalists for Full Employment Newsletter,* #3, Fall 1976, p. 1

[21]*Jobs And Energy,* Grossman, Richard, and Daneker, Gail, Environmentalists for Full Employment, 1977, available from 1101 Vermont Ave., Washington, D.C.

[22]Harris Survey, *Washington Post,* May 23, 1977

[23]*The Conserver Society: A Blueprint for the Future?* K. Valaskakis, GAMMA, University of Montreal, 1976, Montreal

[24]Callenbach, Ernest, *Ecotopia,* Banyan Tree Books, Berkeley, Calif. 1975

EPILOGUE

In our hardware-loving industrial cultures it is exciting to see new "software" emerging. These vital software technologies include institutional redesign, more open political processes, conflict resolution mechanisms, world-order modeling, daring scientific speculation and new hypotheses. Besides, we need a new appreciation for the importance of psychic structures, myths, taboos and other internalized methods of behavior harmonization and self-regulation and for the role of self-expression in art, crafts and production as the best paths to inner, personal growth and the development of evolutionary imagination

So let us not look backwards at our exhausted, confused cultures, but learn to scan them with new eyes and imagination for all their signs that the old instrumental "yang" is now turning into a reemergence of the subtler "yin," intuitive consciousness, to restore the balance. As L.S. Stavrianos states in *The Promise of the Coming Dark Age:*

> The Dark Age following the collapse of Rome was anything but dark. Rather, it was an age of epochal creativity, when values and institutions were evolved that constituted the bedrock foundation of modern civilization. It is true that this creativity was preceded by imperial disintegration—by the shrinkage of commerce and cities, the disappearance of bureaucracies and standing armies, and the crumbling of roads and aqueducts and palaces. This

imperial wreckage explains, but scarcely justifies, the traditional characterization of the early medieval period as "dark." It was an age of birth, as well as death, and to concentrate on the latter is to miss the dynamism and significance of a seminal phase of human history.

Stavrianos points out that the crumbling of the imperial order was due in large part to its particular form of technological stagnation, since slavery devalued the notion of work and blunted the impulse to ingenuity and innovation. The devolution of this slave- and conquest-based imperial system created new frontier conditions, which, in turn, stimulated the invention of labor-saving devices and endowed manual labor with the respect it had lost. Thus, he adds, the Dark Age's advances included the "three-field" system of rotation farming, the heavy-wheeled plough, a new harness that multiplied five-fold the tractive performance of the horse, and the all-important windmill and water mill. These mill technologies were known in Greco-Roman times but were little used due to the abundance of slaves, whereas England's Domesday Book of 1086 documents 5,000 mills, one for every 50 households, sufficient to raise living standards substantially.

Now the wheel of time turns again. It is said that Minerva's owl only flies at dusk, and we only see the age in which we have lived at its twilight. We are now beginning to see more clearly the brief 200-year era of industrialism, its myths, ideologies, intellectual paraphernalia and emotional themes. In order to transcend our industrial perspectives and learn the new survival skills of reimagining and reconceptualizing, we must remember that all humans, including the most "rigorous" of our scientists, in a sense, create their own reality. To use a poetic metaphor based on physicist John Everett's "many worlds" interpretation of quantum mechanics, we all "keyword" our way through the universal computer, "Tarzans of rationality" as poet Ira Einhorn has written, "swinging along, creating our own universes as we go." We must never underestimate the power of the human mind and our unexplored capabilities.

So as I have talked of breakdown and the exhaustion of industrial cultures' metaphysics of "progress," reductionism, compulsive quantification and its profane, manipulative view of nature, I have tried also to celebrate what is being born: the overlooked shoots of a more benign technology, less materialistic goals, more holistic, reintegrated vision, a gentler metaphysic, a humbler, more realistic view of ourselves and an acceptance of our own finiteness and physical death.

But in our new strivings, we must avoid losing our balance when faced with inevitable frustrations. We must avoid the dual traps of escapism or becoming destructively apocalyptic. Today's escapists are dreaming of pristine new paradises in space, where the "high frontier" of space colonies can accommodate our expansionism and competitive spirit. I have commented on these gleaming evasions in the *Co-Evolution Quarterly* of Spring, 1976. We already live on a spaceship more wondrous than we know, and our present task, it seems to me, is first to tune in to its operating principles, which are peacefulness, humility, honesty, cooperation and love. Only then will we be a species fit to be loosed into the universe. The other trap is that of the apocalypse. In our frustration, there is a tendency (like a child losing a chess game who sweeps the pieces off the board) almost to welcome the awful catharsis of apocalypse—where the Gordian knot is cut with a swift blow and the deck is swept clean, as by the mighty, cleansing tidal wave, in Lawrence Ferlinghetti's chilling, beautiful poem, *Wild Dreams of a New Beginning*. Both of these traps are derailments of what I believe must be a long and faithful march, understanding that we, each of us, are bearers of a small, unique package of evolutionary potential, but, all the while, remaining humble actors, mindful of our brief little appearance in the grand drama of human emergence.

The incipient expectation I feel at the birth of planetary consciousness is that it augurs the possibility of another step in human evolution toward the eternal vision of our species: a planetary culture where we humans are in harmony with each other and with our ecosystems. This image is now breaking through again and helping relax old rigid, quantitative assumptions in many disciplines and is leading to magnificent new syntheses, such as that of chemists Lynn Margulis and James Lovelock and their adventurous "Gaia" hypothesis: that the entire biosphere may be one living organism. This image of a planetary "garden" is as old as Eden and is reborn with every generation.

My personal commitment, which I know many of us share, is to do what we can, to work wherever we are, in whatever institutions we find ourselves enmeshed, to seek out others and continue on the path to expanded awareness and the cultural mutation that now must come. In a sense, we must all become educators. More than ever, we all need to teach values for human development and justice and ecological harmony, rather than meaningless, academic, reductionist technique. Most of all, citizens and educators must teach a broader, more realistic definition of self-interest: as coterminous with group and species interest. Garrett Hardin's "Tragedy of the Commons" is the key riddle of our new, interdependent age. And Kenneth Boulding reminds us that the Commons cannot be managed without fostering community and mastering many modes of conflict resolution.

We humans are self-organizing systems, we do not have to tell our hearts when to beat. We have, in our collective history, also developed many examples of stable, self-organizing communities, based on psychic structures, concepts of reverence, transcendence and the sacred, that permitted voluntary internalizing of restraint in our behavior for the sake of others and the good of the whole community. This ability to self-organize is encoded in our DNA. We *know* how it could be—we have *always* known: the vision of empathy between humans and their harmony with the ecosystem. This vision is our commonest myth: the Garden of Eden, the Kingdom of Heaven, Nirvana, the Elysian Fields, the Great Oneness. We also know the hologram: "Do As You Would Be Done By," the Golden Rule. We can now understand, for the first time in our history, the teachings of our great spiritual leaders (who have always been the real futurists), the edict to serve the people, the values of love, caring, sharing, tolerance and humility. For now, at last, we see our true situation on this interdependent planet. It has been said that ethics is merely the acceptance of human interdependence. Morality, in fact, has, at last, become pragmatic.

HAZEL HENDERSON is an internationally-published futurist as well as an activist and founder of many public interest organizations. She co-directs the independent, non-affiliated Princeton Center for Alternative Futures, Inc., a deliberately small, private think tank for exploring alternative futures in a planetary context of human interdependence. She holds an Honorary Doctorate of Science from Worcester Polytechnic Institute for her work in alternative economics and technology. She is a director of the Council on Economic Priorities and the Worldwatch Institute, a member of the U.S. Association for the Club of Rome; an advisor to the Cousteau Society and the Environmental Action Foundation and a member of the Advisory Council of the U.S. Congress Office of Technology Assessment.